New Approaches, Methods, and Tools in Urban E-Planning

Carlos Nunes Silva
University of Lisbon, Portugal

A volume in the Advances in Civil and Industrial Engineering (ACIE) Book Series

IGI Global
DISSEMINATOR OF KNOWLEDGE

Published in the United States of America by
IGI Global
Engineering Science Reference (an imprint of IGI Global)
701 E. Chocolate Avenue
Hershey PA, USA 17033
Tel: 717-533-8845
Fax: 717-533-8661
E-mail: cust@igi-global.com
Web site: http://www.igi-global.com

Library of Congress Cataloging-in-Publication Data

Names: Silva, Carlos Nunes, editor.
Title: New approaches, methods, and tools in urban e-planning / Carlos Nunes Silva, editor.
Description: Hershey, PA : Engineering Science Reference, [2018]
Identifiers: LCCN 2017059207| ISBN 9781522559993 (hardcover) | ISBN 9781522560005 (ebook)
Subjects: LCSH: City planning--Citizen participation. | City planning--Technological innovations. | Urban policy--Citizen participation.
Classification: LCC HT166 .N435 2018 | DDC 307.1/216--dc23 LC record available at https://lccn.loc.gov/2017059207

This book is published in the IGI Global book series Advances in Civil and Industrial Engineering (ACIE) (ISSN: 2326-6139; eISSN: 2326-6155)

British Cataloguing in Publication Data
A Cataloguing in Publication record for this book is available from the British Library.

Advances in Civil and Industrial Engineering (ACIE) Book Series

ISSN:2326-6139
EISSN:2326-6155

Editor-in-Chief: Ioan Constantin Dima, University Valahia of Târgovişte, Romania

MISSION

Private and public sector infrastructures begin to age, or require change in the face of developing technologies, the fields of civil and industrial engineering have become increasingly important as a method to mitigate and manage these changes. As governments and the public at large begin to grapple with climate change and growing populations, civil engineering has become more interdisciplinary and the need for publications that discuss the rapid changes and advancements in the field have become more in-demand. Additionally, private corporations and companies are facing similar changes and challenges, with the pressure for new and innovative methods being placed on those involved in industrial engineering.

The **Advances in Civil and Industrial Engineering (ACIE) Book Series** aims to present research and methodology that will provide solutions and discussions to meet such needs. The latest methodologies, applications, tools, and analysis will be published through the books included in **ACIE** in order to keep the available research in civil and industrial engineering as current and timely as possible.

COVERAGE

- Urban Engineering
- Structural Engineering
- Ergonomics
- Productivity
- Construction Engineering
- Transportation Engineering
- Hydraulic Engineering
- Materials Management
- Quality Engineering
- Earthquake engineering

IGI Global is currently accepting manuscripts for publication within this series. To submit a proposal for a volume in this series, please contact our Acquisition Editors at Acquisitions@igi-global.com or visit: http://www.igi-global.com/publish/.

The Advances in Civil and Industrial Engineering (ACIE) Book Series (ISSN 2326-6139) is published by IGI Global, 701 E. Chocolate Avenue, Hershey, PA 17033-1240, USA, www.igi-global.com. This series is composed of titles available for purchase individually; each title is edited to be contextually exclusive from any other title within the series. For pricing and ordering information please visit http://www.igi-global.com/book-series/advances-civil-industrial-engineering/73673. Postmaster: Send all address changes to above address.

Titles in this Series

For a list of additional titles in this series, please visit:
https://www.igi-global.com/book-series/advances-civil-industrial-engineering/73673

Handbook of Research on Perception-Driven Approaches to Urban Assessment and Design
Francesco Aletta (University of Sheffield, UK) and Jieling Xiao (Birmingham City University, UK)
Engineering Science Reference • ©2018 • 400pp • H/C (ISBN: 9781522536376) • US $295.00

Convective Heat and Mass Transfer in the Free Atmosphere Emerging Research ...
Kaliyeva Kulyash (Lorraine University, France)
Engineering Science Reference • ©2018 • 127pp • H/C (ISBN: 9781522530503) • US $135.00

Innovative Applications of Big Data in the Railway Industry
Shruti Kohli (University of Birmingham, UK) A.V. Senthil Kumar (Hindusthan College of Arts and Science, India) John M. Easton (University of Birmingham, UK) and Clive Roberts (University of Birmingham, UK)
Engineering Science Reference • ©2018 • 395pp • H/C (ISBN: 9781522531760) • US $235.00

Dynamic Stability of Hydraulic Gates and Engineering for Flood Prevention
Noriaki Ishii (Osaka Electro-Communication University, Japan) Keiko Anami (Osaka Electro-Communication University, Japan) and Charles W. Knisely (Bucknell University, USA)
Engineering Science Reference • ©2018 • 660pp • H/C (ISBN: 9781522530794) • US $245.00

Handbook of Research on Trends and Digital Advances in Engineering Geology
Nurcihan Ceryan (Balikesir University, Turkey)
Engineering Science Reference • ©2018 • 765pp • H/C (ISBN: 9781522527091) • US $345.00

Intelligent Vehicles and Materials Transportation in the Manufacturing Sector ...
Susmita Bandyopadhyay (University of Burdwan, India)
Engineering Science Reference • ©2018 • 230pp • H/C (ISBN: 9781522530640) • US $155.00

Recent Advances in Applied Thermal Imaging for Industrial Applications
V. Santhi (VIT University, India)
Engineering Science Reference • ©2017 • 306pp • H/C (ISBN: 9781522524236) • US $205.00

For an entire list of titles in this series, please visit:
https://www.igi-global.com/book-series/advances-civil-industrial-engineering/73673

Table of Contents

Detailed Table of Contents

Austin, Texas, a university-based research team designed a municipal web tool called the Austin Historical Survey Wiki. This chapter offers insights into these prior efforts to augment planning processes with "digitized memory," web-based technology, and public engagement.

During large-scale manmade or natural disasters, such as Superstorm Sandy and Hurricanes Harvey and Irma, collaborations among government agencies, NGOs, and businesses need to be coordinated to provide necessary resources to respond to emergency events. However, resources from citizens themselves are underutilized, such as their equipment or expertise. The citizen participation via social media enhanced the situational awareness, but the response management is still mainly handled by the government or government-sanctioned partners. By harnessing the power of citizen crowdsourcing, government agencies can create enhanced disaster situation awareness and facilitate effective utilization of resources provided by citizen volunteers, resulting in more effective disaster responses. This chapter presents a public engagement in emergency response (PEER) framework that provides an online and mobile crowdsourcing platform for incident reporting and citizens' resource volunteering as well as an intelligent recommender system to match-make citizen resources with emergency tasks.

This chapter explores the theme of collaborative construction of territorial knowledge through the use of ICTs, proposing a new approach to spatial representation based on semantic ontologies. The theoretical perspective is applied in a recent experimentation conducted in collaboration between the Politecnico di Torino and the Università degli Studi di Torino, called OnToMap – Community Maps 3.0. The chapter is divided into two parts. The first part aims to define the theoretical framework of collaborative mapping practices and VGI, social semantic mapping, GIScience, and Web 3.0 applications, specifying which approach to participation is assumed as conceptual background. The second part focuses on the web-based application

Residential energy and water consumption depend on dwelling structure and the behaviour of residents. Aspects of residential behaviour can be derived from census data. Dwelling information is harder to obtain. Using both aerial and street-level views from Google mapping products, exterior dwelling characteristics were captured in each of 40 postal areas in and around Melbourne, Australia. This approach saved the time and cost of travelling to the widely spread suburbs and provided data not otherwise available. The census and dwelling data were compared with resource usage statistics in linear regression models. It was found that energy and water use are highly correlated, with socio-economic variables better explaining water consumption and dwelling structure factors better explaining energy consumption. Nevertheless, the proportions of households that include a couple with children and have a swimming pool provided useful models of variations in both energy and water use. Applications to planning through spatially explicit scenario testing were developed in ArcGIS ModelBuilder.

This chapter builds on research into Australian free Wi-Fi initiatives conducted in late 2012 and early 2013. It tours through a range of global developments in wireless internet delivery, focusing on how these influence the character of public spatiality, participation, and social inclusion. While there have been numerous technical and commercial advances, the authors argue that free public services narrowly focus on constructing public spaces of consumption and spectacle, and valorising public activities through increasingly granular sensor surveillance. The authors offer an expanded conception of what it means to value public space and to participate socially, culturally, and politically in public. The chapter concludes with the concerning gap between small scale projects that experiment with these concepts and the large-scale institutions that ignore them.

Preface

NEW CHALLENGES AND OPPORTUNITIES IN THE GOVERNANCE OF CITIES: INNOVATIVE APPROACHES, METHODS, AND TOOLS IN URBAN E-PLANNING

The growing and wide-ranging use of information and communication technologies in all tiers of government and its improvements in recent years is changing the form public policy is carried out in all policy sectors and across all tiers of public administration (Silva, 2010, 2013, 2015). With the main economies moving into the fourth Industrial Revolution, bringing together innovative combinations in digital, biological, and physical technologies, coupled with the widespread use of artificial intelligence, Big Data analytics, and automation, we may be at the dawn of even more profound changes in the form government, both central and local, will deal with urban planning, equipped as it will be with the infrastructures of the ongoing digital revolution. As the recent UN e-Government Survey suggests, in 2016 people no longer go online, they are online. This idea, highlighting the increasingly ubiquitous, fast, mobile, and content-rich Internet, expresses well the ongoing radical changes in the way citizens live, on how they communicate and work, which together have profound impacts not only in each person' way of life but also on all sorts of organizations, including government in all its tiers.

The UN 2016 e-Government survey points the growing engagement of citizens in public policy decision-making processes as one of the recent trends in the e-government field, associated with the expansion of people-driven services. This citizen-centric approach instead of the initial technology-oriented approaches to e-government and e-planning crosses the debate on smart-city policies, as Capra (2016) refers in the Amsterdam case, Joss and Dayot (2017) for the UK, or as is shown in the literature review on smart cities presented in Mora, Bolici, and Deakin (2017) and in Kummitha et al. (2017). This expanding access and use of information and communication technology, in a world that is getting increasingly connected, with all 193 UN member countries having an online presence since 2014, when a decade before 10% of countries were without any online presence, has impacts also in local government. It is responsible for significant changes in local policy approaches, in the tools and methods used in each stage of the urban planning process (Silva, 2012), and in the way citizens and communities engage in the local policy development, as is shown in recent research on citizen e-participation in urban planning (Silva, 2013; Fredericks, Tomitsch, & Stewart, 2017; Graziano, 2017; Filippi, 2017), on the new models of public sector innovation, namely through living labs (Papadopoulou, Chrysaida-Aliki, & Maria Giaoutzi, 2017; Gascó, 2017), on the efforts to improve public accountability and transparency through new civic technologies in a myriad of local policy issues (for crime mapping, see Scassa, 2016), on online data portals (Bishop et al., 2017), through municipal open data initiatives (Johnson, 2016), on social media (Nummi, 2017), or through the use of new indicators (Stratigea, 2017). In other words, the widespread use of information and communication technologies resulted in a gradual shift, although geographically uneven, from conventional paper-based urban planning to urban e-planning in the last two decades (Silva, 2010; 2015), a move that needs to be sensitive to context (Afzalan, Sanchez, & Evans-Cowley, 2017; Damurski, 2016). This gradual and uneven shift towards e-planning is strongly correlated with growing levels of maturity experienced in the provision of government services and information through the Internet, even if with significant differences among cities, as highlight by Fietkiewicz et al. (2017), and correlated also with the increasing use of social media technologies by municipalities, as examined by Bennett and Manoharan (2017) for the US case.

Also important to note is the worldwide increase in mobile-broadband subscriptions as well as increases in the speed in broadband connections, repeatedly reported in all world surveys of ICT and e-Government development published in recent years, as those published by the ITU and by the UN, with 95% of the world population now living in areas covered by mobile-cellular signal, which has been directly responsible for a higher growth in ICT use. In fact, according to the most recent report on mobile communications worldwide (GAMA, 2016), 2015 saw the adoption of 4G and a continued growth in the number of mobile connections, reaching almost 4,7 billion unique subscribers and more than 1 billion connections, in 151 countries, representing a penetration rate of 63%, expected to reach 72% by 2020, although unevenly distributed, with lower rates in sub-Saharan Africa (43%) and higher ones in Europe (85%). This generalized growth in mobile-cellular penetration, even in less developed countries, namely in the African continent, has been responsible for greater ICT use and may have in the medium term a significant impact in the expansion of urban e-planning and on the diversification of the myriad of its web-based digital technologies even in the less developed countries. Nonetheless, it should also be pointed out that we may be on the eve of a change in this continuous growth pattern registered in recent years, with growth rates expected to slow due to saturation in developed countries and due also to low affordability in the less developed countries.

However, while the access and use of information and communication technologies continue to expand around the world, persistent digital divides continue to exist between more and less developed countries, within countries, between urban and rural areas, in a way that over and over again overlaps and reinforces other pre-existent economic and social divides. According to the ITU (2017), it is clearly the youth who is using mostly the Internet, with youngsters representing almost one-fourth of the total number of persons using the Internet worldwide. In 104 countries analysed by ITU, more than 80% of the youth population are online. This figure rises in developed countries with 94% of the persons aged 15-24 connected to the Internet but is lower on less developed countries (67% in developing countries and 30% in the least developed countries; 9 out of 10 youngsters not using the Internet are in Africa, Asia and Pacific). In other words, despite the improvements in the ICT Development Index in nearly all countries examined (ITU, 2017), compared to previous years, great disparities continue to exist between more and less developed countries with the range between countries remaining almost unchanged year after year, as the e-Government development Index repeatedly shows. With European countries continuing to lead the ranking of the most ICT developed countries, the data available for 2016 continues to highlight the existence of a strong correlation between economic development and ICT development, from which it seems plausible to derive a similar ranking of urban e-planning development. In addition, the ITU data for 2016 also confirms, once again, that investment in ICT and innovations occurs mainly in developed and high-income countries, with a population not only with high income but also with the skills required for an extensive and effective use of ICT and therefore of urban e-planning tools. As expected, the least developed countries are in this case also in a clearly disadvantaged position, in what regards adoption and use of these new planning approaches and web-based digital technologies. Particularly alarming is the fact that the gap between less and more developed countries, which inevitably affects the ability for local governments to incorporate web-based tools in their current planning practices, continues almost unchanged if not worse in some aspects. A similar picture of the impact of ICT on the competitiveness of 139 economies is provided by the World Economic Forum' Network Readiness Index (Baller, Dutta, & Lanvin, 2017).

In sum, these trends confront municipalities with new challenges and opportunities in the governance of cities. If in 2017 more than half of the world population have access to the Internet, the disparities among countries continue to exist. The offline population is, according to UN (2017), mainly female, elderly, less educated, lower income and rural. In developed countries, the proportion of households with Internet access at home is twice that of developing countries. And if most of these citizens already have access, at work or in other places with public connection, the fact is that this low connectivity at home is an important constraint to the ability of citizens to fully participate in the online public policy processes. With these persistent digital divides between and within countries, the development of urban e-planning will require higher growth rates in Internet access, in both fixed and mobile broadband, in order to bridge the divide that currently exists between developed and less developed countries and along pre-existent social divides within countries and cities. Even so, numerous local examples around the world, associated with specific participatory technologies (Gagliardi, 2017), as is the case of PPGIS and Participatory Mapping, show how these sort of constraints can be overcome, as numerous cases examined in the works included in the bibliography compiled by McCall (2017) exemplify, although there are clearly a need for more inclusive civic participation platforms as shown and argued, among others, in Pak, Chua, and Moere (2017).

Confronted with this background of a fast but geographically uneven development in the access and use of information and communication technology, and a similar unequal development of e-government, overlapping several other economic and social divides, and the corresponding new challenges that the theory and ethics of urban e-planning has to address, as well as the new challenges in the organization, and in the technologies and methods used in the local planning system, and in each of its functional sectors, the book aims to present and to discuss recent innovations in urban e-planning: in citizen e-participation, in the use of social media, and in new forms of data collection and idea generation for urban planning, made possible by the ongoing digital transformations in the public policy sphere. The book examines some of these innovations in case studies located in six countries with a high rank in the 2016 e-Government Development Index (Australia -2nd; Finland - 5th; USA - 12th; Canada - 14th; Israel - 20th; and Italy - 22th), and with a high rank position in the UN 2016 e-Participation Index (Australia -2nd; Canada - 8th; Finland - 8th; Italy - 8th; USA - 12th; Israel - 17th). It is organized into three main sections: Citizens as Sensors and e-Participation in Urban Planning; Creativity and Innovation in Data Collection; and Wireless and Social Media.

The first section of the book - Citizens as Sensors and e-Participation in Urban Planning - comprises six chapters and deals in different ways with citizen engagement in the local policy process, with citizens as sensors, with new technological approaches to open digital mapping, and how these e-tools augment participatory planning processes. This section concurs with findings of other studies on the potential uses of social media and community mapping in public policy, namely in urban planning and local governance, with the aim of supporting and enhancing civic engagement in the local policy process, in the different stages of e-participation: e-information; e-consultation; and e-decision-making. The expansion of e-government and the improvements registered in its maturity levels in recent years is important and critical in relation to citizen engagement. If with e-participation citizens could connect more easily to their government and be consulted in urban planning decision-making processes, even those living in more remote areas, with the increase in mobile communication it is expected in the coming future greater digital inclusion of those social groups still unconnected, which will also help to deliver innovative new services and apps, namely in the field of urban e-planning.

In the first chapter of this section, 'Exploring Citizen-Centric E-Government Using a Democratic Theories Framework', Sean Mossey, Aroon P. Manoharan, both from the University of Massachusetts Boston, and Lamar Vernon Bennett, from the Long Island University-Brooklyn, USA, examine how governments engage their citizens online, based on previous studies on e-government and democratic theories. The authors consider three democratic theories - representative, pluralist, and direct – and look at two research questions, using data from 200 American local governments: What factors drive governments to employ each theory? Which theory predominates in the implementation of e-government? In answering these two questions the authors provide also suggestions on how to move forward with the current efforts in e-government and in m-government.

Nili Steinfeld and Azi Lev-On, both from Ariel University, Israel, discuss in the next chapter, 'Writings on the City Wall: An Analysis of Popular Municipal Facebook Pages Posts', the role of municipal Facebook pages, seen as significant social media arenas for maintaining contact between councillors and local citizens. The analysis is based on data from around 24,000 posts collected from the Facebook pages of all Israeli municipalities during a six-month period and combines both a purely automatic linguistic analysis of the texts of all posts considered with a manual coding of a sample of the most popular posts, offering in that way insights into the character of the discourse expressed in the pages examined and on what in fact produces greater citizen participation.

In Chapter 3, 'Enhancing e-Participation in Urban Planning Competitions', Pilvi Nummi, Susa Eräranta, and Maarit Kahila-Tani, from Aalto University, Finland, present and discuss planning competitions, seen as an innovative way to determine alternatives and promote innovative solutions in the early phase of the urban planning process, thus complementing the traditional jury-based evaluation process which faces significant opposition in Finland. These web-based participation tools described and examined in the chapter are used in urban planning competitions to register public opinion, adding therefore new insights to those provided by the expert view of the jury, a view that according to the authors does not consider always adequately the views of local residents. The chapter is therefore focused on how public participation can be organized in these competition processes, how the information produced is used in the following stages of the planning process in that particular area. The authors illustrates this, based on two Finnish case studies, and show that these web-based tools can in fact augment public participation in planning and influence the following stages of the planning process.

Jennifer Minner, from Cornell University, Andrea Roberts, from Texas A & M University, Michael Holleran and Joshua Conrad, both from University of Texas at Austin, USA, in the following chapter - 'A Smart City Remembers its Past: Citizens as Sensors in Survey and Mapping of Historic Places' - explore a neglected area in the existing literature on smart cities, the gathering and sharing of information about historic places within municipal information systems. The authors offer insights into attempts to augment planning processes with digitized memory, web-based technology, and public engagement, and do this through three case studies of technologically augmented planning processes that incorporated citizens as sensors of memory and local knowledge about historic places: SurveyLA, Motor City Mapping, and Austin Historical Survey Wiki. The first case deals with the efforts of the City of Los Angeles preservation professionals and volunteers in surveying over 880,000 parcels of historic places, the second attempts to identify the condition of buildings in neighbourhoods devastated by decades of disinvestment, being the third case a municipal web tool called the Austin Historical Survey Wiki designed by a university-based research team.

Next, in Chapter 5, 'Citizen to Citizen Resource Sharing in Emergency Response', Soon Ae Chun, Jaideep Vaidya, Vijay Atluri, Basit Shafiq, and Nabil R. Adam, from the City University of New York, Rutgers University and Lahore University of Management Sciences, explore the underutilization of citizens resources, equipment or expertise, in the response to emergency events, namely in natural or in large scale man-made disasters, and present a framework, which the authors name PEER - Public Engagement in Emergency Response, that provides an online and mobile crowdsourcing platform for situation reporting, for citizens' resource volunteering and an intelligent recommendation system that intends to match-make the citizen resources to emergency tasks. As argued in the chapter, the expansion of social media creates opportunities for citizens to share information and through that to increase the awareness about specific emergency situations, which can potentially produce more effective disaster responses, even if the response management continues to be mainly in the hands of government or government' partners.

In the last chapter in this section, 'Sharing Semantic Territorial Interpretation. The Contribution of 'OnToMap Project to Spatial Planning' - Angioletta Voghera and Luigi La Riccia, both from Politecnico di Torino, explore forms of collaborative construction of territorial knowledge through the use of information and communication technologies. Voghera and La Riccia propose new approaches to spatial representation based on semantic ontologies, and illustrated with a recent experimentation called OnToMap – Community Maps 3.0, which aims to encourage the construction of collective and shared knowledge of places, seen as an approach that can and should be used to enhance citizen empowerment in the processes of urban e-planning.

The following section, Creativity and Innovation in Data Collection, with three chapters, deals with different aspects of data generation and data collection in the planning process. The first chapter in this section discusses the use of brainstorming in idea generation processes, the second explores the potential of a new tool employed in the early stages of the planning and design process, and the third shows how Google mapping products can easily be used with great advantage compared to other methods in terms of time and cost. These are all new planning digital tools that require, to be effective, easy access to the Internet and to web-based technologies, which is far from being the common situation in numerous countries and cities, characterized by deep socio-economic inequalities, as referred before.

In Chapter 7, 'Enhancing Creativity in E-Planning: Recommendations from a Collaborative Creativity Perspective', Paul B. Paulus, from University of Texas at Arlington, and Jonali Baruah, from Tarleton State University, USA, explore innovative ways to share and evaluate ideas efficiently and effectively in groups, namely through computer based group decision support, in order to overcome the limitations of face-to-face meetings. Although findings provided by the authors about collaborative creativity and their implications for effective e-planning do point to beneficial outcomes, the evidence provided also suggests, as the authors argue, that these electronic meetings need to be employed effectively in order to take full advantage of the intellectual and creative potential of groups. In the following chapter, 'SketchBoard: Design and Evaluation of Interactive Tools for the Design Review Process', Rojin Vishkaie, from Ball State University, USA, and Richard Levy, from University of Calgary, Canada, starting from the assumption that there is a gap in the evaluation of how aspects of the design review process can be performed with digital media, offer an exploratory study whose aim is to investigate the design review process and its embedment in digital media, namely on mobile interactive surfaces. The study combines empirical and conceptual design components in the evaluation of a new communication medium called SketchBoard that uses interactive surfaces to perform selected tasks of the design review process. The third and last chapter in this section, 'Residential Resource Consumption: A New Approach to Collection of Dwelling Data', Maryam Saydi and Ian D. Bishop, both from the University of Melbourne, Australia, offer an innovative way of collecting data, for urban planning and for city governance more generally, related to urban dwellings. As the authors argue, if residential energy and water consumption depend on the structure of the dwelling and on the behaviour of the residents and if these variables can be derived with no special difficulty from census data, that is not always the case with dwelling data. The approach proposed and tested in this study uses both aerial and street level views from Google mapping products. From these data sources the authors captured exterior dwelling characteristics for 50 random dwellings in each of 40 Postal Areas in and around Melbourne, Australia. Among other aspects, this approach proved to be capable to save time and cost of travelling by providing data that could not be easily obtained by other means, in particular in highly dispersed urban areas as the suburban areas considered in this case study in Melbourne. Based on this data, the authors estimated the influence of dwelling and behaviour features on water and energy consumption, with socio-economic variables better correlated with water consumption, and dwelling characteristics with energy consumption, and developed applications for the elaboration of spatial planning scenarios, which may well be replicated in other research studies on cities in other parts of the world.

The two chapters in the third and final section, Wireless and Social Media, deal directly with the effects of the use of social media, first in the context of free Wi-Fi services, and in the second case through the use of social media to enhance sustainability approaches in an international scale. In the first of these two chapters, 'Public Space and the Development of Wireless Media', Alex Lambert, from Monash University, Scott McQuire and Nikos Papastergiadis, from University of Melbourne, Australia, explore how wireless internet influence the character of public spatiality, participation and social inclusion, based on the research conducted into the Australian free Wi-Fi initiatives. The authors argue that despite the technical and commercial progress, free public services are in general narrowly focused on constructing public space in terms of consumption and spectacle. And this seems particularly so in projects of major institutions compared to small scale projects, offering their view of what it means to value public space and to participate socially, culturally and politically in public. In the last chapter, 'The Role of Social Media in Public Involvement: Pushing for Sustainability in International Planning and Development', Tooran Alizadeh, from the University of Sydney, Reza Farid and Laura Willems, both from Griffith University, Australia, explore the potential of social media to enhance public participation in sustainable practices on an international scale through planning and development projects. For that the authors examine the official websites, Facebook pages, and Twitter accounts of four international institutions - World Bank, UN-HABITAT, Unilever and World Business Council for Sustainable Development - suggesting the existence of a strong public move forwards for sustainability in the field of international development, made possible through the use of social media, which goes well beyond the physical geographical borders that traditionally limit the public sphere of each local community.

In sum, in order to offer to more citizens and communities the benefits intrinsic in these new digital technologies, methodologies, and planning approaches, in particular in the less developed countries, regions or cities, it is necessary to bring more people online, which requires a substantial reduction of the current socio-economic inequalities, substantial increases in revenues, and improvements in digital literacy and digital skills, since these are key determinants for people to use or not the Internet, namely the web-based planning tools that each chapter in the book so well examines and discusses. Other barriers, such as Internet prices, even if with a tendency to decrease, as all reports seem to indicate consistently year after year, remain important constraints to the use of the Internet, preventing large parts of the community of getting the benefits from the Internet' full potential in the field of urban e-planning.

And as we can easily derive from the cases presented in the following chapters, urban e-planning, as e-government in general, has positive effects on better urban governance, on accountability and transparency, and has therefore the potential to help support the implementation of the 2030 Agenda, and its 17 sustainable development goals, acting as an enabler for sustainable development in cities of developing and less developed countries. For that reason it is mandatory to expand ICT infrastructures and connectivity, to improve affordability and skills, and to increase the use of web-based digital technologies in urban planning and related urban services, in particular mobile web-based technologies, considering the high mobile phone penetration in less developed countries compared to fixed technologies.

To do this with equity requires action to reach universal and affordable access to technology through the reduction of socio-economic inequalities, eradication of poverty and other inequalities, in particular in less developed countries. The existence of a clear strategy of e-government and e-planning, at both national and local levels, political commitment over the years, a consistent legal framework, adequate financial and human resources, a capacity building initiative at the local government level, the expansion of ICT literacy among the most disadvantaged social groups, and the connection of these changes with the reform of local government structures and procedures, are conditions that ought to be put into practice if cities and common citizens are to take full advantage of the digital technologies applied in the urban planning process and in local policies. Otherwise, the benefits of urban e-planning may well be captured by the elites, reinforcing existing socio-economic inequalities in cities around the world, in particular in developing and less developed countries. Also imperative is the upscaling of what has been so far mostly pilot projects, as is the case of numerous smart cities initiatives, as shown in Winden and Buuse (2017) and in Buck and While (2017), among other research works on these issues.

The examples explored and discussed in the book, taken from countries with a high rank in the E-Government Development Index and in the E-Participation Index, illustrate well the potential these digital technologies have for the development of better and more inclusive urban governance and urban planning, through the provision, by central and local governments, of information and services through the Internet. This is being done namely through open government data approaches, improving transparency, accountability, effectiveness and citizen engagement in the urban e-planning process, being also implicit in these essays issues such as privacy and cyber security that ought to be considered carefully in any process of urban e-planning development. Other challenges lay ahead, as is the case of the choice between technology-oriented approaches in urban e-planning and more citizen-centric ones.

Carlos Nunes Silva
University of Lisbon, Portugal

REFERENCES

Afzalan, N., Sanchez, T. W., & Evans-Cowley, J. (2017, July). Creating smarter cities: Considerations for selecting online participatory tools. *Cities (London, England)*, *67*, 21–30. doi:10.1016/j.cities.2017.04.002

Baller, S., Dutta, S., & Lanvin, B. (Eds.). (2017). *The Global Information Technology Report 2016. Innovating in the Digital Economy*. Geneva: World Economic Forum.

Bennett, L. V., & Manoharan, A. P. (2017). The Use of Social Media Policies by US Municipalities. *International Journal of Public Administration*, *40*(4), 317–328. doi:10.1080/01900692.2015.1113182

Bishop, I. D., Eagleson, S., Pettit, C. J., Rajabifard, A., Badland, H., Day, J. E., ... White, M. (2017). Using an Online Data Portal and Prototype Analysis Tools in an Investigation of Spatial Livability Planning. *International Journal of E-Planning Research*, *6*(2), 1–21. doi:10.4018/IJEPR.2017040101

Buck, N. T., & While, A. (2017). Competitive urbanism and the limits to smart city innovation: The UK Future Cities initiative. *Urban Studies (Edinburgh, Scotland)*, *54*(2), 501–519. doi:10.1177/0042098015597162

Capra, C. F. (2016). The Smart City and its Citizens: Governance and Citizen Participation in Amsterdam Smart City. *International Journal of E-Planning Research*, *5*(1), 20–38. doi:10.4018/IJEPR.2016010102

Damurski, L. (2016). Smart City, Integrated Planning, and Multilevel Governance: A Conceptual Framework for e-Planning in Europe. *International Journal of E-Planning Research*, *5*(4), 41–53. doi:10.4018/IJEPR.2016100103

Fietkiewicz, K. J., Mainka, A., & Stock, W. G. (2017). e-Government in cities of the knowledge society. An empirical investigation of Smart Cities' governmental websites. *Government Information Quarterly*, *34*(1), 75–83. doi:10.1016/j.giq.2016.08.003

Filippi, F. D., Coscia, C., & Guido, R. (2017). How Technologies Can Enhance Open Policy Making and Citizen-Responsive Urban Planning: MiraMap - A Governing Tool for the Mirafiori Sud District in Turin (Italy). *International Journal of E-Planning Research*, *6*(1), 23–42. doi:10.4018/IJEPR.2017010102

Fredericks, Tomitsch, & Stewart. (2017). Design Patterns for Integrating Digitally Augmented Pop-ups with Community Engagement. *International Journal of E-Planning Research, 6*(3), 19-41.

Gagliardi, D., Schina, L., Sarcinella, M. L., Mangialardi, G., Niglia, F., & Corallo, A. (2017). Information and communication technologies and public participation: Interactive maps and value added for citizens. *Government Information Quarterly*, *34*(1), 153–166. doi:10.1016/j.giq.2016.09.002

Gascó, M. (2017). Living labs: Implementing open innovation in the public sector. *Government Information Quarterly*, *34*(1), 90–98. doi:10.1016/j.giq.2016.09.003

Graziano, T. (2017). Citizen e-Participation in Urban Planning: Achievements and Future Challenges in a Mediterranean City. *International Journal of E-Planning Research*, *6*(3), 1–18. doi:10.4018/IJEPR.2017070101

GSMA. (2016). *The Mobile Economy 2016*. London: GSMA.

ITU. (2016). *Measuring the Information Society Report*. Geneva: International Telecommunication Union.

Johnson, P. A. (2016). Reflecting on the Success of Open Data: How Municipal Government Evaluates their Open Data Programs. *International Journal of E-Planning Research*, *5*(3), 1–12. doi:10.4018/IJEPR.2016070101

Joss, S., Cook, M., & Dayot, Y. (2017). Smart Cities: Towards a New Citizenship Regime? A Discourse Analysis of the British Smart City Standard. *Journal of Urban Technology*, *24*(4), 29–49. doi:10.1080/10630732.2017.1336027

Kummitha, R. K. R., & Crutzen, N. (2017). How do we understand smart cities? An evolutionary perspective. *Cities (London, England)*, *67*(July), 43–52. doi:10.1016/j.cities.2017.04.010

McCall, M. K. (2017). Applying Participatory GIS and Participatory Mapping to Participatory Spatial Planning and Local–level Land & Resources Management utilizing Local & Indigenous Spatial Knowledge. Morelia: UNAM, CIGA.

Mora, L., Bolici, R., & Deakin, M. (2017). The First Two Decades of Smart-City Research: A Bibliometric Analysis. *Journal of Urban Technology*, *24*(1), 1, 3–27. doi:10.1080/10630732.2017.1285123

Nummi, P. (2017). Social Media Data Analysis in Urban e-Planning. *International Journal of E-Planning Research*, *6*(4), 18–31. doi:10.4018/IJEPR.2017100102

Pak, B., Chua, A., & Vande Moere, A. (2017). FixMyStreet Brussels: Socio-Demographic Inequality in Crowdsourced Civic Participation. *Journal of Urban Technology*, *24*(2), 65–87. doi:10.1080/10630732.2016.1270047

Papadopoulou, C.-A., & Giaoutzi, M. (2017). Crowdsourcing and Living Labs in Support of Smart Cities' Development. *International Journal of E-Planning Research*, *6*(2), 22–38. doi:10.4018/IJEPR.2017040102

Scassa, T. (2016). Police Service Crime Mapping as Civic Technology: A Critical Assessment. *International Journal of E-Planning Research*, *5*(3), 13–26. doi:10.4018/IJEPR.2016070102

Silva, C. N. (2010). *Handbook of Research on E-Planning. ICTs for Urban Development and Monitoring*. Hershey, PA: IGI Global. doi:10.4018/978-1-61520-929-3

Silva, C. N. (2012). *Online Research Methods in Urban and Planning Studies. Design and Outcomes*. Hershey, PA: IGI Global. doi:10.4018/978-1-4666-0074-4

Silva, C. N. (2013). *Citizen E-Participation in Urban Governance. Crowdsourcing and Collaborative Creativity*. Hershey, PA: IGI Global. doi:10.4018/978-1-4666-4169-3

Silva, C. N. (2015). *Emerging Issues, Challenges, and Opportunities in Urban E-Planning*. Hershey, PA: IGI Global. doi:10.4018/978-1-4666-8150-7

Stratigea, A., Leka, A., & Panagiotopoulou, M. (2017). In Search of Indicators for Assessing Smart and Sustainable Cities and Communities' Performance. *International Journal of E-Planning Research*, *6*(1), 43–73. doi:10.4018/IJEPR.2017010103

UN. (2016). *UN E-Government Survey 2016: E-Government in Support of Sustainable Development*. New York: United Nations.

van Winden, W., & van den Buuse, D. (2017). Smart City Pilot Projects: Exploring the Dimensions and Conditions of Scaling Up. *Journal of Urban Technology*, *24*(4), 51–72. doi:10.1080/10630732.2017.1348884

Section 1

Citizens as Sensors and E-Participation in Urban Planning

Chapter 1
Exploring Citizen-Centric E-Government Using a Democratic Theories Framework

Sean Mossey
University of Massachusetts – Boston, USA

Aroon P Manoharan
University of Massachusetts – Boston, USA

Lamar Vernon Bennett
Long Island University – Brooklyn, USA

ABSTRACT

In this chapter, the authors draw on Scott's work on e-government and democratic theories to examine how governments engage their citizens online. The three theories they focus on—representative, pluralist, and direct—are the most prominent in the democratic theory literature. Using data from 200 U.S. local governments, the authors examine two research questions: What factors drive governments to employ each theory? Which theory predominates in the implementation of e-government? The assumption is that providing answers to these two questions will help set the stage for future research linking e-government and democratic theory. The authors also explore this theory in e-government amidst the rise of m-government, whereby users access e-government services via mobile devices. They suggest as well what governments can do to move forward with their e-government and m-government efforts based on these theories.

DOI: 10.4018/978-1-5225-5999-3.ch001

1. INTRODUCTION

Over the last 20 years e-government has been at the forefront of the management reform movement. Around the world, governments have launched websites and other Internet technologies to enhance government services and engage citizens in the decision-making process. Further, government continue to expand upon existing e-government portals to implement the newest technologies in order to provide a multitude of services to their constituents. E-government refers to

the use of technology to enhance the access to and delivery of government services to benefit citizens, business partners and employees. It has the power to create a new mode of public service where all public organizations deliver a modernized, integrated and seamless service for their citizens (Silcook, 2001).

According to Garson (2006), the adoption of e-government transforms the relationship between government and businesses and the resulting new improved transaction process leads to substantial savings for the government. Moreover, increased use of technology promises the reversal of the declining social capital in the United States.

Research on e-government has recognized the dual components of e-government and e-governance, where e-government refers to a one-way communication between government and its citizens, while e-governance pertains to a two-way communication. According to Garson (2006, p.19), the phenomenon of e-government represents the *provision of government services by electronic means, usually over the Internet, while e-governance points to a vision of changing the nature of the state.* E-governance moves beyond e-government; it represents

the co-evolution of the information and communication technologies with the political institutions, taking in particular into account how these political institutions and the state more precisely are evolving in the context of globalization and by doing so, crystallizing all other relevant function (Rossel & Finger, 2007).

E-governance itself is composed of digital government (delivery of public service) and digital democracy (citizen participation in governance). Citizens, advocacy groups, businesses, and other organizations can connect with government to influence decision making processes. There are greater levels of collaboration among stakeholders, then, who seek to influence the policy making process and their respective government (Eger, 1997). Democratic theories also advance the notion that governments which implement channels by which citizens can influence the government and participate in the policy making process benefit from these greater levels of engagement (Milakovich, 2010). Subsequently, with greater levels of civic participation decisions can be made which are more likely to be accepted by constituents as they represent their majority will (Heberlein, 1976).

The e-governance perspective also involves e-democracy, which will allow for greater government transparency and openness, which in turn leads to a better-informed citizenry. The phenomenon of e-democracy also denotes the potential for information and communication technology (ICTs) to improve the degree and quality of citizen participation in government decision-making. Although e-democracy has not been widely adopted by government in the United States and government worldwide, research shows that its use has the potential to create a more engaged citizenry. For instance, Garrett and Jensen (2010) found that local government officials who use the Internet to communicate with various stakeholders are more engaged with a more diverse array of stakeholders, which speak to the potential of e-democracy to improve civil discourse. What's more is that ICTs can enable direct democracy on a large scale, allowing for greater government transparency and openness, resulting in a better-informed citizenry.

In this chapter, we draw on Scott's (2006) work on e-government and democratic theory to examine how democratic theory relates to how governments engage their citizens on the Internet. We focus on the three most prominent theories in the democratic theory literature: representative, pluralist, and direct (Tercheck & Conte, 2001). Using data from the Rutgers-Newark E-Governance Survey, we examine two research questions: What factors drives governments to employ each theory? What theory predominates in implementation of e-government? Our assumption is that providing answers to these two questions will help set the stage for future research linking e-government and democratic theory.

2. BACKGROUND

Scholars have worked to provide empirical support about how e-government works. The vast majority of this research has focused on trust and confidence in government (Tolbert & Mossberger, 2006; Kim & Lee, 2012; Morgeson, Van Amburg & Mithas, 2011); adoption (Schwester, 2009; Schwester, 2010; Moon 2002; Manoharan, 2013); innovation (Tolbert, Mossberger, & McNeal, 2008), changes and challenges faced by government officials (Carrizales, 2008; Dawes, 2008), IT pedagogy (Ganpati & Reddick, 2016; Manoharan & McQuiston, 2016), and social equity (Helbig, Gil-Garcia, & Ferro, 2008; Manoharan & Carrizales, 2011). Much of this research relies on a similar model of e-government development. Coursey and Norris (2008) lay out five models of e-government (Layne & Lee, 2001; Baum & DMaio, 2000;

Hiller & Belanger, 2001; Westcott, 2001). In the interest of space we will layout the common elements of the models discussed in their work.

E-government development is a five-stage linear process beginning with an emerging web presence where the website provides rudimentary information such as contact information, frequently asked questions, and service offerings. During the second stage of enhanced web presence, the website is periodically updated with links to other agency websites. It also allows users to send messages to government officials. In the third stage of interactive web presence, the website enables users to have access to government documents by searching specialized databases. The fourth stage, transactional web presence, allows users to complete some transactions (paying parking tickets and Renewing drivers licenses) on-line. In this stage, users will also be able to customize the site to meet their individual needs. In the fifth stage of fully integrated web presence, governments provide access to all public services through a single web portal. A government with a fully integrated web presence allows users to complete all government transactions (monetary and non-monetary) online.

Although this model has been the basis of most e-government research, Coursey et al.,(2008, p. 533) are skeptical about its viability. They argue that,

the models are purely speculative. They were not models per se but guesses about what e-government might be and how it may develop.

While their research provides insights regarding e-government, it fails to address one critical question: How does e-government fit into democratic theories of governance? Their examination of the five models of e-governance delves into online service offerings, barriers to implementation, adoption, and impact on governmental operations, but there is little discussion of how this relates to democratic theories of governance. One goal of e-government is to maintain citizens' trust and confidence in government. This can only be realized if governments rely on democratic theories in the implementation of e-government. Coursey et al.'s (2008) analysis focuses primarily on one-way interactions and does not sufficiently address how citizens can be more involved in decision-making processes. One way to address this question is to investigate how units of government apply democratic theory to their websites.

A 10-year retrospective study of the stages model, however, sees them as more complicated and the presence of new technologies has continued to expand the role of e-governance beyond the website both toward more functionality and toward greater levels of citizen input and greater levels of collaboration (Lee, 2010). The stages now contain themes centralized around citizen and service perspective and operation/technology perspectives that gives way to 5 stages: 1) The presenting stage; 2) the assimilation stage; 3) the reforming stage; 4) the morphing stage; and

5) e-governance stage. The latter two stages are characterized by participation and involvement on the part of the citizens in the governance process.

Citizen participation and involvement has been shown to lead to positive overall outcomes for the people and institutions within society (Cialdini & Goldstein, 2004). Over time, the implementation of greater levels of citizen participation has also been shown to lead to greater levels of overall social well–being as the policies implemented are accepted at greater levels and are believed to strengthen and benefit the constituents who advocated for them (Mannarin, Fedi, & Trippetti., 2010). The main issue is the fact that citizen participation venues of the e-government portal are given low priority than portals that enhance internal agency efficiency despite the fact that successful e-service efforts depend on citizen engagement (Axelsson, Melin, & Lindgren, 2013). Many e-government projects have not delivered on their promise of truly engaging the public (Sæbø, Rose & Flak, 2008).

Further, new technologies provide means to greater integrate aspects of democratic theory in their processes and have changed the relationship between governments and citizens (Dutil, Howard, Langford, & Roy, 2007). Tseng, Yen, Hung, and Wang (2008) note that e-government extends to all ICT devices that utilize the internet to access governmental services. With the new age of information technology being defined by new ICT technologies such as web 2.0 technologies, cloud computing, and user generated content there is a greater push for citizen participation in e-governance processes and greater linkage between democratic theories and e-government. Further, cities are increasingly moving toward the classification as "smart cities" (Shapiro, 2006; Hollands, 2008) whereby the city becomes a platform for creativity in application development, user generated content, and empowerment of citizens in policymaking (Mainka, Hartman, Meschede, & Stock 2015).

Mobile Government (m-government), specifically, is a subset of e-government that encompasses the use of mobile devices such as mobile phones, smartphones, and other personal data assistants (PDA's) to interact with the government. Though early studies (Kushchu, & Kuscu, 2003) suggested that m-government was in it's earliest stages of development, there have been a number of more current arguments made that mobile government provides a more equitable, cost effective, efficient, and accessible (Manoharan, Bennett, & Carrizales, 2012). For users mobile phones have been shown to let them more easily organize to carry out political action and carry out civic engagement (Pierskalla & Hollenbach, 2013; Peng & Choi, 2013; Rotberg & Aker, 2013). Further, m-government greatly enhances the fifth stage of e-government by providing additional channels through which citizens can interact with the government. As Hoffman, Nocak, and Schlosser (2001) noted early on in e-government studies, socioeconomic factors (education level, income, race, gender) can lead to a loss in equity in access to government services for those "haves" and "have nots" of technology. The mobile phone has increasingly been seen as a way to bring economically and politically disadvantaged groups to the government arena and allow them better access to government services over traditional ICT technologies as they circumvent many obstacles presented by computer based public infrastructures to create additional channels to strengthen the voice of these groups (Chen 2015; Ndlovu & Mbenga, 2013; Yuan, Raubal, & Liu, 2012).

3. DEMOCRATIC THEORY

3.1 E-Government and Democratic Theory

Scott (2006) highlights the importance of democratic theory in the study of e-government and citizen engagement. He argues that much of the research on e-government has found most websites do little to facilitate citizen involvement. A recent example is Christensen's (2013) study of e-democracy in 188 Finnish municipalities. The author argues that government use of the Internet has the potential to enhance democratic participation. His findings show, however, that most of Finnish municipalities included in the study use the Internet to provide services and information and not to fully engage the public in the decision-making process. Another example is Chadwick's (2003) study of e-government where he argues that most government websites focus on managerialism at the expense of civic engagement. The solution he offers, which Scott echoes, is e-democracy in which government provides opportunities for authentic citizen participation. What is missing from Chadwick's (2003) prescriptions of e-democracy are theory-based indicators of citizen participation, which can help guide the study of public involvement on the Internet. The key to unpacking the concept of participation is to fully define what website components are aligned with each theory, which, in Scott's and our view, is lacking in prior e-government research.

Norris (2003) highlights the various definitions of e-democracy and provides a theoretical framework by broadly categorizing these features into pluralist democracy, representative democracy, and direct democracy. While the representative and pluralist democratic theories emphasize the importance of elections and interest group bargaining, direct democracy theory stresses direct participation of the public in decision-making (Scott, 2006).

The representative theory arises from the perspective that decision-making necessitates specific skills and expertise that citizens do not posses and so they depend on their elected representatives to facilitate this process. Citizens in turn hold these elected officials responsible for their decisions through the conduct of free and fair elections (Norris, 2003). However citizens need adequate information on the performance of their representatives, and the new technologies provide the much-needed information as well as channels of communication with elected officials. Accordingly, Norris has identified three website public involvement indicators in this category: 1) information about the elected officials; 2) communication with elected officials, and 3) direct access to official government records.

The pluralist theory places emphasis on civil society as the foundation of e-democracy and the recognition of the role and interaction of various interest groups in shaping this phenomenon. These can be measured by 1) the website's ability to provide services to specific groups such as residents, visitors, and businesses; 2) the availability of personalized content; 3) e-comment forms; and 4) facilitation of e-transactions.

Finally, direct democracy theory, also called deliberative democracy, calls for efforts beyond elections and e-transactions—a democracy that is directly reflective of citizen involvement. By using new, innovative technologies, direct democracy aims to reach out to all sections of society, thus overcoming space and time requirements and providing the option of anonymity. Accordingly, this perspective involves the previous measures in addition to 1) social networking features; 2) chat rooms or discussion forums; and 3) e-consultation.

Scott's study draws on Norris's (2003) research on e-government and democracy. Scott parses local government websites into the same three categories that relate to democratic theory: representative, pluralist, and direct. He departs from the traditional models of e-government by viewing these theories as dependent but not hieratical.

While the theories are dependent on one another, governments do not have to move though each in a staged process. For example, Mossey and Manoharan (2017) show in sea steads, floating communities with new and emerging government structures, how various theories can take hold more easily due to limited government rules and regulations, geographic concerns, and fiscal savings. Specifically, each iteration of the various democratic theories brings potential with sea steads each having certain aspects of their theory incorporated in their e-government goals.

Scott's findings provide evidence that a majority of websites were linked to representative theory with much smaller percentages of websites linked to pluralist and direct theories. Scott's findings not only helped the field better understand the relationship between e-government and democratic theory, but also increased our understanding of civic engagement. The major question left largely unanswered: What factors influence the mode of civic engagement governments uses? With this information we can propose how governments can increase the opportunities they provide citizens to participate in the decision-making process. In the next section we define each theory-based indicator within the context of m-government.

3.2 M-Government and Democratic Theory

M-government itself can also be referred to as ubiquitous government or u-government (Belanger Carter, & Schaupp, 2005), which is a name indicative of mobile government's spread globally and its widespread proliferation. Mobile phones allow for greater interaction between users with such interactions allowing them to more easily organize and act as political actors and civically engage (Pierskalla, J. H., & Hollenbach, F. M., 2013; Peng, Y., & Choi, S. Y., 2013; Rotberg, R. I., & Aker, J. C., 2013). As such citizens can save time, energy, and money by assessing networks through these mobile portals (Kumar & Sinha, 2007).

For this reason, the success of m-governance is noticeably similar to that of e-governance as it strives for greater levels of two-way citizen participation and greater levels of e-democracy. Amidst the democratic theories Scott (2006) outlines governments move toward greater levels of classification as "smart cities" (Shapiro, 2006; Hollands, 2008) whereby the citizen is empowered in the political realm and m-government becomes a necessary step in the e-governance chain as it seeks to facilitate co-production (Linders, 2012). Mossey et al. (2017) also note how m-government carries with it immense potential in increasing the capabilities of governments in advancing the three democratic theories. The inherent delivery models of e-governance are therefore transformed into models of m-governance:

Government-to-Citizens (G2C) → M Government-to-Citizens (MG2C)

Government-to-Government (G2G) → M Government-to-Government (MG2G)

Government-to-Business (G2B) → M Government-to-Business (MG2B)

Government-to-employees (G2E) → M Government-to-employees (MG2E)

Amidst the overarching goal to increase democratic efforts for many governments over the past 30 years, the internet and e-government has provided a means by which public participation can be facilitated at far greater rates (Lee, Tan, Trimi, 2006). Still, as time passes and mobile phone proliferation continues to outperform computer proliferation mobile users will look to operational m-government channels to interact with government (Lee, S. M., Tan, X., & Trimi, S., 2006). This vision of e-democracy achievement through mobile channels has largely been ignored by governments both in their initial e-government efforts and the subsequent m-government efforts (Mossberger, Wu, & Crawford, 2013; Thiel, Reisinger, Röderer, & Fröhlich, 2016).

Though there are many reasons citizens may not participate in e-democracy efforts, the availability of channels has been demonstrated to greatly limit participation. It has been shown that the mobile phones lead to increased levels of political participation by opening up a new avenue whereby citizens can interact with government (Bratton, 2013; Gil de Zúñiga, Copeland, Bimber, 2014; Lee, Kwak, Campbell, & Ling, 2014). Through such channels e-democracy efforts can prevail. Further,the self-efficacy of the user increases through the use of such a device and confidence in the ability to perform civic engagement functions associated with ICT technology (Cegarra-Navarro, Garcia-Perez, & Moreno-Cegarra, 2014; Kim & Chen, 2015; Kim, Kavanaugh, & Hult, 2011).

3.3 Representative Theory

Representative theory is based on J.S. Mill's seminal work, *On Representative Government,* which holds that:

representatives are chosen from among people who possess the benefits of superior intellect, education credentials, business or superior intellect, and who have shown long dedication and practical discipline to the specialized task of governing (Warner, 2001, p. 407).

Representative theorists argue that society is best served when highly capable individuals are elected by citizens to act on their behalf. This quote highlights the core of the American political system, wherein elected representatives serve the interests of their constituency. It also highlights the tension within public administration between competence and public participation (Warner, 2001). Bureaucrats, the unelected experts, have the requisite knowledge to be able to determine what is in the public's best interest. In the bureaucrat's view, citizen involvement in decision-making will open the process to those who may not have the skills or abilities to participate in a substantive way, rendering the decision-making process inefficient.

Representative theorists would also argue that, to improve the quality of government, elected officials and bureaucrats must provide citizens with the necessary information so they can make informed choices in administrative encounters and the electoral process. Government websites that employ representative theory have four features: information about government activities and services; opportunities for users to interact with these officials directly through e-mail or comments; access to public meeting minutes and agendas; and a calendar of public events. What is noticeably lacking from such websites is indicative of later stages of e-governance and citizen co-production (Linders, 2012). Citizens are not given channels whereby they can affect the policy making realm, or whereby they can provide input into the current state of affairs to direct changes for their governments. Public opinion forums, citizen satisfaction surveys, and online forums are likely lacking in such a system. Instead such decisions are left up to elected officials with the channels of communication directly between the elected officials and the citizens. This could manifest in interactions on the social media presence of the official or through integrated e-mail or contact forms on the website.

3.4 Pluralist Theory

The first sentence of Robert Dahl's (2005) book *Who Governs* poses a question that captures the spirit of pluralist theory:

In a political system where nearly every adult may vote but where knowledge, wealth, social position, access to officials, and other resources are unequally distributed, who actually governs? (Dahl, 2005, p.1).

Unlike representative theory, where elected officials and bureaucrats act on citizens' behalf, pluralist theory takes into account the power of groups to shape the democratic process. Pluralists believe that when interest groups work to bargain for specific policy objectives they provide a check on the power of elected officials. This often gives individuals associated with various interest groups greater power to influence the decisions of elected officials and bureaucrats. From an e-government perspective, pluralist theory increases the availability of information on the web and reduces its cost, which makes interest group mobilization possible. Interest groups are now able to get information from the web and disseminate it to constituents so they can act collectively. In addition to the representative components, websites that engage citizens using a pluralist approach will have the following features: services for multiple client groups, personalized content, external links to civic organizations, and on-line transactions (Scott, 2006).

Lacking again would be the direct communication between the government and the public in the form of public opinion forums, citizen satisfaction surveys, and online forums. However, such features may be present on the external websites of the various organizations in question. Social media components likely would play a powerful role on the part of these interest groups as well as they seek to influence the policy arena and engage constituents on their desires. Tools to engage the government via a pluralist theory approach however would likely be in the hands of the organization rather than the central government website. Here, however, different web 2.0 technologies and m-government platforms such as mobile applications and social media likely would take on a far greater role.

3.5 Direct Theory

Direct theory (Barber, 2004) is when citizens are directly involved in government decision-making. The most popular application of direct democracy is the referendum. Twenty-four states allow citizens to petition to have laws placed on the ballot for citizen approval (Matsusaka, 2005). Most recently Colorado and Washington, which have the most liberal rules regarding referenda, passed a ballot measure to legalize the sale of marijuana for recreational use. Direct democratic theory goes beyond representative and pluralist theory in that it presumes that individual citizens have the skills and competencies to make policy decisions. Scott (2006, p.344) argues that governments that use direct democratic approaches,

aim to deepen democracy by reengaging citizens who are disenchanted with and disengaged from the (electoral) political process.

In addition to the representative and pluralist attributes, websites that employ direct theory will have features such as voluntary opportunities, access to social networking sites, chat rooms, and web consultations (Scott, 2006). Further, such websites would make great use of citizen satisfaction surveys, online forums, and public opinion forums whereby the direction of the government could be influenced by such input. User generated content and participation in the workings of government via a co-production approach (Linders, 2012) would also take route. Cloud computing, user generated content, and integrated services via a Smart City model would continue to move toward greater levels of citizen and government interactions. The tenets of e-democracy by proxy would fully take hold with citizens able to contribute more readily through a variety of ICTs.

Building off Scott (2006) and Norris (2003) work this chapter investigates the factors that influence how local governments use their websites to engage citizens and how these uses are linked to democratic theory. In Scott's study of website features and theories of democracy, he provided evidence that population influences localities to have various features that link to the three democratic theories. His study found that larger cities' websites were more likely to contain features related to each of the three theories. It was his contention that larger cities have more citizens to serve and thus they are more likely to provide multiple options for citizens to engage government. This finding is in line with other studies (Moon & Norris, 2005) of e-government adoption, which provides evidence that larger cities are more likely to adopt e-government as they are under constant financial pressure to find innovative and efficient ways to provide services and information to the public. Beyond population little is said about other factors that influence how local governments use their website to engage citizens. To gain a broader understanding of this issue our study expands Scott's analysis and investigates two other factors that have been found to dictate how governments utilizes e-government to engage citizens: form of government and IT capacity.

In addition, form of government has been found by scholars to play a critical role in government's decision to adopt e-government and offer opportunities for substantive citizen participation. For instance, studies of citizen participation (Ebdon, 2000; Zhang & Yang, 2009) in the budgetary process have consistently found that cities with a council-manager form of government were more likely to provide opportunities for citizens to participate in every phase of the budgetary process. Additionally, studies of e-government adoption and innovation found that cities with a manager-council form of government were more likely to adopt e-government and employ innovative technologies to provide information and deliver services (Moon, 2002; Carrizales, 2008). Taken together, these findings provide ample evidence that cities with a council-manager form of government are more professionalized and have a customer service and efficiency orientation which lends itself more to the utilization of technologies that increase efficiency and service quality.

While these findings are interesting and provide scholars a great deal of insight into how the form of government impacts e-government and citizen participation, little is said about its link to democratic theory. In a city with a council-manager form of government, the city manager (the administrative head) is appointed by the city council. The city manager is, in theory, insulated from the pressures of electoral politics and can focus their energies solely on efficiency and service delivery. One could argue that, in this form of government, democratic values would play a secondary role to the efficient provision of services. For instance, responsiveness - one of the key democratic values - is obscured in council-manager form of government. In a mayor-council form of government the mayor, who is the chief executive of the city, faces electoral pressure and must respond effectively to the needs of citizens and other stakeholders. Hence, it could be argued that in this system of government, the mayor would use the website as a tool to respond to the needs of the residents. The manner in which the mayor uses the city website as a tool would more likely be in line with the theories of democracy discussed above. With this said, it is our feeling that mayor-council form of government will more likely contain features that are in line with the three theories of democracy.

In addition to the form of government, IT capacity has been found to play a critical role in use of e-government. Capacity, in this context, refers to the financial and technical resources necessary to develop and implement e-government. Research (Melitski, 2003) has shown a link between capacity and success of implementing e-government in local governments. One possible solution that is gaining credibility is utilizing mobile technology resources, especially in areas with less resources, as they are more affordable and less reliant on expensive wireless networks. In this sense citizens are able to access government services, participate in decision making, and access transparent government information more easily through mobile devices (Manoharan, A. P., Bennett, L. V., & Carrizales, T. J.,2012). The use of mobile phones enable governments to explore less costly alternatives to traditional e-government channels, but they still often times rely on e-government networks to be in place before implementation. Therefore, despite their cost saving potential more advanced m-government efforts also rely on greater capacity on the part of municipal governments.

As with studies on the form of government and e-government, little has been said about how capacity affects the features that are provided on websites. What is more, this line of research has been nearly silent on how IT capacity relates to theories of democracy in the context of e-government. To bridge this gap, our study will link IT capacity to democratic theory.

4. DATA AND METHODOLOGY

The research examined the official websites of the 200 largest cities, by population, in the United States. The cities were identified from 2010 U.S. Census data and among the 200 cities selected for the survey, 195 had official websites. The research involved a content analysis of the official city websites using a 28-point index classified into the three democratic theory categories, index, as shown in Appendix A. The representative theory category include features related to the information on government agencies/departments and public officials, minutes of public meetings, mission statements of agencies, calendar of events, FAQs, ability to request information and provide comments or feedback to elected officials. The features included in the pluralist theory category include features related to targeted audience links, downloadable documents, online payment of utilities, fines/tickets, file taxes, online application for permits, licenses, provision of social media tools, e-procurement, portal customization, and ability of users to provide comments or feedback to individual departments/agencies through online forms. Finally, the direct theory category consisted of features pertaining to the use of wireless technology; provision of bulletin boards for posting questions to public officials; online discussion forums on policy issues; e-meetings; and tools for online decision-making such as e-petition, electronic citizen juries, or e-referenda. The questions were evaluated using a dichotomous measure (0 or 1), based on the absence or presence of each feature.

In addition to analyzing the presence of various website features, we also divided the cities into a number of categories to facilitate the analysis: population, form of government, and IT capacity. As we mentioned earlier these categories will give us a rudimentary understanding of what drives cities to provide various features on their websites. The cities' population is based on 2010 census data. We refined this category by dividing it into three subcategories: small (less than 160,000 residents), medium (160,000-499,999 residents), and large (greater than 500,000 residents).

The cities are divided into two forms of government: mayor-council and council-manager. Mayor-council forms of government are cities in which the mayor is the chief executive with budgetary and managerial authority and the council has chief legislative authority. In council-manager cities, an appointed manager is the chief executive who manages the day-to-day operation of the agencies. In this form of government, the manager serves at the pleasure of the city council, which is the chief budgetary and legislative authority. To determine the form of government, we conducted an exhaustive search of the city's website and contacted the city (if necessary) to ascertain the form of government.

Our study measures IT capacity by whether or not the city has an independent IT department. This is an imperfect measure of IT capacity, but it provides insight into whether the city has in-house expertise in developing, implementing, and maintaining various website features. We determined whether the cities in the study have an IT department through an analysis of the cities' organizational chart. If that information was not readily available we contacted the city to determine if they had an independent IT department.

5. FINDINGS

5.1 Representative Theory

The findings for the representative theory items appear in Tables 2-4. Representative theorists would argue that government's primary role is to provide information to citizens so that they are better informed and can participate in political and administrative processes. One noteworthy finding is that a majority of cities included in this study, on average, contained most of the online features that are linked to representative theory. This indicates that websites are offering features that contain basic information and services to citizens. Another noteworthy finding is that small cities tended to have more features that are in line with representative theory. Small cities have an average of 7.19 representative theory features. Medium-size cities have an average of 6.76 representative theory features, while large cities have an average of 5.90 features. Digging into the individual features, the findings show that a greater percentage of small cities are more likely or as likely as larger cities to have a mission statement posted on their website, a calendar of events, and allow users to request information online, and send comments to public officials.

Another noteworthy finding is that cities with a manager-council form of government have, on average, more features that relate to representative theory. Cities with a manager-council form of government have an average of 6.93 representative theory features, while cities with a mayor-council form of government have an average of 6.11 representative theory features. This finding provides evidence that the non-political nature of the council-manager form of government is one possible reason why these cities provide more basic information to their citizens than mayor-council cities.

The findings also show that IT capacity plays a role in the presence of representative theory features on municipal websites. As was mentioned earlier, we use the presence of an independent IT department as a measure of IT capacity. Our findings show that municipalities that do not have an IT department (6.80) have, on average, more representative theory features than municipalities that do have an IT department (6.53). This finding appears to be counterintuitive because it was our expectation that municipalities with an IT department would contain more features due to a greater capacity to develop and implement them.

5.2 Pluralist Theory

The findings for the pluralist theory items appear in Tables 5-7. As mentioned earlier, pluralist theory manifests itself in website features that promote or cater to specific constituencies. One noteworthy finding is that cities, regardless of type, contain, on average, less than half of the features linked to pluralist theory. This finding provides evidence that cities are not utilizing all of the tools necessary to engage various constituencies. When we look at cities based on population, form of government, and IT capacity, important differences emerge.

The findings show that larger cities, on average, contain more website features that are associated with pluralist theory. Large cities have the highest number of features associated with pluralist theory (6.69). Medium-size cities have fewer features (5.70), and small cities have the least number of features (4.44). What is most interesting about this finding is that larger cities, and to a lesser degree, medium-size cities are more likely than small cities to contain features that allows residents to access city services online (most notably paying utilities and taxes, applying for permits and licenses, and e-procurement). It is our contention that this link between city size and pluralist features may be because larger cites tend to have more diverse constituencies, which necessitates features that appeal to each group. This may have implications for how larger cities can improve their websites to engage a wide array of groups.

When we look at the role of the form of government on pluralist theory features, the findings show that there is little difference between the presence of features and the form of government. The average number of features is nearly identical for cities with a mayor-council form of government (5.49) and manager-council form of government (5.36). The percentage of cities is also nearly identical for each website feature, with one exception: manager-council cities are more likely to provide feedback to city agencies. It was our expectation that cities with a mayor-council form of government would contain more features associated with pluralist theory than cities with a manager-council form of government. One possible reason for this finding is in mayor-council cities, the mayor is a political actor whose performance is linked to his or her electoral prospects. Given this connection, mayors will be more likely to manage constituencies through bureaucratic action. In this case, mayor-council cities would try to manage constituencies by providing features that cater to each group and its needs.

When we turn to the role of IT capacity, our findings show that cities that have an IT department have, on average, slightly more features (5.63) than cities that do not have an IT department (5.01). With the exception of e-procurement, cities that have an IT department are more likely to have all of the features that relate to pluralist theory. This finding was in line with our expectations that cities with an IT department will have greater capacity to develop and maintain the website features in this category.

5.3 Direct Theory

The findings for the direct theory features are presented in Tables 8-10. A core tenet of direct theory is that it allows citizens to be directly involved in the decision-making process. Notably, few of the cities included in this study contain any features that link to direct theory. The most popular feature, regardless of population, form of government, or IT capacity, is the ability for residents to receive service information or updates on their mobile phones. Although this was the most popular feature, only 15-35% of cities included in this study contain this feature. Both of these findings signify that cities are not using their websites to provide options for citizens participate directly in the political or administrative process.

Looking at these cities based on population, form of government, or IT capacity further underscores this finding. Tables 8-10 show that larger cities have, on average, more features associated with direct theory (.394) than small (.304) or medium-sized cities (.346). Although there are slight differences based on population, the cities in each category have, on average, less than one of the six features that link to direct theory. When it comes to form of government manager-council cities have, on average, slightly fewer features (.314), than mayor-council cities (.360). As with the finding for city population, both categories have, on average, less than one feature that links to direct theory. The findings for IT capacity are quite similar in that cities that did not have an IT department have slightly more features (.400) than cities with an IT department (.304).

Taken together, these findings presents interesting implications for web-based civic engagement. One implication is that cities need to do more to provide features that allow residents to have a direct role in political and administrative processes. The solution may not necessarily reside in including the features that were included in this study, but it may require cities to think about providing features that allow residents more opportunities for direct participation.

6. DISCUSSION OF FINDINGS

The findings of this study present important implications for online citizen participation and the factors affecting their success. While elected officials and bureaucrats may not be directly guided by democratic theory in their day-to-day work, our findings show that city size and structure may play a role in how democratic theory is represented on a city's website. Therefore, larger cities carry with them greater implications for citizen participation development. The results for representative theory present evidence that shows that size (smaller cities), form of government (manager-council) and the presence of IT departments may be the most influential factors. In the case of pluralist theory our findings present evidence that lends support to the idea that size (larger cities), form of government (mayor) and the presence of an IT department are the most influential factors. Finally, population and form of government (manager-council) are related to presence of website features that are in line with the direct theory of democracy.

The variable relating to form of government - council-manager or mayor-council – both seems to be related to online citizen participation, although most research has primarily found council-manager as being more supportive of e-government initiatives. Previous research on e-government has shown that cities with council-manager forms of government have greater rates of e-government adoption (Moon, 2002; Carrizales, 2008), while those cities with mayor-council form of government were found to be less supportive of e-government innovation (Carrizales, 2008; Damanpour & Schnedier, 2009; Krebs & Pelissero, 2010). However, our research expands on this idea by providing evidence to show how different forms of government align with democratic theory. Our findings show that council-manager form of government best align with representative and direct theories of democracy and the mayor-council form aligns with the pluralist theory of democracy. This finding fits well with the day-to-day reality of being a mayor. Mayors are often judged on how effective they are at dealing with multitude of interest groups that exist in large or medium sized cities. This reiterates their potential role as 'institutional catalysts' capable of ensuring that government organizations implement more innovative technologies that directly connect with the public (Ho & Ni, 2004).

Another interesting point of difference between the three theories of democracy is the presence of an IT department. Cities with a dedicated IT department tend to have a greater IT capacity, which has been found to have a positive association with adoption of e-government technologies, which echoes previous research findings in e-government research (Norris & Kraemer, 1996; Teo & Tan, 1998). Our research finds that cities with an IT department tend to be more oriented toward a pluralist view of democracy, while those without IT departments are oriented toward a representative view. Perhaps, the cities without IT departments prefer to depend on IT contracting services, and further research is needed to examine such trends. Another factor is the number of IT employees, which is positively related with a greater e-government score (Schwester, 2009). Finally, population is an important factor in the provision of direct democracy features as those cities serving larger populations tend to have more slack resources, leading to better websites (Chen, 2010); however, greater population does not seem to motivate cities to provide more features related to representative or pluralist forms of democracy.

7. SOLUTIONS AND RECOMMENDATIONS

Key to fostering democracy within municipalities is increasing interactions on government websites which are informative, user friendly, and contain vital points of interest for the citizenry. In addition to a government website being a source of information about community events, leaders must recognize the value of the website in empowering citizen involvement with the governing body. Civic engagement or involvement will also allow for government transparency as well as direct involvement of the people in government and community affairs. E-government has the ability to involve more people in every step of the local policy-making process, whether it is road repairs or the development of a new park. At higher levels of government, ICT technologies allow citizens to keep tabs on elected officials.

Two-way communication and active participation are just as important at the state and federal level, but it is possible that only citizens who are directly affected by the policies in question will choose to participate at this level. This pluralist form of e-democracy at the state and federal level would follow the current pattern of policy making in the "real world".

In addition, cities with larger populations and greater resources should make use of new technologies that encourage citizen co-production efforts (Linders, 2012). Mobile applications that allow citizens to communicate their interests to the city and participate in policy making are first steps compiled with service delivery applications that let citizens contribute to the development of their municipality. These m-government initiatives can jump start lagging e-government initiatives, and enhance current e-government initiatives to better involve citizens and bring about greater levels of overall e-governance. Governments need to recognize the immense power that web 2.0 technologies have in bringing about greater levels of civic engagement and making such engagement beneficial and easy to use for constituents. Amidst social media, government applications that disseminate public opinion forums, service delivery, and information dissemination are all important in keeping up with the latest trends in technology and adhering to citizens' preferred mediums of access. Further, it is time that small-city governments are trained in social media best practices. Governments of all sizes need to recognize and capitalize on social media opportunities available for civic engagement.

8. FUTURE DIRECTIONS

Future research should take measures to examine how the various democratic theories manifest in regard to the most recent technologies across the globe. As discussed in this chapter, m-government is presenting a new way by which governments can engage citizens and foster citizen participation. M-government efforts can be particularly cost effective and powerful in bringing about e-governance efforts among developed and developing nations.

Another pertinent question moving forward is to expand on this study in this realm. Longitudinal studies that show the progression of citizen centric efforts would also be beneficial in realizing how and if governments are improving on these efforts. As more and more "smart cities" continue to arise around the globe, it also becomes important to examine how their efforts foster citizen participation and co-production (Linders, 2012). Examining whether or not two way communication is occurring between government and citizens on official websites, social media, and mobile government platforms is also important in gauging the level of democratic development in e-government efforts.

9. CONCLUSION

In today's society, the computer and the Internet play a key role in the interaction between governments at all levels and the constituents they serve. As a result of ever-evolving technology, there are more ways that people can interact with government. Direct democracy is more prominent at the local level, while higher levels of government tend to focus more on the representative aspects of e-government. Apart from the cost savings, online engagement may be able to involve more members of the jurisdiction than face-to-face engagement, which sometimes tend to be dominated by interest groups. However, participation comes down to the individual interest in online engagement. Nowadays citizens are increasingly participating through non-government electronic communities fueled by social media and mobile technologies. Governments need to recognize these trends and provide links to such forums on their official websites, to encourage more public participation (Calista, Melitski, Holzer, & Manoharan, 2010). As technologies change and improve, so do the demands of the people. Rather than having to choose among the three theories, municipal websites can serve all three viewpoints, with an increasing integration of social media and mobile technologies with their websites.

REFERENCES

Axelsson, K., Melin, U., & Lindgren, I. (2013). Public e-services for agency efficiency and citizen benefit—Findings from a stakeholder centered analysis. *Government Information Quarterly*, *30*(1), 10–22. doi:10.1016/j.giq.2012.08.002

Barber, B. R. (2004). *Strong democracy: Participatory politics for a new age*. University of California Press.

Baum, C., & DiMaio, A. D. (2000). *Gartner's four phases of e-government model*. Retrieved from http://www.gartner.com

Belanger, F., Carter, L. D., & Schaupp, L. C. (2005). U-government: a framework for the evolution of e-government. *Electronic Government, an International Journal, 2*(4), 426-445.

Bratton, M. (Ed.). (2013). *Voting and democratic citizenship in Africa*. Boulder, CO: Lynne Rienner Publishers.

Calista, D. J., Melitski, J., Holzer, M., & Manoharan, A. (2010). Digitized government in worldwide municipalities between 2003 and 2007. *International Journal of Public Administration, 33*(12/13), 588–600. doi:10.1080/01900692.2010.513881

Carrizales, T. (2008). Functions of e-government: A study of municipal practices. *State & Local Government Review*, *40*(1), 12–26. doi:10.1177/0160323X0804000102

Cegarra-Navarro, J. G., Garcia-Perez, A., & Moreno-Cegarra, J. L. (2014). Technology knowledge and governance: Empowering citizen engagement and participation. *Government Information Quarterly*, *31*(4), 660–668. doi:10.1016/j.giq.2014.07.001

Chadwick, A. (2003). Bringing E-Democracy Back In Why it Matters for Future Research on E-Governance. *Social Science Computer Review*, *21*(4), 443–455. doi:10.1177/0894439303256372

Chen, X. (2015). Decentralized computation offloading game for mobile cloud computing. *IEEE Transactions on Parallel and Distributed Systems*, *26*(4), 974–983. doi:10.1109/TPDS.2014.2316834

Chen, Y. (2010). Citizen-centric e-government services: Understanding integrated citizen service information systems. *Social Science Computer Review*, *28*(4), 427–442. doi:10.1177/0894439309359050

Christensen, H. S. (2013). Broadening Democratic Participation? An exploratory study of e-democracy in 188 Finnish municipalities. *Scandinavian Journal of Public Administration*, *17*(3).

Cialdini, R. B., & Goldstein, N. J. (2004). Social influence: Compliance and conformity. *Annual Review of Psychology*, *55*(1), 591–621. doi:10.1146/annurev.psych.55.090902.142015 PMID:14744228

Dahl, R. A. (2005). *Who governs? Democracy and power in an American city*. Yale University Press.

Damanpour, F., & Schneider, M. (2009). Characteristics of innovation and innovation adoption in public organizations: Assessing the role of managers. *Journal of Public Administration: Research and Theory*, *19*(3), 495–522. doi:10.1093/jopart/mun021

Dawes, S. S. (2008, December). The evolution and continuing challenges of e-governance. *Public Administration Review,* 86-102.

Dutil, P., Howard, C., Langford, J., & Roy, J. (2007). 'Re-imagining the service recipient: Customers, clients and citizens in the emerging world of e-government. *Journal of Information Technology & Politics*, *4*(1), 77–90. doi:10.1300/J516v04n01_06

Ebdon, C. (2000). The relationship between citizen involvement in the budget process and city structure and culture. *Public Productivity & Management Review*, 383-393.

Eger, J. (1997). Cyberspace and cyberplace: Building the smart communities of tomorrow. *San Diego Union-Tribune, 26*. empirical assessment. *Public Administration Review*, *68*(3), 523–536.

Ganpati, S., & Reddick, C. (2016). An ostrich burying its head in the sand? The 2009 NASPAA standards and scope of information technology and e-government curricula. *Journal of Public Affairs Education*, *22*(2), 267–286.

Garrett, R. K., & Jensen, M. J. (2011). E-Democracy Writ Small: The impact of the Internet on citizen access to local elected officials. *Information Communication and Society*, *14*(2), 177–197. doi:10.1080/1369118X.2010.490558

Garson, G. D. (2006). *Public information technology and e-governance: Managing the virtual state*. Sudbury, MA: Jones and Bartlett Publishers.

Gil de Zúñiga, H. G., Copeland, L., & Bimber, B. (2014). Political consumerism: Civic engagement and the social media connection. *New Media & Society*, *16*(3), 488–506. doi:10.1177/1461444813487960

Heberlein, T. A. (1976). Some observations on alternative mechanisms for public involvement: The hearing, public opinion poll, the workshop and the quasi-experiment. *Natural Resources Journal*, *16*(1), 197–212.

Helbig, N., Gil-García, J. R., & Ferro, E. (2009). Understanding the complexity of electronic government: Implications from the digital divide literature. *Government Information Quarterly*, *26*(1), 89–97. doi:10.1016/j.giq.2008.05.004

Hiller, J., & Bélanger, H. (2001). *Privacy Strategies for Electronic Government. E-Government Series*. Arlington, VA: PricewaterhouseCoopers Endowment for the Business of Government.

Ho, A. T., & Ni, A. (2004). Explaining the progressiveness of e-government: A case study of Iowa county treasurers' offices. *American Review of Public Administration*, *34*(2), 164–180. doi:10.1177/0275074004264355

Hoffman, D. L., Novak, T. P., & Schlosser, A. E. (2001). The evolution of the digital divide: Examining the relationship of race to Internet access and usage over time. *The digital divide: Facing a crisis or creating a myth*, 47-97.

Hollands, R. G. (2008). Will the real smart city please stand up? Intelligent, progressive or entrepreneurial? *City*, *12*(3), 303–320. doi:10.1080/13604810802479126

Kim, B. J., Kavanaugh, A. L., & Hult, K. M. (2011). Civic engagement and Internet use in local governance: Hierarchical linear models for understanding the role of local community groups. *Administration & Society*, *43*(7), 807–835. doi:10.1177/0095399711413873

Kim, S., & Lee, J. (2012). E-Participation, Transparency, and Trust in Local Government. *Public Administration Review*, *72*(6), 819–828. doi:10.1111/j.1540-6210.2012.02593.x

Kim, Y., & Chen, H. T. (2015). Discussion network heterogeneity matters: Examining a moderated mediation model of social media use and civic engagement. *International Journal of Communication*, *9*, 22.

Krebs, T. B., & Pelissero, J. P. (2010). Urban managers and public policy: Do institutional arrangements influence decisions to initiate policy? *Urban Affairs Review*, *45*(3), 393–411. doi:10.1177/1078087409342693

Kumar, M., & Sinha, O. P. (2007). M-government–mobile technology for e-government. In *International conference on e-government, India* (pp. 294-301). Academic Press.

Kushchu, I., & Kuscu, H. (2003). From E-government to M-government: Facing the Inevitable. In *the 3rd European Conference on e-Government* (pp. 253-260). MCIL Trinity College Dublin Ireland.

Layne, K., & Lee, J. (2001). Developing fully functional e-government–A four stage model. *Government Information Quarterly*, *18*(2), 118–136. doi:10.1016/S0740-624X(01)00066-1

Lee, H., Kwak, N., Campbell, S. W., & Ling, R. (2014). Mobile communication and political participation in South Korea: Examining the intersections between informational and relational uses. *Computers in Human Behavior*, *38*, 85–92. doi:10.1016/j.chb.2014.05.017

Lee, J. (2010). 10year retrospect on stage models of e-Government: A qualitative meta-synthesis. *Government Information Quarterly*, *27*(3), 220–230. doi:10.1016/j.giq.2009.12.009

Lee, S. M., Tan, X., & Trimi, S. (2006). M-government, from rhetoric to reality: learning from leading countries. *Electronic Government, an International Journal*, *3*(2), 113-126.

Linders, D. (2012). From e-government to we-government: Defining a typology for citizen coproduction in the age of social media. *Government Information Quarterly*, *29*(4), 446–454. doi:10.1016/j.giq.2012.06.003

Mainka, A., Hartmann, S., Meschede, C., & Stock, W. G. (2015). Mobile application services based upon open urban government data. *iConference 2015 Proceedings*.

Mannarini, T., Fedi, A., & Trippetti, S. (2010). Public involvement: How to encourage citizen participation. *Journal of Community & Applied Social Psychology*, *20*(4), 262–274. doi:10.1002/casp.1030

Manoharan, A. (2013). A three dimensional assessment of US county e-government. *State & Local Government Review*, *45*(3), 153–162. doi:10.1177/0160323X13494858

Manoharan, A., & Carrizales, T. (2011). Recent trends in e-government: States' and local governments' utilization of Websites. *International Journal of Electronic Governance*, *4*(4), 283–303. doi:10.1504/IJEG.2011.046014

Manoharan, A. P., Bennett, L. V., & Carrizales, T. J. (2012). M-Government: An opportunity for addressing the digital divide. *Citizen 2.0: Public and Governmental Interaction through Web 2.0 Technologies: Public and Governmental Interaction Through Web 2.0 Technologies*, 87.

Manoharan, A. P., & McQuiston, J. (2016). Technology and pedagogy: Information technology competencies in public administration and public policy programs. *Journal of Public Affairs Education*, *22*(2), 175–186.

Matsusaka, J. G. (2005). Direct democracy works. *The Journal of Economic Perspectives*, *19*(2), 185–206. doi:10.1257/0895330054048713

Melitski, J. (2003). Capacity and e-government performance: An analysis based on early adopters of Internet technologies in New Jersey. *Public Performance & Management Review*, *26*(4), 376–390. doi:10.1177/1530957603026004005

Milakovich, M. E. (2010). The Internet and increased citizen participation in government. *JeDEM-eJournal of eDemocracy and Open Government*, *2*(1), 1-9.

Moon, J. M. (2002). The evolution of e-government among municipalities: Rhetoric or reality? *Public Administration Review*, *62*(4), 434–433. doi:10.1111/0033-3352.00196

Morgeson, F. V., VanAmburg, D., & Mithas, S. (2011). Misplaced trust? Exploring the structure of the e-government-citizen trust relationship. *Journal of Public Administration: Research and Theory*, *21*(2), 257–283. doi:10.1093/jopart/muq006

Mossberger, K., Wu, Y., & Crawford, J. (2013). Connecting citizens and local governments? Social media and interactivity in major US cities. *Government Information Quarterly*, *30*(4), 351–358. doi:10.1016/j.giq.2013.05.016

Mossey, S., & Manoharan, A. P. (2017). Seasteads and E-Democracy: The Potential of Mobile the Potential of Mobile Technology. *Seasteads: Opportunities and Challenges for Small New Societies*, 81.

Ndlovu, M., & Mbenga, C. (2013). Facebook, the public sphere and political youth leagues in South Africa. *Journal of African Media Studies*, *5*(2), 169–186. doi:10.1386/jams.5.2.169_1

Norris, D. F., & Kraemer, K. L. (1996). Mainframe and PC computing in American local governments: Myths and realities. *Public Administration Review*, *56*(6), 568–576. doi:10.2307/977255

Norris, D. F., & Moon, M. J. (2005). Advancing E-Government at the Grassroots: Tortoise or Hare? *Public Administration Review*, *65*(1), 64–75. doi:10.1111/j.1540-6210.2005.00431.x

Norris, P. (2003). *Deepening Democracy via E-Governance (Draft chapter for the UN World Public Sector Report)*. Retrieved from http://ksghome.harvard. edu/~. pnorris.shorenstein.ksg/Acrobat/World%20 Public%20Sector%20Report.pdf

Peng, Y., & Choi, S. Y. (2013). Mobile phone use among migrant factory workers in south China: Technologies of power and resistance. *The China Quarterly*, *215*, 553–571. doi:10.1017/S0305741013000738

Pierskalla, J. H., & Hollenbach, F. M. (2013). Technology and collective action: The effect of cell phone coverage on political violence in Africa. *The American Political Science Review*, *107*(2), 207–224. doi:10.1017/S0003055413000075

Rossel, P., & Finger, M. (2007). Conceptualizing E-Governance. In T. Janowaki, & T.A. Pardo (Eds.), *Proceedings of the 1st International Conference on Theory and Practice of Electronic Governance* (pp. 399-407). New York: The Association for Computing Machinery.

Rotberg, R. I., & Aker, J. C. (2013). Mobile phones: Uplifting weak and failed states. *The Washington Quarterly*, *36*(1), 111–125. doi:10.1080/0163660X.2013.751651

Sæbø, Ø., Rose, J., & Flak, L. S. (2008). The shape of eParticipation: Characterizing an emerging research area. *Government Information Quarterly*, *25*(3), 400–428. doi:10.1016/j.giq.2007.04.007

Savas, E. S. (2000). *Privatization and public-private partnerships*. Ithaca, NY: Cornell University Press.

Schwester, R. (2009). Examining the barriers to e-government adoption. *Electronic. Journal of E-Government*, *7*(1), 113–122.

Schwester, R. W. (2010). Socio-Demographic Determinants of E-government Adoption: An Examination of Major US Cities. *Journal of Public Management & Social Policy*, *16*(2).

Scott, J. K. (2006). "E" the people: Do US municipal government Web sites support public involvement? *Public Administration Review*, *66*(3), 341–353. doi:10.1111/j.1540-6210.2006.00593.x

Shapiro, J. M. (2006). Smart cities: Quality of life, productivity, and the growth effects of human capital. *The Review of Economics and Statistics*, *88*(2), 324–335. doi:10.1162/rest.88.2.324

Silcook, R. (2001). What is e-Government? *Parliamentary Affairs*, *54*(1), 88–101. doi:10.1093/pa/54.1.88

Teo, T., & Tan, M. (1998). An empirical study of adopters and non-adopters of the Internet in Singapore. *Information & Management*, *34*(6), 339–345. doi:10.1016/S0378-7206(98)00068-8

Terchek, R. J., & Conte, T. C. (2001). *Theories of Democracy; A Reader*. Lanham, MD: Rowman and Littlefield.

Thiel, S. K., Reisinger, M., Röderer, K., & Fröhlich, P. (2016). Playing (with) Democracy: A Review of Gamified Participation Approaches. *JeDEM-eJournal of eDemocracy and Open Government,* *8*(3), 32-60.

Tolbert, C. J., & Mossberger, K. (2006). The effects of e-government on trust and confidence in government. *Public Administration Review*, *66*(3), 354–369. doi:10.1111/j.1540-6210.2006.00594.x

Tolbert, C. J., Mossberger, K., & McNeal, R. (2008). Institutions, policy innovation, and e-government in the American states. *Public Administration Review*, *68*(3), 549–563. doi:10.1111/j.1540-6210.2008.00890.x

Tseng, P. T., Yen, D. C., Hung, Y. C., & Wang, N. C. (2008). To explore managerial issues and their implications on e-Government deployment in the public sector: Lessons from Taiwan's Bureau of Foreign Trade. *Government Information Quarterly*, *25*(4), 734–756. doi:10.1016/j.giq.2007.06.003

Warner, B. E. (2001). John Stuart Mill's theory of bureaucracy within representative government: Balancing competence and participation. *Public Administration Review*, *61*(4), 403–413. doi:10.1111/0033-3352.00044

Wescott, C. (2001). E-Government in the Asia-Pacific region. *Asian Journal of Political Science*, *9*(2), 1–24. doi:10.1080/02185370108434189

Yuan, Y., Raubal, M., & Liu, Y. (2012). Correlating mobile phone usage and travel behavior–A case study of Harbin, China. *Computers, Environment and Urban Systems*, *36*(2), 118–130. doi:10.1016/j.compenvurbsys.2011.07.003

Zhang, Y., & Yang, K. (2009). Citizen participation in the budget process: The effect of city managers. *Journal of Public Budgeting, Accounting &. Financial Management*, *21*(2), 289.

APPENDIX

Table 1. Website evaluation index

Website Evaluation Index	Y = 1/ N = 0
Representative Theory	
1. Does the website provide information about the location of offices?	
2. Does the website offer contact information for agencies/departments or employees/public officials?	
3. Does the website provide minutes of public meetings?	
4. Does the website contain mission statements for agencies?	
5. Does the website provide budget information online?	
6. Does the website offer a calendar of events?	
7. Does the website allow users to file complaints?	
8. Does the website provide a FAQ (Frequently Asked Questions) corner to guide citizens in city administration?	
9. Does the website allow users to request information?	
10. Does the website allow users to provide comments or feedback to elected officials (e.g. Mayor or city council)	
Pluralist Theory	
1. Targeted audience links: Are targeted audience links available on the homepage? (e.g. general citizens, youths, the old, women, family, citizens in need of social welfare services, businesses, industry, small businesses, public employees, etc.)	
2. Number of downloadable documents for three departments or bureaus in the city government (max 3 departments; examples of departments: budget, social welfare, and transportation; if one of these is not the function of the city, select another department which provides downloadable documents)	
3. Does the website allow users to pay utilities? (State utilities in description: e.g. tap water, sewage, gas, electricity)	
4. Does the website allow users to file or pay taxes?	
5. Does the website allow users to pay fines or tickets?	
6. Does the website allow users to apply for permits (or register)?	
7. Does the website provide online tracking system for the status of permits?	
8. Does the website allow users to apply for licenses?	
9. Does the city provide opportunities for participation through social media?	
10. Does the website allow e-procurement?	
11. Does the website allow users to customize the main city homepage depending on their needs (e.g. portal customization)?	

continued on following page

Table 1. Continued

Website Evaluation Index	Y = 1/ N = 0
12. Does the website allow users to provide comments or feedback to individual departments/ agencies through online forms?	
Direct Theory	
1. Does the website use wireless technology such as messages to a mobile phone, a Palm Pilot, or a PDA to update about applications, events, etc.?	
2. Does the website provide a bulletin board for posting questions about civil applications and answers from public officials?	
3. Does the website have online bulletin board or chat capabilities for gathering citizen input on public issues? ("Online bulletin board" or "chat capabilities" means the city website where any citizens can posts ideas, comments, or opinions without specific discussion topics)	
4. Does the website offer online discussion forum on policy issues? ("Online discussion forum" means the website where the city arranges public consultation on policy issues and citizens participate in discussing those specific topics)	
5. Does the website have scheduled e-meetings for discussion? ("E-meeting" means real-time discussion which takes place at the specific appointed time in a synchronized way so that participants can exchange opinions simultaneously)	
6. Does the website offer tools for online decision-making? (e-petition, electronic citizen juries, e-referenda) Note.E-petition or electronic petition is a formal request to a city council or a government agency, signed by a number of citizens online, to raise issues of concern	
Total Score	

Table 2. Representative theory website features by population

Feature	Small City (n=52)	Medium City (n=107)	Large City (n=33)
Information about the location of offices	47(90%)	100(94%)	28 (85%)
Contact information for city agencies	47(90%)	102(95%)	26 (79%)
Minutes of public hearings or meetings	47(90%)	94(88%)	31(94%)
Mission statement	47(90%)	91(85%)	21(63%)
Budget posted on-line	46(89%)	100(94%)	26(78%)
Calendar of events	47(90%)	92(86%)	26 (78%)
File complaints on-line	35(67%)	85(79%)	23(70%)
Frequently asked questions	30(58%)	53(50%)	19(58%)
On-Line information requests	38(73%)	52(49%)	14(42%)
On-line comments to public officials	31(71%)	55(51%)	9(27%)
Mean score (out of 10 items)	**7.19**	**6.76**	**5.90**

Table 3. Representative theory website features by form of government

Feature	Mayor-Council (n=75)	Manager-Council (n=118)
Information about the location of offices	63(84%)	110 (93%)
Contact information for city agencies	65(87%)	108(92%)
Minutes of public hearings or meetings	63(84%)	107(91%)
Mission statement	56(75%)	101(86%)
Budget posted on-line	63(84%)	107(91%)
Calendar of events	59(79%)	104(88%)
File complaints on-line	50(67%)	91(77%)
Frequently asked questions	36(48%)	64(54%)
On-Line information requests	31(41%)	72(61%)
On-line comments to public officials	35(47%)	64(54%)
Mean score (out of 10 items)	**6.11**	**6.93**

Table 4. Representative theory website features by it capacity

Feature	Has an IT Department (n=128)	Does Not Have an IT Department (n=65)
Information about the location of offices	114(89%)	59 (90%)
Contact information for city agencies	113(88%)	60(92%)
Minutes of public hearings or meetings	115(90%)	55(85%)
Mission statement	100(78%)	57(88%)
Budget posted on-line	112(88%)	58(89%)
Calendar of events	107(84%)	56(86%)
File complaints on-line	95(74%)	47(72%)
Frequently asked questions	65(51%)	36(55%)
On-Line information requests	65(51%)	37(57%)
On-line comments to public officials	64(50%)	36(55%)
Mean Score (out of 10 items)	**6.53**	**6.80**

Table 5. Pluralist theory website features by population

Feature	Small City (n=52)	Medium City (n=107)	Large City (n=33)
Targeted audience links	29(56%)	66(61%)	20(61%)
Download forms	0(0%)	43(40%)	30(91%)
Pay utilities	35(67%)	76(71%)	24(73%)
Pay taxes	13(25%)	44(41%)	23(70%)
Pay fines	25(48%)	73(68%)	21(64%)
Apply for permits	19(37%)	41(38%)	23(70%)
Track permit and license Requests	17(32%)	45(42%)	16(49%)
Apply for licenses	11(21%)	21(20%)	18(55%)
Social media links	15(29%)	75(70%)	29(88%)
E-Procurement	32(61%)	54(51%)	11(33%)
Customized content	5(10%)	24(22%)	2(6%)
Feedback to city agencies	30(58%)	48(45%)	4(12%)
Mean score (out of 12 items)	**4.44**	**5.70**	**6.69**

Table 6. Pluralist theory website features by form of government

Feature	Mayor-Council (n=75)	Manager-Council (n=118)
Targeted audience links	51(68%)	63(53%)
Download forms	37(49%)	36(31%)
Pay utilities	46(61%)	87(74%)
Pay taxes	44(59%)	34(29%)
Pay fines	47(63%)	70(59%)
Apply for permits	31(41%)	52(44%)
Track permit and license Requests	30(40%)	47(40%)
Apply for licenses	19(25%)	30(25%)
Social media links	51(68%)	66(56%)
E-Procurement	26(35%)	69(59%)
Customized content	8(11%)	21(18%)
Feedback to city agencies	22(29%)	58(49%)
Mean score (out of 12 items)	**5.49**	**5.36**

Table 7. Pluralist theory website features by it capacity

Feature	Has an IT Department (n=128)	Does Not Have an IT Department (n=65)
Targeted audience links	77(60%)	37(57%)
Download forms	54(42%)	18(28%)
Pay utilities	91(71%)	42(65%)
Pay taxes	58(45%)	20(31%)
Pay fines	81(63%)	36(56%)
Apply for permits	60(47%)	22(34%)
Track permit and license Requests	51(40%)	26(40%)
Apply for licenses	37(29%)	13(20%)
Social media links	81(63%)	37(57%)
E-Procurement	62(48%)	33(51%)
Customized content	15(12%)	15(23%)
Feedback to city agencies	54(42%)	27(42%)
Mean score (out of 12 items)	**5.63**	**5.01**

Table 8. Direct theory website features by population

Feature	Small City (n=52)	Medium City (n=107)	Large City (n=33)
Mobile phone updates	10(19%)	16(15%)	9(27%)
Bulletin board	3(6%)	1(2%)	1(3%)
Chat room	1(2%)	4(4%)	0(0%)
Discussion forum	1(2%)	2(2%)	1(3%)
E-Meetings	1(2%)	2(2%)	2(6%)
On-line discussion forum	0(0%)	3(3%)	0(0%)
Mean score (out of 6 items)	**.307**	**.346**	**.394**

Table 9. Direct theory website features by form of government

Feature	Mayor-Council (n=75)	Manager-Council (n=118)
Mobile phone updates	14(19%)	19 (16%)
Bulletin board	1(1%)	4(3%)
Chat room	1(1%)	4(3%)
Discussion forum	1(1%)	3(3%)
E-Meetings	1(1%)	4(3%)
On-line discussion forum	0(0%)	3(3%)
Mean score (out of 6 items)	**.360**	**.314**

Table 10. Direct theory website features by it capacity

Feature	Has an IT Department (n=128)	Does Not Have an IT Department (n=65)
Mobile phone updates	21 (16%)	13(20%)
Bulletin board	4(3%)	1(2%)
Chat room	4(3%)	1(2%)
Discussion forum	3(2%)	1(2%)
E-Meetings	5(4%)	0(0%)
On-line discussion forum	2(2%)	1(2%)
Mean score (out of 6 items)	**.304**	**.400**

Chapter 2
Writings on the City Wall:
An Analysis of Popular Municipal Facebook Pages Posts

Nili Steinfeld
Ariel University, Israel

Azi Lev-On
Ariel University, Israel

ABSTRACT

Municipality Facebook pages are significant social media arenas for maintaining contact between representatives and their constituencies. The authors use digital tools to collect and analyze some 24,000 posts from the Facebook pages of all Israeli municipalities in a six-month period. Following a purely automatic linguistic analysis of the texts of all posts published on the pages, this study moved to a manual coding of a sample of the most popular posts in the data in order to gain insight into the character of the discourse in these arenas; the actors; the format, type, and emotionality of the contents that attract the highest levels of engagement.

INTRODUCTION

Democratic thinkers have long argued that the construction of viable arenas for self-governance should start at the workplace and municipal levels (Pateman, 1970). For example, Dahl (1970) argues that the city is,

a unit of human proportions in a world grown giant, demonic, incomprehensible; an optimal site for democracy... an association in which citizens can learn that collective

DOI: 10.4018/978-1-5225-5999-3.ch002

benefits from cooperation and peaceful conflict are so great that rational self-interest must act as a restraint on self-destructive egoism (p. 165; see also Dahl, 1967).

In such environments, people are within physical proximity on a regular basis, they may repeatedly interact with other stakeholders, and it is easy to perceive the link between people's decisions and consequences on the ground. Hence, such environments lend themselves more easily than others to participatory schemes and deliberative programs (Shkabatur, 2010).

Recent years have shown a growing awareness among national, as well as municipal governments to develop new means of communication with citizens, along with a surge in the use of information and communication technologies to promote participatory democratic processes (Coleman & Gotze, 2001; Macintosh & Whyte, 2008; Manosevitch, Steinfeld & Lev-On, 2015). Perhaps, the milestone most associated with the aforementioned phenomenon, is Obama's first presidential memorandum, on his first day in office in 2009, in which he instructed federal agencies to *usher in a new era of open government.* The following key importance to him was to establish a system of transparency, public participation and collaboration. Consequently, executive departments and agencies should use innovative tools, methods and systems to cooperate among themselves, across all levels of government and with non-profit organizations, businesses and individuals in the private sector. In parallel, scholars have begun to develop theoretical models through which they could analyze and evaluate governments' e-transparency, participation and collaboration efforts (Mergel, 2013), as well as the impact of the tools, goals and topics on which they typically focus (Criado, Sandoval-Almazan & Gil-Garcia, 2013). The current chapter is a part of this endeavor, focusing on the topics and formats that are popular in conversations involving governments and citizens, at the municipal level.

Online municipal platforms are novel and attractive arenas for citizens to ask and receive questions, express opinions and communicate with officials. An examination of the interactions between local governments and residents on Facebook is important for several reasons. First, since municipal elections may be determined by a small number of voters, mayors and local candidates may pay significant attention to residents' comments. As a result, a local Facebook page can generate significant incentives and opportunities to initiate discussions and calls to action, especially during election season. Second, participants in municipal forums discuss topics that directly affect their everyday lives (such as education, welfare, and the urban environment). Third, studies demonstrate correlations between municipal website usage and trust in local authorities (Tolbert & Mossberger, 2006).

We studied the prominent concepts and topics of interest in all of the official Israeli municipal Facebook pages from October 2012 to April 2013. The first stage of this research included an automatic linguistic analysis of the texts of all the posts and comments published on the pages during this time scale, and used digital tools to extract the most frequent words and terms in the corpus. The results enabled us to set the ground for improving the service given to citizens in various domains by the city, and understand, for the first time, the interactions between residents and municipalities on Facebook, based on automatic tools. Municipal Facebook pages have emerged from the analysis as a place of reinforcement and positive attitudes, where citizens and cities congratulate, express gratitude and support municipal and citizens' projects and initiatives. Positive terms were significantly more prominent and diverse in the texts than negative terms. The findings further indicate that the discourse tends to be personal in nature, and cities frequently encourage citizens to engage with them on Facebook, suggesting they like content posted by them and let them know what they think, and direct citizens to other online platforms. City pages were also used for public notification of events at both the municipal and national levels, such as emergency drills and wartime notifications; the latter were tied with citizens' expressions of anxiety, confusion, and sense of inability. The pages were also used for discussing issues related to the education system, and for submitting formal requests to the municipality. Finally, despite of the general positive tone, a notable number of posts reflected dissatisfaction and complaints of citizens. The use of such terms was interestingly more prominent in the cities which are ranked higher in socio-economic status (Steinfeld & Lev-On, 2015).

Following this general perspective, we wanted to advance to in-depth analyses of the popular posts, through the use of manual text analysis. We focused on popular contents, i.e. posts that lead in number of Likes, Comments or Shares. Such highly-engaged posts are more prominent and potentially influential than other posts; they are routed to users' newsfeeds, and have greater potential to become viral. Consequently, popular posts that resonate in the public sphere have the greatest potential to reach and involve relevant stakeholders. An in-depth manual coding of the content will enable us a greater understanding of the topics, use of language and interests that most engage citizens.

BACKGROUND

Municipal Facebooking

Studies of municipal e-government typically focus on its scope, and on the factors that affect municipalities' digital presence or absence. These studies show that municipality adoption of the Internet and social networking sites is not uniform, and population size is the most significant factor influencing whether municipalities maintain websites or Facebook pages, and the scope of activity on these sites. The larger the population, the greater the probability that the municipality will maintain some online presence, either a website or a Facebook page, and the more extensive the scope of activity in such online platforms (Lev-On & Steinfeld, 2015; Norris & Reddick, 2013; Reddick & Norris, 2013).

Studies also demonstrate that municipalities rather than residents play a dominant role in uploading contents to Facebook pages (Graham & Avery, 2013; Hofmann, S., Beverungen, D., Räckers, M., & Becker, J., 2013; Magnusson, Bellström & Magnusson, 2012). Municipality websites emphasize static contents such as tenders and information on local events, and place less emphasis on interactions with residents (Mossberger, 2013; Norris & Reddick, 2013). Apparently, this pattern is reproduced on municipal Facebook pages, as municipalities often use their Facebook pages to post materials that also appear on other media such as billboards or mailings (Graham & Avery, 2013; Lovari & Parisi, 2012).

Little research has been conducted on the formats (images, videos, frequent terms and phrases) that municipalities use on their Facebook pages. Municipalities apparently make little use of videos and images (Hofmann, S., Beverungen, D., Räckers, M., & Becker, J., 2013) despite their power to evoke engagement (Hofmann, S., Beverungen, D., Räckers, M., & Becker, J., 2013; Huber, J., Landherr, A., Probst, F., & Reisser, C., 2012).

Israel is a promising site of research on online politics. According to a 2017 report, the number of Israeli Internet users at the end of 2016 was between 5.9 and 6.4 Million – who are 71-80% of the Israeli population (Mann & Lev-On, 2017). With 4.9 Million users (60% of the population), at the beginning of 2016 Facebook was the third most popular application in Israel (83%)- after YouTube (87%) and WhatsApp (84%) (Mann & Lev-On, 2017). A 2011 comScore (2011) study declared Israel as the worldwide leader for the most time spent on social network sites. In October 2011, Israelis spent an average of 11.1 hours a month on social media, almost twice the global average of 5.7 monthly user hours. In addition, Israel is ranked fifteenth in the World Economic Forum's Networked Readiness Index of 2013 (Bilbao-Osorio, Dutta, & Lanvin, 2013). Not only the public but also MPs have a strong presence on Facebook. In 2010, the Director General of the Israeli parliament (the Knesset) advised all MPs to open a Facebook page (Mako, 2010). As of 2015, 106 MPs out of 120 (87.5%) had active Facebook pages (Lev-On & Steinfeld, under review).

In Israel, the few studies that analyzed municipal e-government adoption (Purian-Lukach, 2011; Ravitsch, 2005; Rotem, 2007) have focused on municipality websites. These studies indicate that municipal Internet usage has increased over time, and that small and large local governments in Israel have different adoption and usage patterns. The studies in Israel also indicate that several municipalities, such as municipalities with an Arab or ultra-Orthodox religious majority, are Internet laggards as a result mainly of the cultural makeover of their populations (Lev-On & Steinfeld, 2015).

Manual vs. Digital Analysis of Digital Objects

Internet researchers are involved in recent years in debates over the preferred method to study online phenomena, as more researchers adopt computational research methods and digital tools for their research. Arguments for and against the move to digital research methods are diverse. Some argue that digital objects and phenomena that are "natively digital" are best studied by natively digital methods and tools (Herring, 2009; Rogers, 2009). These tools allow an unprecedented scale and scope of analysis, specifically in social science research (Giles, 2012; Lazer, D., Pentland, A., Adamic, L., Aral, S., Barabási, A., Brewer, D., Christakis, N., Contractor, N., Fowler, J., Gutmann, M., Jebara, T., King, G., Macy, M., Roy, D., & Van Alstyne, M., 2009). A digital approach to digital social research can reveal more than traditional methods may, and enable researchers to see the broad picture more clearly (Boyd & Crawford, 2012; Cioffi-Revilla, 2010; Lazer, D., Pentland, A., Adamic, L., Aral, S., Barabási, A., Brewer, D., Christakis, N., Contractor, N., Fowler, J., Gutmann, M., Jebara, T., King, G., Macy, M., Roy, D., & Van Alstyne, M., 2009). Critics of these new methods of research state the need for a more theory-guided research, where in many cases the research outlines what was found, but lacks some necessary grounded social science theory as its basis (Boyd & Crawford, 2012; Snijders, Matzat & Reips, 2012). The reliance on third party platforms that operate as black boxes for the gathering and analysis of datasets, and the necessary process of "cleaning" the data, can cause bias and sometimes prevent researchers from accounting for the representativeness and accuracy of the data (Bollier & Firestone, 2010; Boyd & Crawford, 2012). A main question to be asked regarding automated, computer-mediated analysis is: What kind of questions can this type of research answer, and where does it fail?

Following a previous study, which used purely automatic processes for textual analysis, the current study uses a manual coding of content and focuses on the popular posts published on municipal Facebook pages, i.e. posts that attract the highest number of Likes, Comments or Shares. Although the literature on content popularity in municipal forums is limited, insights from existing literature in marketing and journalism may be useful for further analysis. For example, studies indicate that one relevant variable in determining user engagement is the frequency of page posts: The more active a Facebook page and the more frequently materials are uploaded, the greater engagement users generate with the contents (Huber, J., Landherr, A., Probst, F., & Reisser, C., 2012).

A post's potential popularity is also affected by the emotional states it evokes. "Arousing" contents that stimulate the nervous system and trigger emotions such as sadness, anger, anxiety, or revulsion have a greater probability of becoming popular (Berger, 2011). A consistent finding in studies about viral transmission of content in diverse settings such as advertising (Eckler & Bolls, 2011), news content (Berger & Milkman, 2012), Twitter posts (Stieglitz & Dang-Xuan, 2013) and Facebook videos (Nelson-Field, Riebe & Newstead, 2013), has indeed been that emotionality predicts popularity of content on social media.

Still, engagement with online content can assume many forms, in the case of Facebook users can, for example, like it, comment on it, share it and more. A few studies demonstrate that different types of content trigger different types of user engagement; for example, Tenenboim and Cohen (2013) found that political contents appear more prominently among the highly commented news items while sensational topics appear more prominently among the heavily clicked items, in a longitudinal study of popular items in a news site; Boczkowski and Mitchelstein (2012) also found that public affairs are more commented on than clicked on or e-mailed. Such studies justify analyzing the popularity of Facebook posts using various parameters, looking at the most prominent forms on engagement with them: Liking them, commenting on them and sharing them.

RESEARCH QUESTIONS AND HYPOTHESES

The current study aims at characterizing the most popular posts in terms of format, author, language, and content. To do so, we drafted several specific research questions. We begin with exploring the entire dataset of all the posts sampled for this study, and then continue with coding and analyzing a sample of popular posts extracted from the larger population. Due to the paucity of research in this field, several research questions are unaccompanied by specific hypotheses and the analysis is more exploratory in nature.

- **RQ1:** What are the most popular posts in the entire dataset?
- **RQ2.** What is the overlap in engagement measures (Like, Comment, and Share)?
- **H1:** In line with Tenenboim and Cohen (2013), we expect to find some overlap in the posts that lead in different engagement measures, but expect a significant proportion of posts to lead in a single engagement measure only.
- **RQ3:** How does engagement with Facebook posts (i.e. shares, likes and comments) correlate with various city's population demographics and level of activity on the city's Facebook page?

- **H2:** Engagement will be correlated with municipality size and the socio-economic status of its population, in line with previous research findings (Lev-On & Steinfeld, 2015; Norris & Reddick, 2013; Reddick & Norris, 2013).
- **H3:** Engagement will be correlated with the number of Facebook fans.
- **H4:** Engagement will correlate with the number of posts, such that engagement increases with the number of page posts (Huber, J., Landherr, A., Probst, F., & Reisser, C., 2012).
- **RQ4:** Which contents stimulate greater engagement: content uploaded by the municipality or by the residents?
- **H5:** Content uploaded by municipalities stimulates greater engagement than content uploaded by residents (Graham & Avery, 2013; Hofmann, S., Beverungen, D., Räckers, M., & Becker, J., 2013; Magnusson, Bellström & Magnusson, 2012).

The following research questions refer to the sample of the most popular posts (the posts that generated the highest engagement measures):

- **RQ5:** What are the themes and the formats that feature most prominently in the most popular posts?
- **RQ6:** How emotional are the most popular posts?
- **H6:** Following previous studies that found that emotional contents evoke greater engagement (Berger & Milkman, 2012; Nelson-Field, Riebe & Newstead, 2013; Stieglitz & Dang-Xuan, 2013; Eckler & Bolls, 2011), we hypothesize that emotional posts will account for a significant proportion of the most popular posts.
- **RQ7:** What proportion of the most popular posts contains expressions of criticism or support?
- **RQ8:** What is the format that is most frequently used in the most popular posts?
- **H7:** We expect to find a greater proportion of rich content (videos and images), which has been found to trigger greater engagement, in the most popular post sample, compared with the study population (Hofmann, S., Beverungen, D., Räckers, M., & Becker, J., 2013).
- **RQ9:** What distinguishes the most popular posts initiated by municipalities from those initiated by residents in terms of themes, format, and criticism/support?

METHODOLOGY

A list of all 75 cities in Israel was created based on data from the Israeli CBS (Central Bureau of Statistics, 2011). To locate the official Facebook page of each city, we searched in Google's search engine and in Facebook's internal search function, entering the name of each city on the list in Hebrew, Arabic, and English, Israel's three official languages. Also, we scanned the municipalities' Websites for possible links to their Facebook pages.

A total of 43 Facebook pages were found and included in the study. Socio-demographic data were collected from the CBS website (Central Bureau of Statistics, 2011). Data includes the city's socio-economic status (SES cluster), population size, age median, education level (percentage of matriculation certificate holders among all twelfth-grade students in the city in the year 2010-2011, the most recent metric for the entire country), and geographic location (periphery cluster). These measures were found to predict Internet usage in previous research (Lev-On, 2012, 2014; Lev-On & Steinfeld, 2015).

To analyze the data, we used a combination of automated and manual content analysis. For the automated analysis, we used Netvizz, a Facebook application developed by Digital Methods Initiative labs, to extract data from Facebook pages. We used the application to extract and then archive all the posts published by the administrators or fans of the official municipality pages in the six-month period from October 22, 2012 to April 22, 2013. No municipal elections were held during this period (elections took place on October 22, 2013). In addition to the posts texts, Netvizz automatically extracted and included in the data the following information for each of the posts: engagement measures (number of Likes, Comments, and Shares) that the post received, post origin (pages or users), and post format (image, video, link, survey, or text status). For pages that were created during this period, we extracted all of the posts from page launch date to April 22, 2013. A total of 23,768 posts were extracted from all the pages. In addition, in June 2013 (shortly after the conclusion of the data collection phase of this study), we recorded the number of fans on each page (this data is not extracted by Netvizz and hence was collected manually). To answer RQ3 and RQ4, all posts were analyzed based solely on the information retrieved from Netvizz and information regarding the municipalities' demographics which was gathered from the CBS website.

To add to the previous analysis of all posts and comments using automatic linguistic tools, we used manual content analysis to analyze, in a higher resolution, a smaller sample of the most popular posts. For this purpose we constructed three lists of popular posts per municipality, based on the data received from Netvizz: A sample of 474 posts that received the greatest number of Likes each month; A second sample of 502 posts that received the greatest number of Comments each month; and a third sample of 510 posts that received the greatest number of Shares each month. All lists were generated per municipality- so that we gathered the most popular posts per month from each of the municipalities separately (comparing engagement rates between posts of the same municipality only). In this manner we ensured that all municipal pages received equal consideration, and that the samples were free of bias in favor of the most active municipality Facebook pages. For municipality Facebook pages that were established during the data collection period, the samples include the leading posts each month, from the Facebook page establishment month to April 2013.

The three lists of the most popular posts were coded manually by three raters who were trained by the researchers. We conducted repeated training sessions for categories that achieved less than 90% reliability, and repeated the coding procedure until inter-rater agreement of 90% was achieved. Raters coded the posts for the following categories:

- Post length (number of words).
- Topic (each post could be classified as belonging to several topics): emergencies (example: "One hour ago, an explosion was heard [...]"); holiday or Sabbath (examples: "Giant Street party next Tuesday"); sports (example: "City's night run begins tonight"); calls to online or offline action (such as a call for donations for a local cause or to participate in a specific action, such as "Share this post," "Come to the library for a lecture on..."); congratulations and commendations (example: "New Year's greeting to everyone," "Birthday wishes to ...," "Congratulations to The Voice contenders from our city"); public service announcements; security-related posts (example: "Please unlock the doors to bomb shelters in condominium buildings"); education (announcements related to the school system, such as "Another local school has joined the road safety Project"); and hazards ("Drainage problems after the rains endanger pedestrians," "Public bathrooms are filthy").
- Criticism or support (for example, the post "It's a pity that citizens have to do the municipality's job" was classified as "criticism," while the post "City residents bring honor to the city" was classified as "support").
- The object of criticism or support (municipality, mayor, other municipality official, resident/s, organization/non-profit, Federation of Local Authorities in Israel, or other).
- Emotionality. Posts were coded "emotional" if the text included words that describe emotions or contained exclamation points (example: "Extending condolences on the passing of….." or "Dear (and costly) municipality!!!").
- Personal angle. Posts that included a personal story or perspective (example: "Yesterday I fell on the sidewalk opposite the mall").

Several categories were determined on the basis of previous studies.

Additionally, a central source for thinking of relevant categories for coding was the results of the preliminary automatic linguistic analysis which found prominent terms, pointing to prominent topics in the posts. Some of the prominent terms found in the previous study were related to education (schools, kindergartens, ministry of education etc.), some expressed greetings (good morning, congratulations, feel better) or related to holidays/ Sabbath (good Sabbath, happy holiday). Some of the prominent terms suggested criticism (shame on you, I disapprove) or support (well done). Others related to emergencies (the alarm) or to official town-hall messages (please contact, your request has been forwarded) etc. On the basis of these findings several categories for posts topics were drafted.

FINDINGS

General

An average of 553 posts were published on each page in the six-months data collection period. Posts attracted an average of 23.69 Likes, 3.09 Comments, and 2.1 Shares per post. Pages had an average of 7,654 fans, with results ranging from 45 to 67,413 fans.

Of the 23,768 posts in the dataset, 9,930 (41.7%) were posted by the municipal pages (248 posts per page, on average), and 13,838 (58.3%) by fans (419 posts per page, on average). Images are the most prevalent content type, accounting for 10,117 posts (43%). Text-only statuses, links, and videos account for 9,071 posts (38%), 3,341 (14%), and 1,216 (5%) of the posts, respectively.

The Most Popular Posts in the Dataset

The post that attracted the highest number of Likes and Comments in the dataset was published by Kfar-Saba municipality and stated,

Have you ever thought of how you'll celebrate a 100th birthday? Greta, a senior Kfar Saba resident, a wonderful and funny woman, is celebrating her 100th birthday today!!! She deserves at least 100 Likes! Give her a Like!:)

Accompanying the text is a photo of Greta holding a sign saying "I'm celebrating 100. Give me a Like!" This post received 93,676 Likes, 1,742 Comments, and 415 Shares (8th place in the Share count).

Figure 1. The most popular post in the dataset: Greta's 100th birthday post Kfar Saba Municipality

The second most engaging post was posted by a user on the Rishon Lezion page. The post refers to an old man who owns a tire shop and experiences financial difficulties. The user attached a photo of the man working in his tire shop, and asked people to help him by using his services, and sharing the post. This post attracted 2,935 Shares (2nd in the Share count), 22,605 Likes (2nd in the Like count), and 422 Comments (6th in the Comment count).

The third most engaged post is a complaint posted by a user and addressed to Tel-Aviv municipality. The user's car was towed for parking in a no-parking zone, but a luxury car parked right next to his did not get towed. The clerk in the parking lot plainly told him that the city doesn't tow luxury cars because they prefer not to get involved in law suits or pay for damages if the cars are scratched. The user attached a photo of the luxury car parked alone in a no-parking zone. This post received 18,159 Likes (3rd in the Likes count), 1,457 Comments (2nd most commented-on post) and 2,734 Shares (3rd most Shared post).

The fourth most popular post is also a resident's complaint related to a parking ticket, addressed to the municipality of Rishon-Lezion and to its mayor in person.

The user describes how she witnessed a parking ticket being issued to an elderly couple, when the husband was sitting in the car and his wife went to the pharmacy for drugs for his cancer treatment. The user describes the parking official's disgraceful refusal to cancel the ticket, and added a photo of the elderly man in his car. This post received 15,915 Likes (4th in the Like count), 1,319 Comments (3rd in the Comments count) and 4,270 Shares (most Shared post).

The fifth most popular post is a complaint regarding violence in Haifa, posted by an 18 year-old Arab man, addressed to the mayor of Haifa personally. The poster writes about the rising violence in the city, and how he has been hearing about anti-violence city-funded projects for years but hasn't seen any change. He added a photo of a youngster who was killed in Haifa. This post received 5,073 Likes (6th in the Like count), 554 Comments (4th in the Comments count), but only 231 Shares (this post was not among the ten most Shared posts).

The next five most popular posts include four posts by municipalities (a post announcing the city of Jerusalem's plan to build one of the biggest skating rinks in Israel; two posts from Tel-Aviv municipality about the Tel-Aviv Marathon, including links to registration; and one by the city of Ashkelon thanking the Air Force soldiers and Iron Dome system for defending the city's residents during the military operation "Pillar of Defense," in November 2012), and one post by a resident of Tel-Aviv complaining about a parking ticket.

Studying the 10 most popular posts reveals that highly engaged-with posts are usually either complaints by users, or posts initiated by municipalities which deal with major events. Note that receiving high levels of engagement to posts initiated by users is much more difficult in comparison to posts initiated by the municipality, due to the design of Facebook pages: At the time this study was conducted, Facebook page fans automatically receive updates made by the page to their News Feed, but view users' posts to the page only when they visit the page itself or if other users are their friends. For this reason, user-initiated posts are inherently less visible to other users. In addition, it is not surprising that the most popular posts are usually posts published on the pages of the biggest cities in Israel (Tel-Aviv, Haifa, Rishon-Lezion and Jerusalem) with the exception of Kfar Saba and Ashkelon's posts.

Moving on to RQ2, another interesting point is that the ten most popular posts tend to be among the leaders in all engagement measures (Likes, Comments, and Shares). This overlap is especially strong between Likes and Comments: Eight posts featured in both the 10 most Liked list and the 10 most Commented list. Posts that make the most shared list seem to be a bit different: of the 10 most Shared posts, five did not appear on either the 10 most Liked or the 10 most Commented lists. Of these five posts, four are emergency or public announcements. These findings support H1.

Correlations of Engagement Variables With City and Facebook Variables

Table 1 includes the correlations between all the variables included in the study.

A Pearson test was used to measure the relationship between population size and socio-demographic variables, for the various Facebook engagement measures indicating a post's popularity. The only variable that significantly correlated with any engagement measure was population size. There is a significant and strong correlation between population size and mean Shares per post (r=0.65, p<.001); a significant moderate correlation between population size and average engagement (r=.32, p<.05), mean Comments per post (r=0.4, p<.01), and a weak correlation between population size and mean Likes per post (r=0.26, p<.05). All correlations are positive, implying that engagement increases with city population size. No significant correlation was found between the Facebook engagement indices and any other socio-demographic indicators: education, peripherality, socio-economic status, and median age. These findings partially support H2.

Pearson correlations indicate a significant positive correlation between the number of page fans and average Likes per post (r=0.4; p<.01), average Comments per post (r=0.34; p<.05), and average Shares per post (r=0.69; p<.001). The greater the number of fans per page, users' engagement with the posts increases. These findings support H3.

A significant positive correlation was also found between the number of posts per page and the engagement measures with the posts: average Likes per post (r=0.44; p<.01), average Comments per post (r=0.34; p<.05), and average Shares per post (r=0.69; p<.011). These findings support H4.

Posts by Municipalities and Users

Moving on to RQ4, the findings demonstrate that posts uploaded by the municipalities generate significantly greater engagement and are more popular than posts by fans, supporting H5.

The number of Likes per municipality-uploaded post (M=44.99, SD=946.98) was significantly higher than the number of Likes per user-uploaded post (M=8.41, SD=286.82) ($t_{(11243.06)}$=3.73, p<.001). The number of Comments per municipality post (M=4.48, SD=21.86) is also significantly higher than the number of Comments per user post (M=2.1, SD=18.7) ($t_{(19280.78)}$=8.78, p<.001). In addition, fans rarely share posts uploaded by other fans, in contrast to Shares to posts by pages; the number of Shares per municipality-uploaded post is significantly higher (M=3.73, SD=19.09) than the number of Shares per user-uploaded post (M=0.94, SD=50.11)

Table 1. Correlations between the City, Facebook and Engagement Variables

	1	2	3	4	5	6	7	8	9	10	11	12	13
1. Population Size	1.00	.07	.4***	-.03	.13	.59***	.08	.6***	.6***	.26*	.40**	.65***	.32*
2. Age Median		1.00	.12	.72***	.75***	.22	.28	.14	.17	.09	.16	.05	.10
3. Peripherality			1.00	.19	.41***	.31*	-.13	.37*	.19	.09	.24	.11	.12
4. Education				1.00	.81***	.14	-.09	.18	.11	.17	.22	-.01	.18
5. Socio-Economic Cluster					1.00	.28	-.12	.34*	.25	.17	.22	.11	.19
6. Total Number of Posts						1.00	.36*	.95***	.86***	.44**	.34*	.69***	-.00
7. Number of Posts by Pages							1.00	.06	.86***	.44**	.34*	.69***	-.00
8. Number of Posts by Users								1.00	.86***	.4**	.36*	.63***	.04
9. Fans									1.00	.4**	.34*	.69***	.02
10. Total Post Likes										1.00	.48**	.55***	.08
11. Total Post Comments											1.00	.4**	.81***
12. Total Post Shares												1.00	-.01
13. Total Overall Engagement													1.00

$^*p < .05$ $^{**}p < .01$ $^{***}p < .001$

($t_{(18922.5)}$=5.96, p<.001). Lastly, municipality-uploaded posts have a mean overall engagement score of 56.89, and user-uploaded posts have a mean overall engagement score of a mere 13.37.

Interestingly, when analyzing exclusively the posts by the municipalities, the correlation between number of posts and engagement becomes insignificant- with the exception of the correlation between posts per page and average Shares per post (r=0.36; p<.05). But when analyzing the posts by fans, all correlations between number of posts and the various engagement measures are significant: average Likes per post (r=0.4; p<.001), average Comments per post (r=0.36; p<.005), and average Shares per post (r=0.63; p<.001).

Findings: Sample of the Most Popular Posts

The sample of the most-Liked posts includes 474 posts that received an average of 453.9 Likes (Median=49) each. The sample of most-Shared posts includes 510 posts that received an average of 44.61 Shares (Median=6) each. The sample of most-Commented posts includes 502 posts that received an average of 36.32 Comments (Median=12) each. Table 2 summarizes the distribution of posts by engagement measures.

Themes of the Most Popular Posts (RQ5)

The most popular posts are about various distinct topics:

- The main topics of the most-Liked posts were cultural events (24.3%), holidays/Sabbath (14.7%), public announcements (13.2%), congratulations (10.6%), and education (10.1%).
- In the most-Commented sample, 22.9% of the posts concerned cultural events, 19.8% concerned public announcements, 10.6% concerned congratulations and commendations, and 10.4% concerned calls to action.
- The main topics of the most-Shared posts were cultural events (34.8%), public announcements (21.5%), calls to action (19.4%), and holidays/Sabbath (14.1%).

Emotionality of the Most Popular Posts (RQ6)

Posts containing an emotional tone account for 8.8% of the most-Liked posts, 10.4% of the most-Commented posts, and a mere 2.3% of the most-Shared posts. These low figures are not unexpected, as municipality-originated contents are typically matter-of-fact and avoid emotional language. However, even the small amount of

Table 2. Features of the Most Popular Posts by Pages and Fans

	Most Commented	Most Liked	Most Shared
Number of Posts	Total: 502 posts	Total: 474 posts	Total: 503 posts
	Posted by pages: 411 (81.9%)	Posted by pages: 412 (86.7%)	Posted by pages: 468 (93%)
	Posted by users: 91 (18.1%)	Posted by users: 62 (13.1%)	Posted by users: 35 (7%)
Length of Posts	Average= 59.96 words (Median=30)	Average= 54.57 words (Median=24)	Average= 44.44 words (Median=23)
	Pages' posts: $M = 50.65$, $SD = 69.86$ **	Pages' posts: $M = 45.94$, $SD = 76.04$ **	Pages' posts: $M = 42.02$, $SD = 61.36$
	Users' posts: $M = 99.88$, $SD = 139.1$ ** $t_{(100.83)} = 3.28$	Users' posts: $M = 109.98$, $SD = 140.69$ ** $t_{(66.66)} = 3.51$	Users' posts: $M = 76.65$, $SD = 107.79$
Format of Posts	Photos Lead: 61.8% Status: 26.1% Links: 7.6% Video: 4.2% Survey: 0.4%	Photos Lead: 71.3% Status: 17.7% Links: 7.2% Video: 3.8%	Photos lead: 73.8% Links: 10.7% Video: 8% Status: 7.6%
	Pages' posts***: Photos: 69.3% Status: 17.5% Links: 8.5% Video: 4.1% Survey: 0.5%	Pages' posts***: Photos: 76.5% Status: 13.1% Links: 6.3% Video: 4.1%	Pages' posts: Photos: 72.4% Links: 11.3% Video: 8.1% Status: 8.1%
	Users' posts***: Status: 64.8% Photos: 27.5% Video: 4.4% Links: 3.3% Survey: 0% χ^2=88.21; Cramer's V=0.42	Users' posts***: Status: 48.4% Photos: 37.1% Video: 1.6% Links: 12.9% χ^2=53.72; Cramer's V=0.34	Users' posts: Photos: 91.4% Links: 2.9% Video: 5.7% Status: 0%
Topics:			
Emergency	Total: 3.3%	Total: 5.9%	Total: 5.3%
	By pages: 3.3%	By pages: 6.6%	By pages: 5.7%
	By users: 3.3%	By users: 1.6%	By users: 0%
Holiday	Total: 11.2%	Total: 14.7%	Total: 14.1%
	By pages**: 13.3%	By pages**: 16.8%	By pages: 15.6%
	By users**: 2.2% χ^2=9.18	By users**: 1.6% χ^2=9.8	By users: 6.1%
Cultural Events (including community events)	Total: 22.9%	Total: 24.3%	Total: 34.8%
	By pages***: 26.2%	By pages***: 27.1%	By pages*: 36.1%
	By users***: 8.8% χ^2=12.68	By users***: 6.5% χ^2=12.41	By users*: 17.6% χ^2=4.75
Sports	Total: 2.9%	Total: 3.7%	Total: 4%
	By pages: 3.1%	By pages: 4.3%	By pages: 4.3%
	By users: 2.2%	By users: 0%	By users: 0%

continued on following page

Table 2. Continued

	Most Commented	Most Liked	Most Shared
Mobilization/ Call to Action	Total: 10.4%	Total: 8.4%	Total: 19.4%
	By pages: 10.3%	By pages: 7.9%	By pages*: 18.4%
	By users: 11%	By users: 11.3%	By users*: 33.3% χ^2=4.4
Greetings and Commendations	Total: 10.6%	Total: 10.6%	Total: 8.6%
	By pages**: 12.9%	By pages***: 24.7%	By pages: 9.3%
	By users**: 1.1% χ^2=10.73	By users***: 1.6% χ^2=16.92	By users: 0%
Public Announcements	Total: 19.8%	Total: 13.2%	Total: 21.5%
	By pages***: 22.8%	By pages**: 15.3%	By pages: 21.8%
	By users***: 6.6% χ^2=12.26	By users**: 0% χ^2=10.94	By users: 18.2%
Education	Total: 8.9%	Total: 10.1%	Total: 7.4%
	By pages: 8.5%	By pages: 10.5%	By pages: 7.7%
	By users: 11%	By users: 8.1%	By users: 3%
Hazards	Total: 7.5%	Total: 2.6%	Total: 2.3%
	By pages***: 3.6%	By pages***: 1%	By pages**: 1.8%
	By users***: 24.2% χ^2=45.01	By users***: 12.9% χ^2=29.37	By users**: 9.1% χ^2=7.17
Criticism/ Support	Total: Criticism=12.5% Support=9.4% Neutral=78.1%	Total: Criticism=8.4% Support=12.6% Neutral=79.1%	Total: Criticism=3.8% Support=5.7% Neutral=90.5%
	By pages***: Criticism=0.8% Support=11.1% Neutral=88.2%	By pages***: Criticism=0.5% Support=13.8% Neutral=85.7%	By pages***: Criticism=0.5% Support=5.9% Neutral=93.7%
	By users***: Criticism=62.6% Support=2.2% Neutral=35.2%	By users***: Criticism=58.1% Support=4.8% Neutral=37.1%	By users***: Criticism=48.5% Support=3% Neutral=48.5%
Emotional/rational text	Total: 10.4%	Total: 8.8%	Total: 2.3%
	By pages***: 4.1%	By pages***: 4.1%	By pages***: 3.2%
	By users***: 37.4% χ^2=87.37	By users***: 38.7% χ^2=79.9	By users***: 45.5% χ^2=95.55
Personal Angle	Total: 9.4%	Total: 6.6%	Total: 3.6%
	By pages***: 1.8%	By pages***: 1.3%	By pages***: 1.6%
	By users***: 41.8% χ^2=138.6	By users***: 40.3% χ^2=132.26	By users***: 30.3% χ^2=73.22

$*p < .05$ $**p < .01$ $***p < .001$

emotional posts did seem to attract much user engagement. In the most-Liked sample, emotional posts received an average of 2981.51 likes, while non-emotional posts received an average of 222.84 likes. In the most-Commented sample, emotional posts received an average of 120.16 comments, while non-emotional posts received 27.79 comments on average. And in the most-Shared sample, emotional posts received an average of 268.07 shares while non-emotional posts received 99.89 shares on average. These findings suggest that although emotional posts are not as prevalent among the most popular posts as H6 assumed, they are nonetheless among the most engaged-with posts in the sample.

Criticism and Support in the Most Popular Posts (RQ7)

Posts expressing criticism of a target (the municipality, a resident, or outside entity) account for 12.5% of the most-Commented posts, 8.4% of the most-Liked posts, and a mere 3.8% of the most-Shared posts. Posts that express support account for 9.4% of the most-Commented posts, 12.4% of the most-Liked posts, and 5.4% of the most-Shared posts.

Formats of the Most Popular Posts (RQ8)

All three most-popular samples are dominated by posts containing images. Images account for the majority of the leading posts in all categories: Likes (71.3%), Comments (61.8%), and Shares (73.8%). Among the most-Commented and most-Liked posts, text statuses account for the second largest group of posts, and these are followed by links, and lastly- videos. Among the most-Shared posts, links constitute the second largest group of posts, followed by videos, and finally, text statuses. These findings support H7, especially with respect to images.

Popular Posts by Users and Municipalities (RQ9)

The most popular posts by fans and pages differ on many levels (see Table 1). While among the entire dataset of posts user posts accounted for a larger share of the posts than did posts uploaded by municipalities, the reverse occurs among the most popular posts sample, in which posts by municipalities constitute 86.7% of the most-Liked posts, 81.9% of the most-Commented posts, and 93% of the most-Shared posts. Here too, it is important to note that the structural features of Facebook may have affected these patterns: As page fans automatically receive updates made by the page to their News Feed, but view other users' contents only when they visit the page itself or if other users are their Friends, it is not surprising that users engage more strongly with contents uploaded by municipality pages to which they are passively exposed and which do not require users to actively access the municipality page.

Fan posts are also longer than municipality posts: In all three most popular samples, users upload longer posts than the municipalities. The average post length of posts uploaded by users in the most-Liked sample was 109.98 words, while municipalities' posts in the sample were 45.94 words on average. In the most-Commented sample, users' posts were 99.88 words on average, while posts uploaded by the municipalities averaged on 50.65 words. And in the most-Shared sample users' posts were 76.65 words on average, while municipalities' posts averaged on 42.02 words.

The most popular user posts are more textual: In contrast to previous findings that indicate that posts including images generate greater engagement than posts in other formats, findings of the most popular samples show that the most popular user posts are actually comprised mostly of text. In the most Liked and most Commented user post samples, the leading format was textual statuses (64.8% of the most-Commented posts and 48.4% of the most-Liked posts), followed by image-based posts. The majority of municipality posts are images (69.3% in the most-Commented posts and 76.5% of the most-Liked posts), followed by statuses. The difference in post format used by users and municipalities is significant in both samples (most-Commented posts: χ^2=88.21; Cramer's V=0.42; most-Liked posts: χ^2=53.72; Cramer's V=0.34). A different picture emerges from the Most-Shared post sample, where images dominate municipality posts (72.4%) and constitute almost the sole format in user posts (91.4%).

Users and municipalities also address different themes, and significant differences were found in the most-Commented and most-Liked posts. Specifically, in these two samples, municipalities communicate on holidays, congratulations and commendations, and public service announcements more frequently than users. In all three most popular post samples, municipalities communicate on cultural events more frequently than do users. In contrast, in all three most popular post samples, users communicate on hazards more frequently than do municipalities, and in the sample of the most-Shared posts, users communicate calls for action more frequently than do municipalities.

Users are more emotional than municipalities: In all three most-popular post samples, user posts contain a significantly more emotional tone than municipality-uploaded posts. While the most popular municipality posts hardly contain any emotional tone (4.1%, in the most-Commented and most-Liked samples), more than one-third of all user posts in each of the samples are written in an emotional tone.

Users criticize municipalities, while municipalities express support for their residents: In the three samples, user posts are significantly more critical than municipality posts, and the latter contain significantly more expressions of support than do user posts. Furthermore, a significant correlation was found in all samples between criticism/support and the object of criticism/support: In the most-Liked and most-Shared post samples, criticism was directed mainly at the municipality (65.8% in the most-Liked samples and 55.6% in the most Shared sample), followed by criticism that was directed at the mayor (15.8% and 22.2%, respectively). Support is mainly directed at the residents (49.1% and 40.7% in the most-Liked and most-Shared samples, respectively). In the most-Commented sample, criticism is directed mainly at the municipality (56.7%), followed by the mayor (13.3%), but support is mainly directed at the municipality (37.8%), followed by expressions of support for the residents (20%).

Finally, posts by users include personal perspectives (30.3% of the most-Shared user posts and 41.8% of the most-Commented user posts, respectively), significantly more frequently than municipality posts, which contain very few personal perspectives (1.3% of the most-Liked municipality posts and 1.8% of the most-Commented municipality posts).

CONCLUSION

The chapter contributes to scholars' efforts to develop theoretical models through which they could analyze and evaluate governments' e-transparency, participation and collaboration efforts (Mergel, 2013), as well as the impact of the tools, goals and topics on which they typically focus (Criado, Sandoval-Almazan & Gil-Garcia, 2013). Online social media offer governments novel and convenient platforms for communicating with the public, and the municipal sphere is an important site to study these patterns of communication. This chapter describes such social media activities, based on a thorough analysis of the contents uploaded to the Facebook pages of local governments in Israel in a six-month period, and focuses on the characteristics of popular posts in terms of Likes, Comments and Shares.

The general findings demonstrate significant correlations between municipality size and the scope of user engagement with posts on the municipality page, which suggests that the municipal arenas of the major local governments benefit from economies of scale, offer greater communicative potential and function as vibrant public spheres.

Other population-related variables do not correlate with level of user engagement. The number of posts on a page and page engagement measures correlate, but this finding does not necessarily indicate that municipalities that regularly update their Facebook pages attract higher levels of residents' engagement; instead, the findings suggest that on pages where residents are active- they are active in all levels of activity: The number of posts uploaded by residents to a page significantly correlates with all engagement indices, suggesting that when residents are more active by uploading posts to a page- they are also more active in engaging with posts uploaded to the page (by both residents and municipalities). The three measures of engagement overlap significantly, but not perfectly (Tenenboim & Cohen, 2013).

However, and in line with previous findings, municipalities are the dominant players on municipal Facebook pages, which is reflected in the number of posts they upload, and the engagement levels that their posts attract. While the number of posts uploaded by users to the page correlates with users' engagement with posts in general, the users Like, Comment, and especially Share contents uploaded by municipalities more frequently than they do with contents uploaded by other users.

Combining manual and automatic textual analyses enabled us to reach higher resolutions in characterizing popular topics discussed on municipal Facebook pages. The use of automatic tools is effective for studying social big data, as in Facebook, due to the increasing amounts of content uploaded to the platform (Cambria, E., Rajagopal, D., Olsher, D., & Das, D., 2013; Manovich, 2011; Orgaz-Bello, Jung & Camacho, 2016). Automatic tools enable an unprecedented scale and depth of analysis, specifically in social sciences research (Giles, 2012; Lazer, D., Pentland, A., Adamic, L., Aral, S., Barabási, A., Brewer, D., Christakis, N., Contractor, N., Fowler, J., Gutmann, M., Jebara, T., King, G., Macy, M., Roy, D., & Van Alstyne, M., 2009), in an attempt to extract the most important information (Javed & Muralidhara, 2015). Thus, this methodology reveals more than traditional methods may (Boyd & Crawford, 2012); automated text mining may effectively point to citizens' general perceptions on policies and services (Reddick, Takeoka Chatfield & Ojo, 2016). More specifically, automated tools are highly productive in linguistic analysis of uses, discourses and interactions between citizens-users and municipalities-pages. Finally, human coding data often faces problems of validity, reliability and bias, due to attributing different meaning to the same text (Mikhaylov, Laver & Benoit, 2012). Thus, at first stage and using purely automatic methods and tools we were able to get some clear sense of what occupies citizens in Israel on the municipal level (Steinfeld & Lev-On, 2015).

Manual text analysis, on the other hand, and in the current study- through the focus on the most popular posts- can reveal more in-depth patterns and uses of municipal Facebook pages by citizens and by municipalities. While looking at words and terms frequencies only, the depth of analysis was quite limited. Without knowing the cultural, municipal and practical context, the automatic findings had limited meaning for the research (Steinfeld & Lev-On, 2015). In addition, an analysis that was designed and conducted by the researchers, based on direct access and examination of selected raw data, freed us from reliance on software tools that operate to some extent as black boxes, which we cannot completely account for (Bollier & Firestone, 2010; Boyd & Crawford, 2012).

The findings from the manual content analysis of popular posts portray an image of the topics, forms of expression, and targets of discussions that mostly engage citizens on municipal Facebook pages. They indicate that information concerning cultural events dominates popular municipality Facebook pages. Leading posts by users and municipalities typically concern different topics: Municipalities focus on informative contents while users upload contents that are emotional in tone, a fact that may account for the popularity of users' posts (Berger & Milkman, 2012; Nelson-Field, Riebe & Newstead, 2013; Stieglitz & Dang-Xuan, 2013; Eckler & Bolls, 2011). Municipality posts contain little emotional tone, while one third of the most popular user-uploaded contents contain emotionality. Content by users also includes a significantly higher percentage of personal posts than posts by municipalities. Critical opinions, which are more frequently found in user posts rather than in municipality posts, trigger Comments on the topic in question, while supportive contents trigger more Likes.

Another aspect of activity on municipality pages is their dominant visual aspect. Images are not merely the most frequently used format, they also attract significant engagement levels.

To conclude, our analysis of the characteristics and indicators of post popularity on municipal Facebook pages reveals that several general variables such as the city population and general activity of the users on the city page correlate with posts' popularity. Municipal posts in general attract more user engagement, but user posts which manage to trigger much engagement are different than municipal posts in content (less informative, more critical) and in form (more emotional, using more personal tone).

FUTURE RESEARCH DIRECTIONS

Future studies should focus not only on automated and quantitative content analyses, but also include visual analysis which may add another layer to conventional analyses by providing information on the images that are posted, such as which images attract more Likes, Shares, and Comments.

Future studies might also incorporate interviews with page administrators and users who post to municipal pages, in order to gain insight into their views on Facebook page communications, strategies, and activity. Another direction for future research may compare the current findings with findings from other municipal public spheres such as Facebook pages or groups managed by residents, or Facebook pages of mayors, to determine the type of contents posted on these platforms that generate engagement. To extend a comparative analysis further, it would also be interesting to compare the activities on these platforms to activities on Facebook pages of other public institutions and figures, such as government ministries or parliament members, to provide a more comprehensive understanding of the character of the relationship between the public and its representatives in diverse social media arenas.

ACKNOWLEDGMENT

The study was supported by the Center for the Study of New Media, Politics and Society at Ariel University. The authors thank Valerie Aronov, Veronika Makogon, Avital Mandel, Bar Ram, Michael Barmatz, Yinon Sarusi and Nadav Gelbert for their assistance in data collection and coding, and Nomy Bitman for assisting with manuscript preparation.

REFERENCES

Berger, J. (2011). Arousal increases social transmission of information. *Psychological Science*, *22*(7), 891–893. doi:10.1177/0956797611413294 PMID:21690315

Berger, J., & Milkman, K. L. (2012). What makes online content viral? *JMR, Journal of Marketing Research*, *49*(2), 192–205. doi:10.1509/jmr.10.0353

Bilbao-Osorio, B., Dutta, S., & Lanvin, X. B. (2013). *The Global Information Technology Report 2013: Growth and jobs in a hyperconnected world*. Retrieved from http://www3.weforum.org/docs/WEF_GITR_Report_2013.pdf

Boczkowski, P. J., & Mitchelstein, E. (2012). How users take advantage of different forms of interactivity on online news sites: Clicking, E-Mailing, and commenting. *Human Communication Research*, *38*(1), 1–22. doi:10.1111/j.1468-2958.2011.01418.x

Bollier, D., & Firestone, C. M. (2010). *The Promise and Peril of Big Data*. Washington, DC: Aspen Institute.

Boyd, D., & Crawford, K. (2012). Critical Questions for Big Data: Provocations for a Cultural, Technological, and Scholarly Phenomenon. *Information Communication and Society*, *15*(5), 662–679. doi:10.1080/1369118X.2012.678878

Cambria, E., Rajagopal, D., Olsher, D., & Das, D. (2013). Big social data analysis. In R. Akerkar (Ed.), *Big Data Computer* (pp. 401–414). Boca Raton, FL: CRC Press. doi:10.1201/b16014-19

Central Bureau of Statistics. (2011). *Local Authorities in Israel 2011*. Retrieved from http://www.cbs.gov.il/publications/p_libud_11.xls

Cioffi-Revilla, C. (2010). Computational Social Science. *Wiley Interdisciplinary Reviews: Computational Statistics*, *2*(3), 259–271. doi:10.1002/wics.95

Coleman, S., & Gotze, J. (2001). *Bowling together: Online public engagement in policy deliberation* (pp. 39–50). London: Hansard Society.

ComScore. (2011). *It's a social world: Top 10 need-to-knows about social networking and where it's headed*. Retrieved from www.comscore.com/it_is_a_social_world

Criado, J. I., Sandoval-Almazan, R., & Gil-Garcia, J. R. (2013). Government innovation through social media. *Government Information Quarterly*, *30*(4), 319–326. doi:10.1016/j.giq.2013.10.003

Dahl, R. A. (1967). The city in the future of democracy. *The American Political Science Review*, *61*(04), 953–970. doi:10.2307/1953398

Dahl, R. A. (1970). *After the revolution? Authority in a good society*. New Haven, CT: Yale University Press.

Eckler, P., & Bolls, P. (2011). Spreading the virus: Emotional tone of viral advertising and its effect on forwarding intentions and attitudes. *Journal of Interactive Advertising*, *11*(2), 1–11. doi:10.1080/15252019.2011.10722180

Giles, G. (2012). Computational Social Science: Making the Links. *NATNews*, *488*, 448–450. Retrieved from http://www.nature.com/news/computational-social-science-making-the-links-1.11243 PMID:22914149

Graham, M., & Avery, E. (2013). Government public relations and social media: An analysis of the perceptions and trends of social media use at the local government level. *The Public Relations Journal*, *7*(4), 1–21.

Herring, S. C. (2009). Web Content Analysis: Expanding the Paradigm. In J. Hunsinger, L. Klastrup, & M. M. Allen (Eds.), *International Handbook of Internet Research* (pp. 233–249). New York: Springer. doi:10.1007/978-1-4020-9789-8_14

Hofmann, S., Beverungen, D., Räckers, M., & Becker, J. (2013). What makes local governments' online communications successful? Insights from a multi-method analysis of Facebook. *Government Information Quarterly*, *30*(4), 387–396. doi:10.1016/j.giq.2013.05.013

Huber, J., Landherr, A., Probst, F., & Reisser, C. (2012). Stimulating user activity on company fan pages in online social networks. ECIS 2012 Proceedings, Paper 188.

Javed, N., & Muralidhara, B. L. (2015). Framework for Participative and Collaborative Governance using Social Media Mining Techniques. *International Journal of Computers and Applications*, *123*(8), 36–41. doi:10.5120/ijca2015905435

Lazer, D., Pentland, A., Adamic, L., Aral, S., Barabási, A., Brewer, D., ... Van Alstyne, M. (2009). Life in the Network: The Coming Age of Computational Social Science. *Science*, *323*, 721–723. doi:10.1126cience.1167742 PMID:19197046

Lev-On, A. (2012). YouTube usage in low-visibility political campaigns. *Journal of Information Technology & Politics*, *9*(2), 205–216. doi:10.1080/19331681.2011.645360

Lev-On, A. (2014). Campaigning online, locally. *International Journal of E-Politics*, *5*(3), 14–28. doi:10.4018/ijep.2014070102

Lev-On, A., & Steinfeld, N. (2015). Local engagement online: Municipal Facebook pages as hubs of interaction. *Government Information Quarterly*, *32*(3), 299–307. doi:10.1016/j.giq.2015.05.007

Lev-On, A., & Steinfeld, N. (Under Review). *MPs on Facebook: An analysis of politicians' Facebook pages*.

Lovari, A., & Parisi, L. (2012). Public administrations and citizens 2.0: Exploring digital public communication strategies and civic interaction within Italian municipality pages on Facebook. In F. Comunello (Ed.), *Networked sociability and individualism: Technology for personal and professional relationships* (pp. 238–263). Hershey, PA: Information Science Reference. doi:10.4018/978-1-61350-338-6.ch012

Macintosh, A., & Whyte, A. (2008). Towards an evaluation framework for eParticipation. *Transforming Government: People. Process and Policy*, *2*(1), 16–30.

Magnusson, M., Bellström, P., & Magnusson, M. (2012). Facebook usage in government: A case study of information content. In *Proceedings of the 18th Americas Conference on Information Systems 2012* (pp. 353–362). Red Hook, NY: Curran Associates.

Mako. (2010). *Coming soon: A Facebook page for all MKs*. Retrieved from: http://www.mako.co.il/news-money/tech/Article-f89c1d2216b4721004.htm

Mann, R., & Lev-On, A. (2017). *Annual Report: The Media in Israel 2016. Agendas, Uses and Trends. Ariel*. Ariel University. (in Hebrew)

Manosevitch, I., Steinfeld, N., & Lev-On, A. (2014). Promoting online deliberation quality: Cognitive cues matter. *Information Communication and Society*, *17*(10), 1177–1195. doi:10.1080/1369118X.2014.899610

Manovich, L. (2011). Trending: the promises and the challenges of big social data. In M. K. Gold (Ed.), *Debates in Digital Humanity* (pp. 460–475). Minneapolis, MN: University of Minnesota Press.

Mergel, I. (2013). A framework for interpreting social media interactions in the public sector. *Government Information Quarterly*, *30*(4), 327–334. doi:10.1016/j.giq.2013.05.015

Mikhaylov, S., Laver, M., & Benoit, K. R. (2012). Coder Reliability and Misclassification in the Human Coding of Party Manifestos. *Political Analysis*, *20*(01), 78–91. doi:10.1093/pan/mpr047

Mossberger, K., Wu, Y., & Crawford, J. (2013). Connecting citizens and local governments? Social media and interactivity in major US cities. *Government Information Quarterly*, *30*(4), 351–358. doi:10.1016/j.giq.2013.05.016

Nelson-Field, K., Riebe, E., & Newstead, K. (2013). The emotions that drive viral video. *Australasian Marketing Journal*, *21*(4), 205–211. doi:10.1016/j.ausmj.2013.07.003

Norris, D. F., & Reddick, C. G. (2013). Local e-government in the United States: Transformation or incremental change? *Public Administration Review*, *73*(1), 165–175. doi:10.1111/j.1540-6210.2012.02647.x

Orgaz-Bello, G., Jung, J., & Camachoa, D. (2016). Social big data: Recent achievements and new challenges. *Information Fusion*, *28*, 45–59. doi:10.1016/j.inffus.2015.08.005

Pateman, C. (1970). *Participation and democratic theory*. London: Cambridge University Press. doi:10.1017/CBO9780511720444

Purian-Lukach, R. (2011). Municipal e-government: Comparative index, Success factors and models for management and evaluation. In E. Cohen & A. Lev-On (Eds.), *Connected: politics, Technology and society in Israel*. Israeli Political Science Association Press. (in Hebrew)

Ravitsch, P. (2005). *Difference between Israeli municipal websites, 2005: To what extent the inventory of online services fulfills the potentials of the medium and meets public expectations?* (MA thesis). Bar-Ilan University, Department of Political Science. (in Hebrew)

Reddick, C. G., & Norris, D. F. (2013). Social media adoption at the American grass roots: Web 2.0 or 1.5? *Government Information Quarterly*, *30*(4), 498–507. doi:10.1016/j.giq.2013.05.011

Reddick, C. G., Takeoka Chatfield, A., & Ojo, A. (2016). A social media text analytics framework for double-loop learning for citizen-centric public services: A case study of a local government Facebook use. *Government Information Quarterly*, *34*(1), 110–125. doi:10.1016/j.giq.2016.11.001

Rogers, R. (2009). *The End of the Virtual – Digital Methods*. Text prepared for the Inaugural Speech, Chair, New Media & Digital Culture, University of Amsterdam. Retrieved from http://www.govcom.org/rogers_oratie.pdf

Rotem, E. (2007). *Municipal e-government in Israel: A comparative perspective*. Hartog School for Governance and Policy. (in Hebrew)

Shkabatur, J. (2010). Cities at crossroads: Digital technology and local democracy in America. *Brooklyn Law Review*, *76*, 1413–1485.

Snijders, C., Matzat, U., & Reips, U. D. (2012). "Big Data": Big Knowledge Gaps in the Field of Internet Science. *International Journal of Internet Science*, *7*(1), 1–5.

Steinfeld, N., & Lev-On, A. (2015). 'Well-Done, Mr. Mayor!': Linguistic Analysis of Municipal Facebook Pages. *International Journal of E-Planning Research*, *4*(2), 26–38. doi:10.4018/ijepr.2015040102

Stieglitz, S., & Dang-Xuan, L. (2013). Emotions and information diffusion in social media: Sentiment of microblogs and sharing behavior. *Journal of Management Information Systems*, *29*(4), 217–248. doi:10.2753/MIS0742-1222290408

Tenenboim, O., & Cohen, A. A. (2013). What prompts users to click and comment: A longitudinal study of online news. *Journalism*, *20*(10), 1–20.

Tolbert, C. J., & Mossberger, K. (2006). The effects of e-government on trust and confidence in government. *Public Administration Review*, *66*(3), 354–369. doi:10.1111/j.1540-6210.2006.00594.x

Chapter 3
Enhancing E-Participation in Urban Planning Competitions

Pilvi Nummi
Aalto University, Finland

Susa Eräranta
Aalto University, Finland

Maarit Kahila-Tani
Aalto University, Finland

ABSTRACT

Planning competitions are used as a way to determine alternatives and promote innovative solutions in the early phase of urban planning. However, the traditional jury-based evaluation process is encountering significant opposition, as it does not consider the views of local residents. This chapter describes how web-based public participation tools are utilized in urban planning competitions to register public opinion alongside the expert view given by the jury. The research focus of this chapter is on studying how public participation can be arranged in competition processes, how the contestants use the information produced, and how it has been utilized in further planning of the area. Based on two Finnish case studies, this study indicates that web-based tools can augment public participation in the competition process. However, the results indicate that the impact of participation on selecting the winner is weak. Instead, in further planning of the area, the public opinions are valuable.

DOI: 10.4018/978-1-5225-5999-3.ch003

INTRODUCTION

In Nordic countries and in Europe, urban planning and design competitions play a significant role as planning instruments in the field of urban planning. Indeed, such competitions are often used as a way to identify and promote new ideas and alternatives in relation to various urban design projects or even as a starting point for an actual plan. In the context of urban planning, competitions are used, for example, in the planning of new housing areas and in urban renewal projects in areas where changes in land use are envisaged.

In Finland, municipalities have a strong mandate for public urban planning. The land use system is hierarchical, from the regional level steered by the unions of counties to the municipal level of planning. Though the demand for participatory planning is acknowledged in the Finnish Land Use and Building Act (132/1999) and the role of participation has strengthened in recent years, the form of participation varies across Finnish cities (Bäcklund & Mäntysalo, 2010). This has especially been the case in planning competitions, where participation is not taken into account in the Finnish rules for competitions (SAFA, 2008). Even though competitions are used as planning instruments aiming to develop urban areas, participation during the competition process is not mandatory. The official participation process usually only begins after municipal planning has commenced. In this chapter, we study how web-based tools are used in urban planning competitions to enhance public participation in planning.

Various academic studies have been conducted in the field of public participation and collaboration of urban planning. These studies have focused either on the activities in which people are engaged or on the broader purpose of participation (Schlossberg & Shuford, 2005). In very recent years, there has been an increasing research interest in architectural and urban planning competitions (Andersson, Zettersten, & Rönn, 2016). Despite this, research cases concerning public participation in planning competitions represent something of an exception, with relatively little previous work being recorded. The need for enhancing participation in the competition process has recently been identified and studied. For example, a dialogue-based architectural competition model presented by Kreiner, Jacobsen, and Jensen (2011) adds interaction between the contestants and the competition jury; Volker (2010) has studied design competitions from the client perspective; and Grangaard and Ryhl (2016) have approached the topic from the perspective of universal design. However, public participation in urban planning competitions is a sparsely studied area, and there is an evident gap in research of web-based public participation in this context. This study aims to fill this gap by providing insight into how *e*-participation can be implemented in urban planning competitions.

European urban planning has traditionally been based on rational planning thinking, which arose in Europe in the 1960s. In the rational planning process, planning solutions can be based on scientific reasoning; that way, the environment can be planned to improve human well-being (Taylor, 1998). Currently, participatory and communicative planning approaches are on the ride and are challenging the rational planning paradigm. The rational process itself is not in conflict with participatory planning, but, in the tradition of rational planning, there is no space for residents' views, as opinions are not considered to be scientific information (Rantanen & Kahila, 2008). In Finland, the move away from the rational planning paradigm towards a communicative and participative planning practice remains an ongoing process.

At the same time, the use of the Internet and social media has reached a significantly high level in Finland. In 2015, it was estimated that 92% of Finns aged 16-74 used the Internet, and 84% did so on a daily basis. Apart from that, more than half of the population is registered on at least one social media service (OSF, 2016). In recent years, the wide use of the Internet and social media has offered new possibilities to involve people in urban planning processes via the web. The electronic participation tools that aim to support more diverse preparation of plans, enable more efficient and democratic processes, and produce more comprehensive and reliable follow-up are now well accepted (*Evaluation of the Land Use and Building Act in practice*, 2014). In order to become an essential part of future competition processes, web-based participation (*e*-participation) practices should be further developed, and their impact on the planning process should be evaluated.

This chapter contributes to the research of *e*-participation and is organized as follows: the next section presents the background for this study by describing aspects of public participation in urban planning, the tradition of urban planning competitions in Finland, the concept of *e*-participation, and recent research related to the topic. Before entering into the two case studies, the research methods and questions are presented. In this chapter, two case studies are explored. In Vaasa, we have analyzed the ideas competition in the Strait of Palosaari and, in Sipoo, the River Valley of Sipoo. Results derived from these case studies are formulated to demonstrate solutions and provide recommendations. At the end of the chapter, we suggest a few new avenues for future research and draw conclusions.

BACKGROUND

Public Participation in Urban Planning

The rational planning tradition has long been dominant in Finland (Roininen, Horelli, & Wallin, 2003). However, recent decades have seen residents and other local actors acquire a more active attitude towards planning. Indeed, citizen activism and bottom-up participation in planning issues is currently rising (Graziano, 2017). This has created a need to develop planning practices regarding communicative and collaborative planning. Compared to rational planning, communicative and collaborative planning both recognize the diversity of local values and interests and emphasize the role of different stakeholders in the practice underlining the process and the content of planning (Healey, 1997). It is also viewed as the most influential theory among the alternatives to the rational planning paradigm (Silva, 2010).

The development of new approaches to communication and collaborative decision-making are important for engaging citizens in contemporary urban communities (Anttiroiko, 2012). Evidently, different approaches or orientations exist regarding the idea of participation (Schlossberg & Shuford, 2005). When Arnstein (1969) formulated the 'ladders' of public involvement, public participation was framed along a spectrum from manipulation to citizen control. In this definition, participation is framed from the perspective of the level of citizen power. The approaches followed by Arnstein (1969) often frame participation differently, as in Dorcey's approach (as cited in Schlossberg & Shuford, 2005), where the participation more closely follows the actual planning process, thus allowing the change of participation during the decision-making process.

As Silva (2013) argues, public participation is described as having a limited impact on the urban planning process. Participatory planning has been criticized for neglecting the issue of how information might shape the output of planning processes and the type and form of knowledge needed (Brown & Weber, 2012). Rantanen and Kahila (2008) note that resident knowledge has sometimes been dismissed in planning processes on the grounds that it is 'soft' and 'subjective,' while priority is given to 'hard' data. In practice, this has led to a situation where only residents with professional skills have the ability and, thus, the opportunity to influence the process (Staffans, 2004). Though planners are willing to both use and value experience-based data, their competence in the analysis and utilization of such data often varies significantly (Sanvig-Knudsen & Kahila, 2012). As such, this can lead to understating of the meaning and value of resident information in planning and decision-making processes (Eräranta, 2013).

According to the Finnish Land Use and Building Act (Finlex, 1999), everyone has the right to participate in urban planning. However, this applies only to official planning projects. As such, public participation is not mandatory in planning competitions. Moreover, opposition to the ways in which traditional planning competitions are conducted, particularly in relation to their lack of public participation, is clearly growing (Rönn, 2009).

Furthermore, other scholars agree that the traditional competition process does not engage citizens in evaluation of the entries or in the selection of the winner. Although

Volker (2010) points out several participatory elements in the competition process, she also admits that the potential of public participation cannot be exploited, due to the power of the jury. It is evident that new participatory competition processes are needed. Dialogue-based architectural competition (DAC) is one example of a new model, where interaction between stakeholders is arranged before and during the competition process. The aim in the DAC process is to clarify the clients' needs and preferences during the competition in order to gain mutual understanding about the planning task (Kreiner, Jacobsen, & Jensen, 2011). The organizers of participation in architectural competitions have also learned from the experiences of how the design industry perceives and engages end users in the development process (Våland, 2009). Grangaard and Ryhl (2016) argue that understanding the users is a key factor in a user-centered planning competition. They present a case study where universal design approach was applied in an architectural competition of the Vandhalla sports center. In that process, knowledge about user needs was proven to be an eye-opener and success factor for the contestants.

Urban Planning Competition Tradition in Finland

The tradition of planning competitions is more than a hundred years old (Rönn, 2011). In Finland, each year there are approximately 5 to 10 open competitions and 10 to 30 invitational competitions organized (SAFA, 2011). Thus, urban planning competitions are a significant institution that shapes our living environment and the way it is planned.

The General Council of the Finnish Association of Architects (SAFA) approves the official competition rules, which determine the process of organizing a competition. SAFA is usually invited to be a partner in the arranging of such competitions. The Finnish competition tradition can be categorized, for example, as open or invitational competitions and planning/design or ideas competitions. Open competitions are open to anyone meeting the basic criteria set out in the competition conditions; invitational competitions, on the other hand, are only open to participants who have been pre-selected on an invitation-only basis (SAFA, 2008). Ideas competitions can be utilized to map out broad solutions for further development. It should also be noted that such competitions do not necessarily lead to a design commission being granted, while planning and design competitions generally aim at finding a designer and a basic solution for further development of the project (SAFA, 2008).

A competition brief sets out the goals, the requested competition material, the evaluation criteria, and the information with respect to the display of the entries. The brief is important, as it contains all the necessary information required by the contestants, who must follow the program carefully. It is evident that the competition brief substantially directs the outcomes of the competition, and a well-designed competition process and brief are crucial for success. Volker (2010) argues that there is *no one formula for success*; every competition is a design challenge with specific goals and challenges. Kreiner, Jacobsen, and Jensen (2011) emphasize that there are possibilities of misunderstanding the competition goals. Therefore, they propose a dialogue-based architectural competition (DAC) as a method for a participatory competition process that enables early clarification of the needs and preferences presented in the brief.

For every competition, a jury is appointed to approve the competition conditions, answer questions about the competition, evaluate entries and decide upon the winner. The jury has an important role to play during the competition process; the competition rules state that at least one-third of the jury's members have to be professionals in a relevant field (SAFA, 2008). Jury members can include representatives from the organizing municipality and also municipal decision-makers. In order to guarantee competition secrecy and the impartial evaluation of the entries, contestants must ensure anonymity when submitting their entries. The anonymous competition entries are presented to the jury, which discusses each entry on the basis of the evaluation criteria confirmed in the competition brief. The traditional jury process is closed and autonomous, with public discussion of the entries usually only initiated after the results have been announced (Svensson, 2009). Accordingly, Volker (2010) argues that,

competitions do have a lot of participation potential, but (that) the power of a jury panel does not leave a lot of room for other stakeholders to have decisive rights and therefore influence the final outcome (p. 9).

New Tools for eParticipation in Urban Planning

The Internet and social media represent a challenge to contemporary urban planning practices and to expert-led processes in particular. The development of information and communication technologies (ICT) has created a demand for planning and participation processes to enhance participation methods. Using web-based methods in urban planning participation is often termed *e*-planning or *e*-participation. According to Anttiroiko (2012), the impact of the Internet on urban planning is so strong that it can be conceived of as a paradigm change.

Here, *e*-participation is defined as,

the sociocultural, ethical, and political practice in which people take part online and offline in the overlapping phases of the urban planning and decision-making cycle (Horelli & Wallin, 2010, p. 3).

The ways in which participation, administration, and decision-making are conducted are clearly now changing (Wallin, Amati, Saad-Sulonen, & Horelli, 2012). This opens up the possibility that *e*-participation practices in planning competitions could also be altered. According to Wallin et al. (2012), the focus of *e*-participation is on new tools, structures, and the timing of participation in the long planning and development processes of urban areas. Despite this, the development process of new tools is vulnerable to various organizational issues, like power relations, unbalanced collaboration, and weak communication (Kahila-Tani, 2015). On the other hand, more often, many geographic information systems (GIS) applications function more top-down, and interaction between the residents and planners is more one-way (Talen, 2000).

The scope of *e*-participation in urban planning varies from conversational to GIS-based methods and from one-phase projects to ongoing processes. For example, typical conversational tools include planning blogs, chat rooms, discussion forums, virtual communities, and mailing lists (Silva, 2010). GIS-based tools, such as PPGIS, are often used at the beginning of the process. On the other hand, PSS (planning support systems) are utilized throughout the planning process. Generally, various collaborative geospatial/geovisual decision support systems sustain collaborative decision-making, using GIS-based methods in participatory democracies (Bailey & Grossardt, 2010). Generally, *e*-participation tools used in planning competitions focus on the evaluation of competition entries. There is, however, a contradiction between the ideal and actual practice of public participation. If public participation do not have a chance to affect the process (planning of the entries and selecting the winner) nor the outcome (plan proposal) it does not fulfill the conditions of impactful public participation. Real involvement in decision-making requires an experience of influencing the process (Roininen, J., Horelli, L., & Wallin, S., 2003).

In recent years, a number of new participatory methods have developed in conjunction with various Finnish planning competitions. These methods include pre-competition workshops and PPGIS tools for gathering local information for assisting the contestants in understanding local values and needs, for appointing resident representatives in the jury panels, and for public evaluation of the competition entries. The development of public evaluation methods has become more popular in competitions and has precipitated further discussion (e.g., Kazemian, R., Rönn, M., & Svensson, C., 2006) of both the challenges faced in the traditional jury process and the timing of public evaluation. As such, there is a need to review the development of participation methods in planning competitions. On the other hand, the notion that Stern, Gudes, and Svoray (2009) posit is valid, both in planning competitions and in planning processes, generally: when supporting public participation with web-based tools, the potential advantages should be studied to clarify whether the tools only complement the set of traditional participation techniques or if the tools are capable of replacing them (Stern, E., Gudes, O., & Svoray, T., 2009).

Related Academic Research

The importance of architectural and planning competitions is reflected in the research field. Since the 1990s, research on this topic has emerged, and the growing numbers of research results indicate that the topic has emerged as an individual research field (Andersson, Zettersten, & Rönn, 2016)

Recently, the knowledge and dialogue behind the judgment of architectural competitions has been acknowledged more in research. For example, Wallhagen, Malmqvist, and Eriksson (2017) have studied professional knowledge in architectural competition processes. Andersson (2016) studied how architectonic visions were converted into a built environment under the influence of Swedish civil administration. Jacobsen, Harty, and Tryggestad (2016) discussed wicked design problems and the use of the dialogue-based competition process in their research. Al-Qaysi et al. (2016) frame judgment in architectural competitions as a communicative, deliberative practice, through acknowledging the diversity of stakeholders involved in the decision process.

Kazemian (2009) and Våland (2009) have pointed out that a significant gap remains in relation to studies of public participation in urban planning competitions. Since then, there have been few studies that fill this gap. For example, when considering an ethnographic case study carried out in the Hobart Waterfront urban design competition in Tasmania, Picken (2013) points out that urban design competitions have the potential to increase citizens' awareness of urban planning and willingness to participate in planning. In that study, local residents are considered to be consumers and urban designs are considered to be products to be consumed. Furthermore, Garde (2014) reported a case study of a public participation process in an international design competition for the design of the Great Park in Irvine, California. In that case study, public participation was arranged before and during the competition. At the beginning, innovative ideas were collected in workshops with local representatives, and in the evaluation phase, citizens were invited to give feedback and rate the entries. Based on the case study, Garde argues that *citizen participation by itself is inadequate in achieving excellence in the product* (p. 114). As a recommendation for functioning public participation, Garde suggests a multi-phased competition model that starts with an open competition aiming to produce a large number of alternatives and continues through voting on an invitational competition. Garde also suggests that the winner could be selected based on public feedback and the jury.

The development of a new participatory competition format is evidently in progress. Dialogue-based architectural competition (DAC) is a model that integrates dialogic interaction in the competition process. The aim of the process is to enable interaction between the client and contestants in order to create mutual understanding of the goals and preferences for the design. Due to the legal requirements, this kind of interactive competition process has to be implemented in two phases: a collaborative starting phase and a closed concluding phase (Jacobsen, Harty, & Tryggestad, 2016; Kreiner, Jacobsen, & Jensen, 2011).

In general, recent studies prove that there are needs for and approaches to the inclusion of participatory actions in planning competitions. It is also evident that there is a clear need for further research on the area, especially focusing on the development of *e*-participation methods within the competition process.

RESEARCH METHOD AND QUESTIONS

This chapter is an enhanced and updated version of the article 'Web-based Public Participation in Urban Planning Competitions' (Eräranta, Kahila-Tani, & Nummi-Sund, 2015). These studies synthesize the results of two case studies, where participation during the competition process was studied. The focus is on the possibility of pragmatically including public participation in the competition process; two cases in which a variety of web-based participation tools were utilized in the city of Vaasa and in the municipality of Sipoo are presented. The Sipoon Jokilaakso (River Valley of Sipoo) planning competition represents a typical planning competition of a new housing area. The Palosaaren salmi (Strait of Palosaari) project in Vaasa is an example of an urban renewal project. In the context of this chapter, we simply refer to these projects as Vaasa and Sipoo. The aim of *e*-participation was to enable the public to participate in the competition process as fully as possible. Internet tools were used to present and evaluate the competition entries. In addition, in the case of Vaasa, public participation geographic information systems (PPGIS) software was used from the outset to collect experience-based information from residents. This chapter presents the use of these tools and their effect on the competition process and further planning of the areas.

In both cases, research was conducted by following the competition process closely from beginning to end. The researchers participated in the meetings of the competition planning phase and studied the material related to the process. The material collected during the observation period composed the key data for the study. In addition, two key interviews were conducted after the competition phase in 2012, and one email interview was conducted again in 2017. The aim of the interviews was to gain understanding of how the public participation succeeded from the administrative perspective and how the information generated by the residents was utilized in the decision-making and further planning of the area. Also, the competition briefs and competition entries, as well as planning material produced after the competition, are used as research data. The case studies are described in the following section.

The research approach is twofold: first, possibilities to arrange public participation are studied, and after that, the utilization of the information produced in the participatory process (i.e., public opinions) is analyzed.

In relation to the first problem, two research questions are answered:

- How can public participation be arranged in a planning competition process?
- What kinds of *e*-participation tools are useful in conducting participation processes in respect to planning competitions?

Regarding the use of the information produced by residents, the following questions are examined:

- Is the information that is produced by residents used in the competition process?
- Is the information used in further planning of the area?
- What issues can influence the utilization of the residents' views?

UNRAVELLING THE STORY OF TWO CASE STUDIES

In this study, we present two case studies. In the city of Vaasa and in the municipality of Sipoo, a variety of web-based participation tools were utilized in ideas competitions. The aim of the cases was to incorporate public views into the competition at the beginning of the process by providing the contestants insight into residents' views. Furthermore, the goal was to give the local community a better chance to participate in the evaluation of the competition entries while testing new web-based methods and tools in an ideas competition. The coastal city of Vaasa, with a population of 60,000, is a lively center of culture, education, and tourism in western Finland. The municipality of Sipoo is in close proximity to Finland's capital city, Helsinki, in a metropolitan area. There are 20,000 inhabitants in Sipoo, and it is currently the fastest growing municipality in Finland (Figure 1).

The research project in Vaasa started in autumn 2010. Initially, resident views were collected with a softGIS method. The ideas competition was organized during the spring and summer of 2011. Public evaluation of the competition entries took place after the jury concluded the official evaluation. In Sipoo, the competition process started in autumn 2011, with public evaluation open from December 2011

Figure 1. Aerial photo of Palosaari (Vaasa) and the Jokilaakso area (Sipoo). © (2011, City of Vaasa and 2010, Municipality of Sipoo). Used with permission.

until January 2012. The jury was able to become familiar with the public's opinions before the evaluation period commenced (Figure 2). The case studies are compared in detail in Table 1.

The Strait of Palosaari in Vaasa

Ideas Competition Phase

The research study in Vaasa started when the planners contacted researchers at Aalto University with a view to utilize the softGIS method developed by the Aalto University Land Use Planning and Urban Studies Group. This initial contact led to a multi-year project, during which the participation process and tools were developed and studied. The softGIS methodology is an advanced example of an Internet-based PPGIS methodology that allows residents to produce localized experiential knowledge that can be analyzed together with the register-based data included in geographic information systems (Kahila & Kyttä, 2009; Kyttä, 2011; Kyttä & Kahila, 2011). The softGIS method has been rigorously tested in empirical research projects (Kahila & Kyttä, 2009). The difference between previous softGIS methods and the one used here was that the first continuously open softGIS portal was developed in Vaasa (Figure 3). Researchers and IT specialists coded and tailored the first generation softGIS tools, especially for each project. Later, a "do it yourself" softGIS service, Maptionnaire, was built (see more at http://maptionnaire.com).

Researchers contributed to the competition brief and, together with the planners, designed the participation process for the invitational competition. The planners wanted to emphasize the role of the residents at the beginning of the process. The researchers and planners drew up the set of questions used in the query. The aim was to create a permanent set of questions to be used in every planning project in Vaasa for the collection of more continuous data on residents' experiences.

Figure 2. Process charts for the Vaasa and Sipoo ideas competitions.

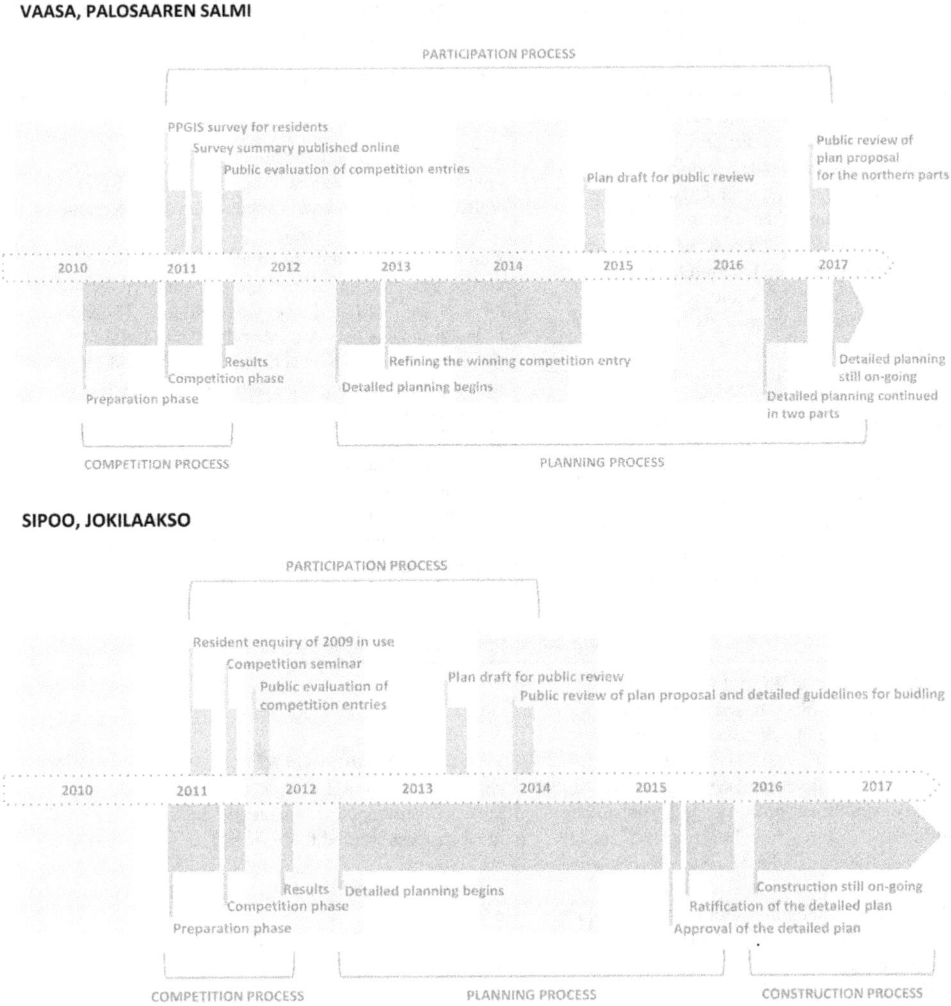

Generally, all feedback received with the softGIS survey was positive and appropriate. Most of the comments originated on the small island of Mansikkasaari (Figure 4). The respondents marked more new houses than preserved places. Areas to be improved were mainly located on Mansikkasaari. Interestingly, residents marked, as new proposals on the map, both high rises and low-rise townhouses. By studying all of the marked locations, it was easy to identify an area that was clearly viewed as a special place by the residents, namely, a small beach in Mansikkasaari. Respondents hoped that this place could be preserved, as it is a very popular beach among residents living nearby. Implacable resident resistance to new construction

Table 1. The comparison of the case studies in Vaasa and Sipoo

THE STRAIT OF PALOSAARI IDEAS COMPETITION IN VAASA	THE RIVER VALLEY OF SIPOO IDEAS COMPETITION IN SIPOO
General Information about the Competition	
• Competition area: Existing 1950's neighborhood of 5,500 inhabitants in Palosaari, Vaasa. • Initiative: 30-year history of opposing views of developing the area. Urban planners and politicians have shared an interest in redevelopment of the Palosaari neighborhood, primarily because of its location next to the city center and its scenic coastline; residents and the resident union have strongly resisted these planning proposals. • Form and participants: Invitational, five offices, among 45 registered offices were selected; three were selected on the basis of references and two by drawing lots. • Competition website: Your Vaasa portal was developed to collect the various tools used during the ideas competition process to inform residents of the arranged competition and of the schedule and to allow the local community to evaluate the competition entries.	• Competition area: New housing area for approximately 1,000 new inhabitants of the River Valley of Sipoo, situated in the administrative center of Sipoo (4,300 inhabitants). • Initiative: The planning area is currently almost completely an undeveloped agricultural area located close to the commercial center of Nikkilä. There seems to be consensus between inhabitants and politicians on the development. However, there are different views on how the new housing areas should be planned. • Form and participants: Invitational, the organizer chose five contestants by drawing lots among 26 registered offices. • Competition website: The competition website was developed to inform the local community of the competition and to allow the local community to evaluate the competition entries.
Pre-Competition Public Participation	
• **Method:** The softGIS tool was used to collect residents' place-based experiences, such as positive and negative characteristics of the physical landscape, social atmosphere, appearance, and functional settings. Residents also located development proposals (e.g., preserved and improved places, new houses, services, and routes) on the map and assessed everyday services used. • **Timing:** In the beginning of competition phase. • **Amount:** 69 respondents with 121 locations marked on the map. • **Utilization:** Collected softGIS data formed part of the competition background material provided to the contestants through an online tool and as a written summary. The material was also delivered to the jury and to city officials. The contestants were required to state how they utilized the data in their competition entries.	• **Method:** Due to a tight schedule, an earlier resident enquiry realized in 2009 was utilized. In addition, a kick-off conference was arranged, in which two local inhabitants presented their views and wishes to the contestants. • **Timing:** In 2009 and in the beginning of competition phase. • **Amount:** 126 respondents in the resident enquiry. • **Utilization:** Resident views of the development of Nikkilä were described in the competition brief and presented in the kick-off conference. Resident views were presented in either written or spoken form. No maps were produced to visualize these viewpoints.

continued on following page

Table 1. Continued

THE STRAIT OF PALOSAARI IDEAS COMPETITION IN VAASA	THE RIVER VALLEY OF SIPOO IDEAS COMPETITION IN SIPOO
Public Evaluation of Competition Entries	
• **Method:** Public evaluation of the competition entries at the competition website. Evaluation was accomplished through a multiple-choice questionnaire covering the issues of physical landscape, social atmosphere, appearance, and functional settings in addition to an open-ended comment section. Also, a general grade was available. • **Timing:** After the jury concluded the official evaluation. When the competition ended, the jury chose the winner. Web pages for public evaluation of the entries were then opened. • **Amount:** 39 open-ended comments • **Utilization:** Residents' comments were compared to the comments received from the jury. The aim is to include these comments into the planning process that starts after the competition phase.	• **Method:** Public evaluation of the competition entries at the competition website. Evaluation tool was a web form, where the user gave grades and verbal feedback. Numerical evaluation was accomplished through a questionnaire covering the issues of the evaluation criteria: energy efficiency and low emission rates, housing solutions, connection to the surroundings, distinctiveness, and functioning traffic arrangements. Also a general grade was available. • **Timing:** Entries were published on the competition website before the jury began their evaluation process. • **Amount:** 181 open-ended comments • **Utilization:** The secretary of the jury delivered copies of the comments to each jury member before the official evaluation. The comments are utilized in the further planning process.
Further Planning of the Area	
• Detailed planning process commenced after the competition ended. • The winning entry was developed further, based on the jury's comments. • Public opinions are not documented in the plan report (City of Vaasa, 2017b). • The detailed planning process is currently still ongoing.	• Detailed planning commenced in cooperation with the winner after the competition ended. • Feedback from the competition phase was analyzed and written in a separate report. • Public opinions were taken into account in the detailed planning process and documented in detail in the plan report (Municipality of Sipoo, 2015). • The detailed plan is approved and ratified. • The construction of the area is currently ongoing.

was not evident from the results. The results showed that residents were more worried about a possible change in the atmosphere and of the loss of 'old charm' associated with the place.

In the competition brief, the following topics were highlighted:

- The goal is to find a high-grade land use solution to support a redevelopment process that takes into consideration the conservation and landscape challenges of the area.
- The above-mentioned challenges should fit into the city structure, architecture, and functional whole of the city and should be realizable.
- The solution should consolidate the existing cityscape and historical identity of the city and enable the development of the shore to become livelier by zoning the area for park and recreational use.

Figure 3. SoftGIS tool in Vaasa. © 2011, City of Vaasa and Aalto University. Used with permission.

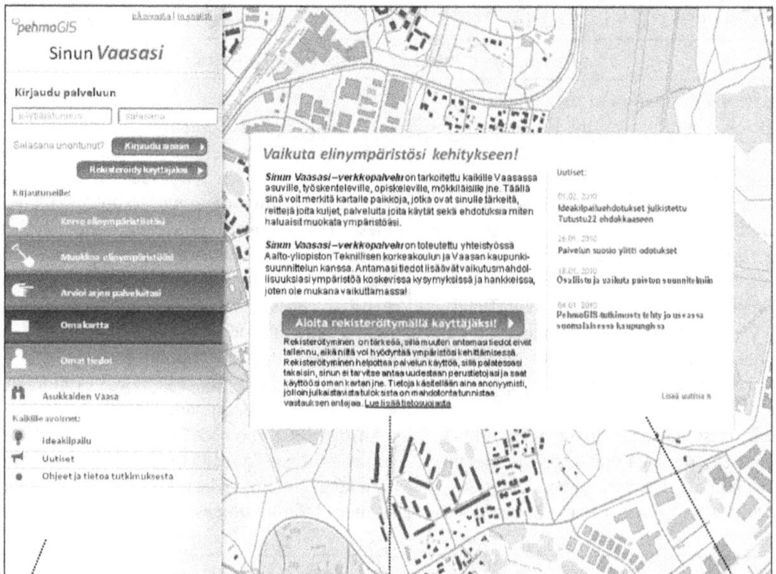

softGIS
Your Vaasa

Log in
User name Password
Forgotten your password? Log in
Register as a user

For logged in users:
Tell about your living environment
Edit your living environment
Evaluate your everyday services
Own map
User information

Residents' Vaasa
Open for everyone:
Idea competition
News
Guide and information of the research

Influence the development of your living environment!

Your Vaasa internet portal is meant for everyone who lives, works, studies, has holiday home etc. In Vaasa. Here you can mark on a map places that are important to you, routes that you use, services you use and suggestions of how you would like to edit your living environment.

Your Vaasa internet portal is realised together with Aalto University and Vaasa city planning department. The information you give will increase your possibilities to influence the development of your living environment, so participate!

Start with registration!

Registration is important, because otherwise your information will not be saved, and thus cannot be used in developing your living environment. Registration will make the use of the service easier, because you don't have to give again your basic information, when coming back to the page, you'll have your own map etc. All information will be treated anonymously, so it is not possible to identify respondents from the published results. Read more about confidentiality.

News:

01.02.2010
Idea competition proposals published
Explore the 22 proposals

26.01.2010
Popularity of the service exceeded all expectations

18.01.2010
Take part and influence park planning

04.01.2010
SoftGIS research has been done already in many Finnish cities

More news >>

Public participation was mentioned in the competition brief as follows:

- The ideas competition aims to utilize the softGIS project of Palosaari. The project enables the citizens to produce information about their living environment and feed this information into future proposals made for the area.
- In their proposals, contestants must clearly state how they plan to utilize data collected with the softGIS tool.

Figure 4. Residents' located views with softGIS-tool. © 2011, City of Vaasa and Aalto University. Used with permission.

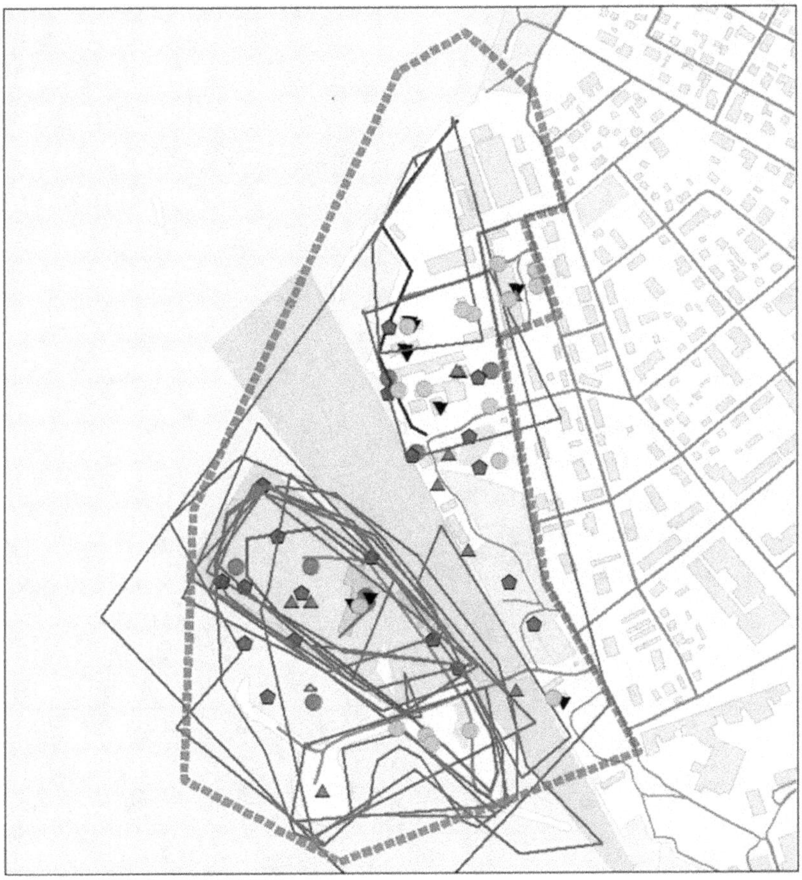

After the competition, the web pages for the public evaluation were opened. In the web pages, the residents were able to get information about the competition and comment on the competition entries (Figure 5).

Further Planning: Detailed Plan for the Area of Palosaaren Salmi

After the competition phase, the winning entry was developed further with the winning designer A detailed plan process was initiated in November 2012, and the planning is currently still in progress. It is evident that there are issues that have delayed the planning process. After the first draft of the detailed plan was published for public

Figure 5. Evaluation pages in Vaasa. © 2011, City of Vaasa and Aalto University. Used with permission.

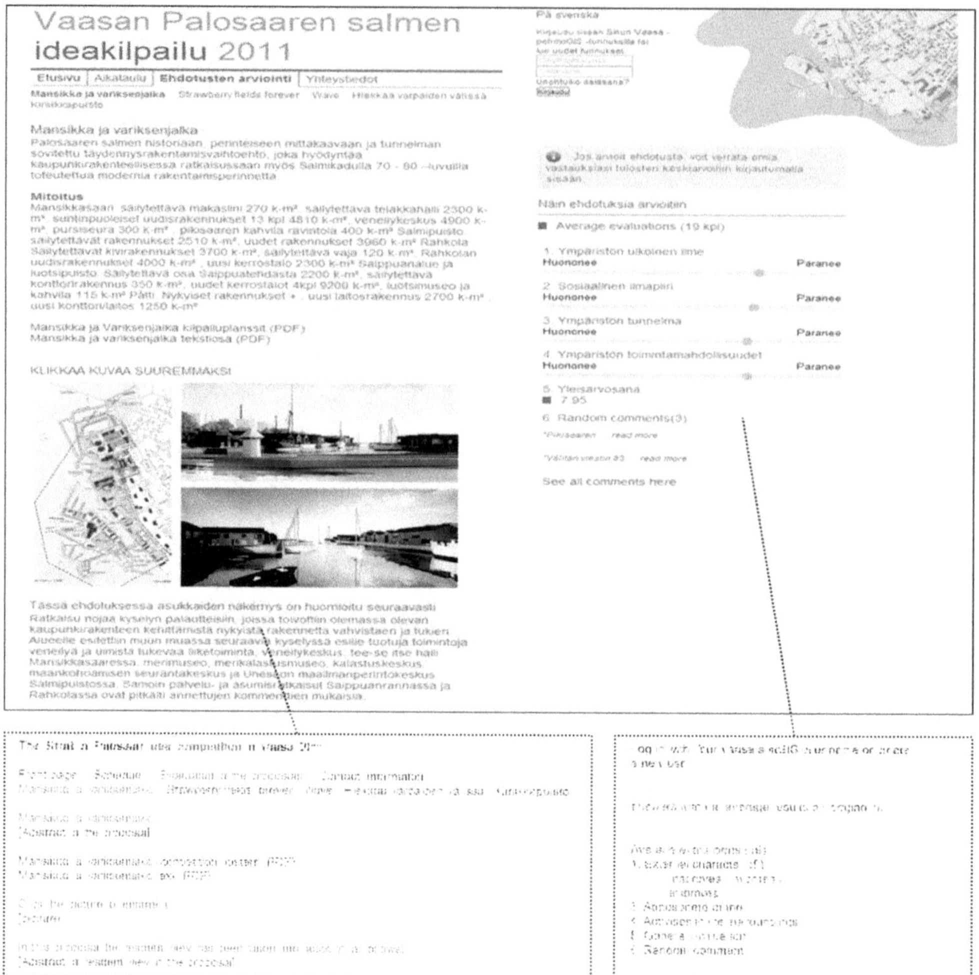

participation in spring 2015, the planning area was divided into two separate detailed planning processes: the northern part of Palosaaren salmi comprises the park of Rahkola-Varisselkä (City of Vaasa, 2017b), and the southern part covers the island of Mansikkasaari and park Salmenpuisto (City of Vaasa, 2017a).

The ideas competition is mentioned in the plan report as one starting point for the planning of Palosaaren salmi in parallel with the general plan of Vaasa. The winning entry and the development goals described in the jury's evaluation protocol are listed as background material for the detailed plan. The winning entry has been developed after the competition, based on the feedback from the jury. It is noteworthy that the information that residents produced during the competition has not been described in the plan report (City of Vaasa, 2017b).

The responsible planner has changed after the competition phase in Vaasa. Consequently, there has been a break in the communication between the researcher and the city planning. Possibly, this has also affected the information flow from the competition phase to further planning of the area.

The River Valley of Sipoo

Ideas Competition Phase

The research project in Sipoo commenced in spring 2011 with the planning of the competition website and the public evaluation tool. The planners in Sipoo carried out the design of the tools. An earlier ideas competition (the Sibbesborg competition for a sustainable community), held in 2010-2011 in Sipoo, provided an important starting point for the River Valley competition project. The Sibbesborg case helped highlight the challenges encountered in respect to the participation methods that helped to shape development for future competitions (Merikoski, Eräranta, & Staffans, 2012). These experiences and the knowledge they generate within the municipality effectively made it possible to implement *e*-participation in the context of the River Valley competition.

Figure 6. Evolution of the plan: the winning entry on the left, developed version of the winning entry in the middle, and illustration of the detailed plan draft on the right. ©2011, Käpy and Simo Paavilainen Architects (the winning entry) and City of Vaasa.
Used with permission

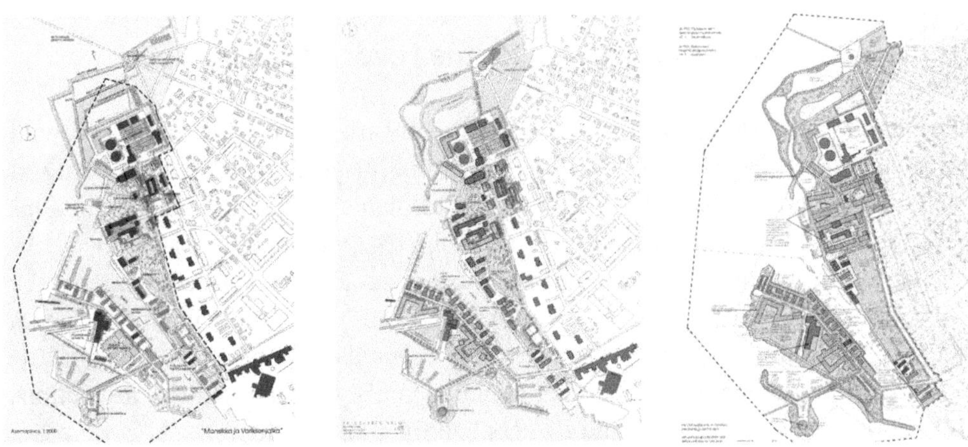

The aim of the competition was to create a charming, livable townhouse-style area for housing and urban functions with personal identity. The following goals were stated in the competition brief:

- The structure of the area shall encourage people to move around on foot or by bike or to use public transport.
- Safe and functional areas will provide an enjoyable living environment for people from different age groups.
- The banks of the river will be developed into an active part of the green area network.
- Eco-efficiency is one of the leading goals in planning and it shall support the community and the personal identity of the area.

As in Vaasa, researchers contributed significantly to the competition brief with relevant issues, such as *e*-participation. The public evaluation of the entries was promised in the competition brief. Instructions were also given to produce a short text and select images to be shown on the competition website.

The results of a survey conducted in 2009 were utilized as background information for the competition brief. In that survey, residents expressed a desire for small-scale buildings, bicycle and pedestrian transport connections, traditional wood construction methods, and new services.

In Sipoo, participation focused on the public evaluation of the entries. The entries were published on the competition website before the jury began their evaluation process (Figure 7). In addition, a comparison table, where entries could be arranged by the average grades of each criterion, was published on the website after the public evaluation (Figure 8). Thus, contrary to the situation in Vaasa, the jury was able to familiarize themselves with the public's opinions before the evaluation period commenced.

Further Planning: Detailed Plan for the River Valley of Sipoo

The planning of the River Valley continued directly after the competition phase with the architects of the winning entry. The planning project was initiated in August 2012, and the preliminary draft of the detailed plan and the building instructions were published for public participation in April 2014 (Municipality of Sipoo, 2015).

The public participation and information generated by residents during the competition is described in great detail in the plan report (Municipality of Sipoo, 2015, p. 52). Public opinions are presented as goals that originated during the planning process in parallel with other planning goals, such as technical restrictions and development aims of the land owners and the municipality.

Figure 7. Evaluation questionnaire on the competition website of the River Valley of Sipoo. © 2012, Municipality of Sipoo. Used with permission.

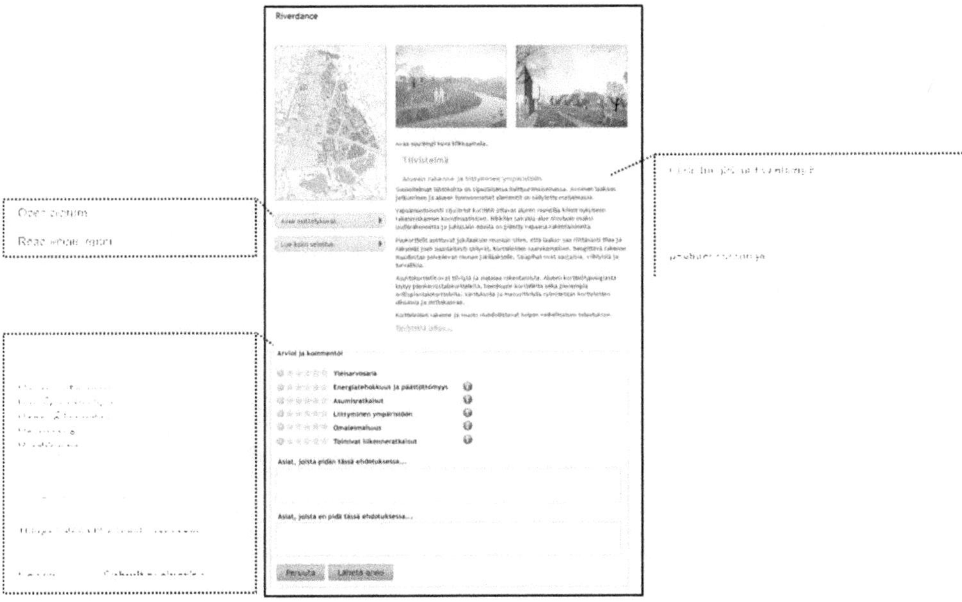

Figure 8. Comparison table in Sipoo competition website. © 2012, Municipality of Sipoo. Used with permission.

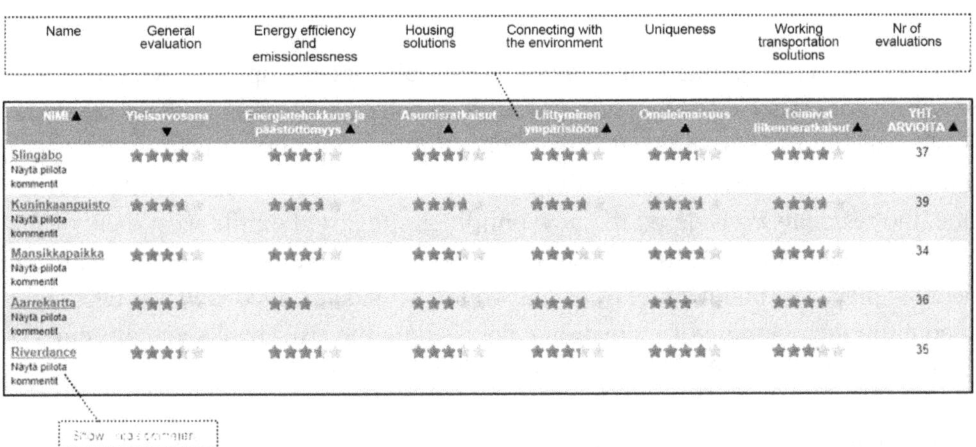

The basic structure and idea of the plan has not changed remarkably since the competition phase, as shown in the illustrations (Figure 9). However, several changes have been made during the planning process. In the email interview, the planner from the Municipality of Sipoo said that the aim was to take the residents' views

Figure 9. The winning entry, "Treasure Map" (Aarrekartta), and illustration of the final detailed plan of the River Valley of Sipoo. © 2011, 2014 Arkkitehtitoimisto AJAK Oy and Municipality of Sipoo.
Used with permission

into account in further planning as much as possible. For example, the winning entry was criticized for the dull and box-like appearance of the buildings. This feedback was passed on to planning so that the detailed plan and building instructions were guided towards more colorful buildings, and no flat roofs were allowed. Apart from that, the residents considered the new buildings suggested on the west bank of the river in the competition entry unacceptable, and thus, that area was left out from the detailed plan. The planner mentioned as well that residents' feedback was taken into account in the planning of recreational areas along the river banks and alignments of the walkways. She described the information as follows:

Many of the comments received from the residents related to tangible aspects concerning everyday life, and those were relatively easy to integrate into the detailed plan (Author's own translation).

The plan was approved by the council of Sipoo in October 2015 and ratified without any complaints. Currently, the infrastructure is built, and the area is under construction, as can be seen from the aerial photo (Figure 10). New residents will move to their new homes in 2017. Thus, the River Valley of Sipoo is an example of a successful participatory planning competition with the winning entry implemented.

RESULTS

Early Phase Participation

In practice, one of the primary challenges to effective public participation is that the residents are involved in the planning process at a stage that is too late to have an impact on the results. However, in the competition processes in Vaasa and Sipoo, the commitment to early phase participation was enshrined at the beginning of the process. In Vaasa, a clear commitment was made in the competition brief, which stated precisely how the residents were to be approached during the competition. The brief took a functional approach by outlining the requirements for the utilization of the data collected from the residents. In Sipoo, the residents' viewpoints were

Figure 10. Aerial photo of the River Valley of Sipoo, June 2017. The infrastructure of the new housing area is being built. © 2017, Lentokuva Vallas Oy / Hannu Vallas. Used with permission

written into the competition brief, based on an earlier study carried out in 2009. This saved costs and time, enabling residents' opinions to be included in the competition from the onset.

Furthermore, it is relevant to consider what kind of information public participation can produce in the competition process. Garde (2014) presents a censorious viewpoint of public participation in architectural competitions, arguing that public participation produces commonplace ideas rather than excellence in urban design. Like in Vaasa, in Garde's case study, participation was also carried out in an early phase, before the competition commenced. The difference is that, in Vaasa, residents were asked to present their preferences for the existing living environment and wishes for the future, while in Great Park, the aim of public participation was to generate innovative ideas. For contestants, resident engagement at the beginning of the process was deemed to be fruitful, as they were made aware of how inhabitants perceived the area.

Though early phase participation played a significant role in both cases, it was not clear how best to engage the residents at the beginning of the process. Residents could have participated in numerous ways, but a survey was viewed as the most efficient approach. At least in inhabited areas, the map-based survey method is valid. Residents already have viewpoints of the area and are easy to approach.

Evaluation Phase

In both cases, participation was highlighted at the end of the competition process, when the residents had an opportunity to evaluate the competition entries online. In Vaasa, the public evaluation occurred after the jury evaluation phase, while in Sipoo, the evaluation was undertaken before the jury made their decision about the winner. In Vaasa, planners were concerned about the possibility that the residents' comments could influence the expert evaluation; planners thought it was important to give the experts the opportunity to form their opinions first. It was emphasized to the public that the jury would select the winner and the public feedback and evaluation would affect further planning of the area. In Sipoo, the opposite approach was taken: the jury was able to follow the public evaluation on the web simultaneously, as they conducted their own evaluation. A summary of the public comments was delivered to the jury members before a final decision on the winner was made. In both cases, the planners emphasized the value and relevance of public opinions in further planning of the area.

The evaluation pages were individually designed and implemented for these projects with the same idea. As a difference, in Vaasa, the system required registration in order to give feedback, while in Sipoo, the evaluation did not require registration. Another difference was that in Vaasa, the respondents were not able to see the other comments instantly, while in Sipoo, the comments were published on the website once a day. In Sipoo, the competition entries were also displayed in an exhibition in the library, where visitors were able to give feedback through traditional paper forms.

In Vaasa, organizers were concerned that the amount of text regarding the competition entries on the evaluation pages was insufficient. In both cases, it was possible to open the whole entry presentation as a PDF file, although most of the respondents formed their opinions on the basis of the material that was found directly on the page. Respondents were not eager to familiarize themselves with the extra material offered. The problem was that these presentations were made with the traditional expert evaluation in mind. As such, they were not suitable for public evaluation or for presentation on the web. In Sipoo, the respondents primarily formed their opinions based on the visual material. Issues like the visual appearance of the area and the color of the houses were quite common. The planner had expected more feedback on the overall solution (e.g., functionality of the solution and traffic systems, housing solutions, and overall uniqueness).

In Vaasa, the residents did not actively evaluate the entries, even though they were widely advertised by the organizer. The winning entry received 19 written comments, while the others received only four comments each. Older people were more likely to participate during the evaluation phase than younger people. All evaluations received were in line with the expert comments on the winner; the comments received were quite positive and constructive. In Sipoo, the experience was different: the residents were active during the evaluation phase. There were 181 responses to the evaluation. Half of these responses were given in the library, with the other half given on the website. During the evaluation period, more than 1,500 visits to the website were recorded. Each of the entries was given diverse feedback, both positive and negative. In Vaasa, the decision to first select the winner and then allow residents to evaluate the entries may have affected residents' willingness to evaluate.

Residents' Viewpoints: Process and Outcomes

In addition to engaging the organizers and contestants, the jury also needed to commit themselves to the participatory process. The jury should have been aware of local needs and residents' viewpoints and should have supervised the data flow from the residents to the competition entries through the competition brief. In Vaasa, the jury was interested in studying the summary of the residents' viewpoints because of the difficult planning history of the area. The jury was willing to include the residents' viewpoints in the process and to ensure a certain level of engagement between the residents' experiences and the competition entries. In Sipoo, four of the jury members were themselves local inhabitants or municipal politicians, and as such, they were interested in the residents' perspectives. The jury discussed resident feedback in the evaluation meetings and pondered whether the viewpoints could be addressed in the context of each entry.

In both cases, residents' viewpoints were taken into account in the competition entries one way or another. In Vaasa, the planner commented that all the contestants took residents' views seriously, and accordingly in Sipoo, the planner said that it was possible to spot that residents' wishes were utilized in every competition entry. Thus, it is arguable that the information gathered from local people did function as a source of knowledge and inspiration for the contestants.

In Vaasa and Sipoo, various attempts were made to determine the contestants' understandings of residents' views. In Vaasa, the contestants were required to stipulate how data were utilized. The aim was to publish a brief text on the evaluation pages so that everyone could read how each contestant had integrated residents' viewpoints into their plans. Even though this was a requirement of the competition brief, only two of the five offices provided the comments. The jury discussed the possibility of rejecting offices that did not provide an explanation. Eventually, organizers themselves decided to produce the missing explanations on behalf of those who had not already done so. In Sipoo, the contestants were not asked to describe how they used the residents' impressions. Although, in every entry, there were elements that could be derived from residents' wishes.

In Vaasa, it was evident that the contestants had both acknowledged the issues the residents identified as important and used the data when it was deemed appropriate to their situation. One illustrative detail was the small beach in Mansikkasaari, which all of the contestants subsequently retained, as the residents had stipulated. In Sipoo, the residents' desires for a functional recreation area by the riverbank, colorful and low-rise buildings, and new services were addressed. It seems that the contestants were willing to use local knowledge provided during the participation, even without being forced to do so. Possibly, knowing that the entries would be published and evaluated publicly on the web may have affected their willingness to address the wishes of the residents in their plans.

Though in each entry the residents' impressions were clear, the entries reflected these notions differently. In Vaasa, some offices took the desires to arrange recreational routes along the coastline seriously; some followed the comments about higher apartment buildings, while others took the desire for low-rise buildings seriously. In Vaasa, the winning entry took the residents' viewpoints most seriously by representing comments about the special history of the place. In Sipoo, the entry that won displayed the resident wishes: the riverbank was redesigned into a functional green area with fruit trees, while the bicycle and pedestrian transport connections to the town center were carefully designed. The winner also received some negative comments from respondents about the visual appearance and the height of the buildings, as well as on the connection to the built cultural heritage and to nature.

It is relevant to notice that, in either Vaasa or Sipoo, the winner was not selected based on public rating. In both cases, the public evaluation included a possibility to rate the entries. In Vaasa, the winner of the competition was rated best by the public as well. In Sipoo, the winning entry was not rated as the best by residents, and the numerical rating was considered problematic by the planners in Sipoo. It is relevant to consider how the results of the public rating or voting are utilized in the jury process; the possibility of rating the entries creates an expectation that the result has an effect on the selection of the winner.

Further Planning

Regardless of the evaluation process, in both cases, planners underlined the fact that public evaluation produced valuable feedback to be used in further planning of the area. In Vaasa, the planner pointed out that the aim is to disseminate the comments forward to the detailed planning process. Similarly, in Sipoo, utilizing the feedback from public evaluation was considered important. Guidelines for further planning were produced from the feedback received in the evaluation phase. These guidelines were then used in the detailed planning process and referenced in detail in the plan report (Municipality of Sipoo, 2015).

In Sipoo, the planning process continued a straight–forward manner, with detailed planning aiming to design the new residential area. The winning architects were hired to continue planning based on the entry. At this stage, the planner carefully studied the comments from the public evaluation, and improvements were made to the plan in collaboration with the architects. The planner mentioned that the architects felt that public feedback made it easier to improve the plan after the competition phase. The planner underlined that utilizing the information produced by residents helped them to achieve better quality in the final plan. She also felt that the residents' attitudes towards the detailed plan were very positive, especially due to the participatory actions.

Based on these results, the two most important factors affecting the utilization of public feedback in further planning are: (1) the motivation and dedication of the planner; and (2) ensuring the information flow throughout the competition and planning processes. In Sipoo, the planner was certainly very motivated to engage residents in the process and was the key person in disseminating the information from the competition process to the final detailed plan. According to her, it was fairly easy to utilize the information, although she admitted that it could have been more systematic in various phases of planning. For example, investing more time in visualization of the information could have been beneficial. However, this would have required more resources in the process.

SOLUTIONS AND RECOMMENDATIONS

Over the last twenty years, public participation in Finland has improved and gained a more stable role in the urban planning process. However, the planning competition as a planning instrument has remained a closed, top-down process, and the demand for public participation has generally been neglected, even though planning competitions play a significant role as planning instruments. This study has revealed four important perspectives in respect of developing participatory competition processes. These findings support the thinking that, in planning competition processes, public participation should be arranged, as in any planning process, while competitions are used as impactful planning instruments in land use planning.

Firstly, it is important to engage residents in the competition process from the outset. Competitions often take place in an existing urban environment, where local residents already have strong and well-defined views. As Healey (1997) points out, collaborative planning is about sorting out the diversity of local knowledge. In planning competitions, this is even more evident, as the offices that participate in competitions are rarely local; this presents significant challenges to their understanding of the local context. In Vaasa and Sipoo, the importance and impact of local insight was reflected in the competition entries. It is arguable that the information was a source of inspiration for the contestants and that local residents' views should be included in the background data of planning competitions.

Secondly, the competition process can be supported and facilitated through the use of different kinds of *e*-participation tools in countries like Finland, where the vast majority of people use the Internet (OSF, 2016). However, the use of *e*-participation tools does contain some level of risk. Such tools need to be ordered from private sector suppliers, since the commissioning cities themselves do not have access to ready-made tools, like competition evaluation pages. The implementation of new tools entails costs and is difficult to organize; furthermore, it is difficult to find the right IT supplier. The so-called 'implementation gap' between the supplier and the customer can also affect the quality of the outcome in respect of the ordered *e*-participation tool (Kahila-Tani, 2015). On the other hand, as also Garde (2014) points out, no single approach to public participation and planning competition works in every situation. Purpose and approach of participation should be stated at the beginning of the process in order to be able to select the right methods and tools. Clearly, there is not one method that would be practical from one planning project to another; instead, a diverse selection of tools is needed. This indicates that partially web-based participation tools could even replace the existing and traditional tools used (Stern, E., Gudes, O., & Svoray, T., 2009).

Thirdly, the resident perspective should be supported through the jury process. If the competition processes are planned to support participation, then the actual practice associated with the competition, as well as the jury action, should be changed accordingly. One possibility would be to include someone on the jury who would act as an advocate for resident insight (Kazemian, R., Rönn, M., & Svensson, C., 2006). However, this approach carries risks, as it assumes that one person has the ability to express multiple and not necessarily consistent views in an articulate way. As stated in this chapter, if the planning competitions are desirable for planning instruments in the future, public participation should be included in these processes in a more meaningful way. This study reveals that, despite strong will and investment in participation, the level of citizen empowerment in competitions remains at the consultancy level when compared to Arnstein's (1969) ladders. This notion should be further elaborated to consider whether the residents could have a stronger role in the future in planning competitions and in planning in general. The dialogue-based competition model presented by Kreiner, Jacobsen, and Jensen (2011) and a participatory planning competition in Great Park, California (Garde, 2014) are the first examples of new interactive competition forms. These examples, as well as results from Vaasa and Sipoo, suggest that there is a need to rethink the traditional competition process when enhancing public participation.

Fourthly, the competition entries, as plan proposals, should always be publicly evaluated. In Finland, this has generally been done after the competition by giving lay people the opportunity to become acquainted with the entries in a special show where all the proposals are presented. The presented cases introduce two different approaches to the residents' role in the jury evaluation process. Taken together, the Vaasa and Sipoo cases show that it is not difficult to integrate lay people into the competition evaluation phase with appropriate *e*-participation tools. Although the entries are difficult and laborious to study, the evaluation allows lay people to react to the envisaged changes in their living environment. This facilitates the start of the official planning process by indicating residents' views to the planner. Nevertheless, the competition process should be steered in this direction with a view towards ensuring that structured public opinion plays a more prominent role. Similarly, Garde (2014) argues that it is possible to engage citizens in the evaluation and selection of the winner.

Even though the residents' role is important, it is hard to ensure a process that reflects their views. One reason is that resident information is not exact or comprehensive and can be contradictory. Planners are more accustomed to dealing with 'hard' data about the physical properties of the area than with 'soft' data on residents' viewpoints and experiences. For future cases, it is important to consider in what form resident feedback is offered to the contestants as well as to urban planners. If the resident views are offered as unprocessed raw data, new tools need to be developed for the analysis of this material. Also Kreiner, Jacobsen, and Jensen

(2011) notice the problem of conflicting feedback. They suggest an open phase in the competition process, where continuous dialogue between the stakeholders is enabled in order to achieve mutual understanding of requirements before the actual closed competition phase.

As always, in public participation cases, emphasis is placed on the organization's willingness to manage participation efforts successfully. In Vaasa and Sipoo, planners had a significant interest in arranging a more transparent process in which the participants could be involved. Residents' viewpoints are important and have an impact, but the question remains regarding how to make this more evident in order to ensure the use of *e*-participation tools in subsequent competitions. Though the competition process should be open and democratic, the ultimate dilemma between democracy and the jury's independence described by Kazemian et al. (2006) remains unresolved.

FUTURE RESEARCH DIRECTIONS

In this study, the participatory competition processes were supported with *e*-participation tools, and the process was followed from the beginning of the competition phase until the planning process commenced in Vaasa and the council in Sipoo approved the detailed plan. In order to evaluate the ultimate impact of public feedback, a post-occupancy evaluation could be carried out after the plan's implementation.

Based on the results, it is necessary to develop tools, not only for the evaluation of the competition entries, but also for disseminating and maintaining feedback throughout further planning processes. As the willingness to utilize public participation from the early competition phase onwards is organization-specific, or even person-specific, it is important to develop tools for managing the information throughout the whole development process. As people may change when planning proceeds from one phase to another, it is meaningful to have the participatory information easily and understandably available to all relevant stakeholders.

Obviously, more research is needed related to participatory competition processes from the public participation perspective. There is especially still a lack of empirical case studies of *e*-participation in urban planning competitions. It is self-evident that new technological solutions provide new possibilities in *e*-participation. Interactive, web-based 3D models or virtual environments are already in everyday use, and immersive augmented reality is already available in an urban planning context. These technologies have the potential to change the way planning competitions are arranged in the future.

CONCLUSION

By emphasizing the current role of public participation in planning competitions, this chapter presents findings that generally offer relevant insight into urban planning studies. Like planning competitions, urban planning projects in their current form are also often bereft of local contextualization. This observation is based on two salient facts. Contestants especially have little or no relationship with the area in which the competition takes place, and the competition can be arranged as an addendum to the local planning process. The arrangement of public participation during competition binds it more closely to the local planning process by allowing the contestants to become familiar with and utilize local insight. Kyttä, Broberg, Tzoulas, and Snabb (2013) describe the definition of contextually sensitive urban planning as something that takes residential environmental quality into consideration. The experiences from the case studies indicate how public participation in the competition process can alleviate the difficult impasses that often emerge in situations where development plans clash with residents' views.

Furthermore, information produced by residents may also become a source of inspiration for the contestants and planners. Grangaard and Ryhl (2016) emphasize the understanding of the users in a design process and argue that involvement of users in a competition process may act as an eye-opener to the planners and architects. This means that a human-centered approach is relevant in place-based planning processes. It is arguable that planning competition practices, as well as urban planning, generally should, in the future, acknowledge the importance of these local aspects, the place, and the people living there by developing more robust public participation processes.

As the technology and *e*-participation tools develop further, it is even more important to consider how the competition practices and processes should change to enhance, as Picken (2013) points out, residents' *willingness to participate* and the outcomes and impact of participation. The rising urban planning activism and citizen self-organizing *e*-participation (Graziano, 2017) challenges municipal authorities in the planning field. Results from this study indicate that active citizens should be considered as a resource by engaging citizen activists in planning competitions. The production of local knowledge and participating with their own plans or designs in the competition is an opportunity to enhance public engagement.

ACKNOWLEDGMENT

This research was supported by the City of Vaasa and Municipality of Sipoo.

REFERENCES

Al-Qaysi, N., Piroozfar, A., Southall, R., & Farr, E. R. P. (2016) Judgment in architectural competitions as communicative deliberative practice In *Proceedings of the ICC 2016: 6th International Competitions Conference: The Competition Mesh*. Leeds Beckett University.

Andersson, J. E. (2016). When all is said and done: An architectural competition, was it a good idea? In ICC 2016. Leeds Beckett University.

Andersson, J. E., Zettersten, G. B., & Rönn, M. (Eds.). (2016). *Architectural Competitions as Institution and Process*. The Royal Institute of Technology and Kulturlandskapet.

Anttiroiko, A.-V. (2012). Urban planning 2.0. *International Journal of E-Planning Research*, *1*(1), 16–30. doi:10.4018/ijepr.2012010103

Arnstein, S. R. (1969). A ladder of citizen participation. *Journal of the American Institute of Planners*, *35*(4), 216–224. doi:10.1080/01944366908977225

Bäcklund, P., & Mäntysalo, R. (2010). Agonism and institutional ambiguity: Ideas on democracy and the role of participation in the development of planning theory and practice – The case of Finland. *Planning Theory*, *9*(4), 333–350. doi:10.1177/1473095210373684

Bailey, K., & Grossardt, T. (2010). Towards structured public involvement: Justice, geography and collaborative geospatial/geovisual decision support systems. *Annals of the Association of American Geographers*, *100*(1), 57–86. doi:10.1080/00045600903364259

Brown, G., & Weber, D. (2012). Measuring change in place values using public participation GIS (PPGIS). *Applied Geography (Sevenoaks, England)*, *34*, 316–324. doi:10.1016/j.apgeog.2011.12.007

City of Vaasa. (2017a). *Ak1000: Palosaaren salmen alue, Mansikkasaari ja Salmenpuisto* [Detailed plan for the area of Palosaaren salmi, The island of Mansikkasaari and park Salmenpuisto]. Retrieved July 13, 2017 at: https://www.vaasa.fi/kaavoitus/ak1000

City of Vaasa. (2017b). *Ak1081: Palosaaren Salmen alue, Rahkola-Varisselänpuisto* [Detailed plan for the area of Palosaaren salmi, The park of Rahkola-Varisselkä]. Retrieved July 13, 2017 at: https://www.vaasa.fi/kaavoitus/ak1081

Eräranta, S. (2013). *Situation awareness in urban planning* (Unpublished Master's Thesis). Aalto University, Espoo, Finland.

Eräranta, S., Kahila-Tani, M., & Nummi-Sund, P. (2015). Web-based public participation in urban planning competitions. *International Journal of E-Planning Research*, *4*(1), 1–18. doi:10.4018/ijepr.2015010101

Evaluation of the Land Use and Building Act in practice 2013 [Arviointi maankäyttö- ja rakennuslain toimivuudesta 2013]. (2014). Suomen ympäristö, Ympäristöministeriö, 1/2014. Helsinki.

Garde, A. (2014). Citizen participation, design competition and the product in urban design: Insights from the Orange County Great Park. *Journal of Urban Design*, *19*(1), 89–118. doi:10.1080/13574809.2013.854697

Grangaard, S., & Ryhl, C. (2016). Vandhalla – A sport centre and a successful example of first-generation universal design. In *Universal Design 2016: Learning from the Past, Designing for the Future: Proceedings of the 3rd International Conference on Universal Design (UD 2016)* (pp. 243-245). Amsterdam: IOS Press.

Graziano, T. (2017). Citizen e-participation in urban planning: Achievements and future challenges in a Mediterranean city. *International Journal of E-Planning Research*, *3*(6), 1–18. doi:10.4018/IJEPR.2017070101

Healey, P. (1997). *Collaborative planning. Shaping places in fragmented societies*. Hampshire, UK: McMillan Press.

Horelli, L., & Wallin, S. (2010). The future-making assessment approach as a tool for e-planning and community development – The case of ubiquitous Helsinki. In C. N. Silva (Ed.), *Handbook of research on e-planning: ICTs for urban development and monitoring* (pp. 58–79). Hershey, PA: IGI Global. doi:10.4018/978-1-61520-929-3.ch004

Jacobsen, P. H., Harty, C., & Tryggestad, K. (2016). (Re)constructing the wicked problem through the visual and the verbal: the case of a dialogue based architectural competition. In *Proceedings 32nd Annual ARCOM Conference*. Manchester UK. Association of Researchers in Construction Management.

Kahila, M., & Kyttä, M. (2009). SoftGIS as a bridge-builder in collaborative urban planning. In S. Geertman & J. Stillwell (Eds.), *Planning support systems best practice and new methods* (pp. 389–411). Springer. doi:10.1007/978-1-4020-8952-7_19

Kahila-Tani, M. (2015). *Reshaping the planning process using local experiences: Utilising PPGIS in participatory urban planning* (Doctoral dissertation). Aalto University Publication Series, 223/2015.

Kazemian, R. (2009). Design interactivity and communicative quality judgment versus urban design competition – A design methodology statement. *Nordic Journal of Architectural Research, 21*(2/3), 68–78.

Kazemian, R., Rönn, M., & Svensson, C. (2006). Arkitekturtävlingar i Finland. *TRITA-ARK-Forskningspublikationer, 2006*, 11.

Kreiner, K., Jacobsen, P. H., & Jensen, D. T. (2011). Dialogues and the problems of knowing: Reinventing the architectural competition. *Scandinavian Journal of Management, 27*(1), 160–166. doi:10.1016/j.scaman.2010.12.003

Kyttä, M. (2011). SoftGIS methods in planning evaluation. In A. Hull, E. R. Alexander, A. Khakee, & J. Woltjer (Eds.), *Evaluation for participatory and sustainable planning* (pp. 334–354). London: Routledge.

Kyttä, M., Broberg, A., Tzoulas, T., & Snabb, K. (2013, May). Towards contextually sensitive urban densification: Location-based softGIS knowledge revealing perceived residential environmental quality. *Landscape and Urban Planning, 113*, 30–46. doi:10.1016/j.landurbplan.2013.01.008

Kyttä, M., & Kahila, M. (2011). SoftGIS methodology – Building bridges in urban planning. *GIM International, 25*(3), 37–41.

Land Use and Building Act, Finlex. (1999). *Maankäyttö- ja rakennuslaki*. Retrieved November 15, 2013 from http://www.finlex.fi/fi/laki/ajantasa/1999/19990132

Merikoski, T., Eräranta, S., & Staffans, A. (2012). *Sibbesborg – Kestävän yhdyskunnan suunnittelukilpailu. Raportti kilpailuhankkeen valmistelusta ja toteutuksesta 2010–2011*. Espoo: Aalto University.

Municipality of Sipoo. (2015). *N 40, Sipoon Jokilaakso, selostus* [Report for the detailed plan of River Valley of Sipoo]. Retrieved July 13, 2017 at: https://www.sipoo.fi/fi/jokilaakso/asemakaava

OSF Official Statistics of Finland. (2016). *Use of information and communications technology by individuals*. Retrieved July 11, 2017 from http://www.stat.fi/til/sutivi/index_en.html

Picken, F. (2013). From designed spaces to designer savvy societies: The potential of ideas competitions in willing participation. *Environment & Planning A, 45*(8), 1963–1976. doi:10.1068/a45460

Rantanen, H., & Kahila, M. (2008). The SoftGIS approach to local knowledge. *Journal of Environmental Management, 90*(6), 1981–1990. doi:10.1016/j.jenvman.2007.08.025 PMID:18556113

Roininen, J., Horelli, L., & Wallin, S. (2003). *Osallistuminen ja vuorovaikutus kaavoituksessa. Seurannan ja arvioinnin viitekehys ja menetelmät. Suomen ympäristö 664*. Helsinki: Ympäristöministeriö.

Rönn, M. (2009). Judgment in the architectural competition – Rules, policies and dilemmas. *Nordic Journal of Architectural Research, 21*(2/3), 52–67.

Rönn, M. (2011). Architectural quality in competitions. A dialogue-based assessment of design proposals. *FORMakademisk, 4*(1), 100–115.

SAFA. (2008). *SAFA Competition Rules 2008*. SAFA.

SAFA. (2011). *Competitions 1972-2011*. Retrieved November 12, 2013 from http://safa.fi/document.php?DOC_ID=1133&SEC=03702b3ae4d10944d643485c927250eb&SID=1#competitions_2011.pdf

Sanvig-Knudsen, A. M., & Kahila, M. (2012). The role of volunteered geographic information in participatory planning: Examples from Denmark and Finland. *Perspektiv, 21*, 35–46.

Schlossberg, E., & Shuford, E. (2005). Delineating "public" and "participation" in PPGIS. *URISA Journal, 16*(2), 15–26.

Silva, C. N. (2010). The *e*-planning paradigm - Theory, methods and tools: An overview. In C. N. Silva (Ed.), *Handbook of research on e-planning: ICTs for urban development and monitoring* (pp. 1–14). Hershey, PA: IGI Global. doi:10.4018/978-1-61520-929-3.ch001

Silva, C. N. (2013). Foreword. In L. Horelli (Ed.), *New approaches to urban planning, insights from participatory communities* (pp. 7–8). Helsinki: Aalto University.

Staffans, A. (2004). *Vaikuttavat asukkaat. Vuorovaikutus ja paikallinen tieto kaupunkisuunnittelun haasteina. Yhdyskuntasuunnittelun tutkimus- ja koulutuskeskuksen julkaisuja A 29* (Doctoral dissertation). Helsinki: Yliopistopaino.

Stern, E., Gudes, O., & Svoray, T. (2009). Web-based and traditional public participation in comprehensive planning: A comparative study. *Environment and Planning. B, Planning & Design, 36*(6), 1067–1085. doi:10.1068/b34113

Svensson, C. (2009). Speaking of architecture. A study of the jury's assessment in an invited competition. *Nordic Journal of Architectural Research, 21*(2/3), 94–107.

Talen, E. (2000). BottomUp GIS. A new tool for individual and group expression in participatory planning. *Journal of the American Planning Association, 66*, 279–294. doi:10.1080/01944360008976107

Taylor, N. (1998). *Urban Planning Theory Since 1945*. London: SAGE Publications Ltd.

Våland, M. S. (2009). End user participation as an input to shape the brief in architectural competitions. A threefold translation process. *Nordic Journal of Architectural Research*, *21*(2/3), 108–122.

Volker, L. (2010). *Designing a design competition: The client perspective*. Paper at the 2010 Design Research Society (DRS) International Conference Design & Complexity, Montreal, Canada.

Wallhagen, M., Malmqvist, T., & Eriksson, O. (2017). Professionals' knowledge and use of environmental assessment in an architectural competition. *Building Research and Information*, *45*(4), 426–442. doi:10.1080/09613218.2015.1118264

Wallin, S., Amati, M., Saad-Sulonen, J., & Horelli, L. (2012). Exploring *e*-planning practices in different contexts: Similarities and differences between Helsinki and Sydney. *International Journal of E-Planning Research*, *1*(3), 17–39. doi:10.4018/ijepr.2012070102

Chapter 4
A Smart City Remembers Its Past:
Citizens as Sensors in Survey and Mapping of Historic Places

Jennifer Minner
Cornell University, USA

Andrea Roberts
Texas A&M University, USA

Michael Holleran
University of Texas at Austin, USA

Joshua Conrad
University of Texas at Austin, USA

ABSTRACT

Integral to some conceptualizations of the "smart city" is the adoption of web-based technology to support civic engagement and improve information systems for local government decision support. Yet there is little to no literature on the "smartness" of gathering information about historic places within municipal information systems. This chapter provides three case studies of technologically augmented planning processes that incorporated citizens as sensors of data about historic places. The first case study is of SurveyLA, a massive effort of the city of Los Angeles to comprehensively survey over 880,000 parcels for historic resources. A second case study involves Motor City Mapping, an effort to identify the condition of buildings in Detroit, Michigan and a parallel historical survey conducted by volunteers. In Austin, Texas, a university-based research team designed a municipal web tool called the Austin Historical Survey Wiki. This chapter offers insights into these prior efforts to augment planning processes with "digitized memory," web-based technology, and public engagement.

DOI: 10.4018/978-1-5225-5999-3.ch004

INTRODUCTION

Local governments around the world aspire to be 'smart cities' where information and communication technologies are applied in innovative ways to increase the efficacy of municipal services in a quest for more efficient, and, in some cases more equitable city (Albino, Berardi, & Dangelico, 2015; Fietkiewicz, Mainka, & Stock, 2016; Glasmeier & Christopherson, 2015; Townsend, 2013). Although the rhetoric of smart cities appears to reach beyond volunteered geographic information and Web 2.0 to autonomous vehicles, artificial intelligence, and robotics (Batty, 2017; Dia, 2017), web-based planning support technologies remain a central vision of 'smartness' (Afzalan, Sanchez, & Evans-Cowley, 2017). Web infrastructure for technology-augmented planning processes, include the use of social media, crowdsourcing platforms, and web-based geographic information systems (GIS) for urban planning and public administration (Townsend, 2013; Evans-Cowley & Hollander, 2010; Seltzer & Mahmoudi, 2013; Gordon & de Souza e Silva, 2011). The aim of many of these efforts is to enhance municipal decision-making by incorporating volunteered time, perspectives, and knowledge.

Some municipal historic preservation programs have expanded the use of digital technologies to serve preservation and urban planning. An example is the award-winning, multi-million-dollar effort of the City of Los Angeles to survey historic resources citywide, including development of specialized GIS tools, a web presence for public outreach and data collection, and a robust public engagement plan to accomplish it (Bernstein, Sun, & Sucre, 2009; (Bernstein & Hansen, 2016; City of Los Angeles, 2017). In Detroit, Michigan, Motor City Mapping and a related effort among preservationists to survey for historic places aimed to provide information for a city losing building stock to decades of population decline and disinvestment (Scola, 2014; Evans, 2014).

In Austin, Texas, a university-based research team created The Austin Historical Survey Wiki (referred to throughout this chapter as the Wiki) as municipal web infrastructure to maintain a cumulative database of historic resources that is open to public contributions. Through this web-based tool, historic resources were intended to be surveyed, documented, and maintained over time by a combination of municipal officials, professional preservationists, and interested members of the public. The Wiki was inspired by visions of advancing municipal decision-making and urban planning. The effort was based on the conviction that public participation, online or otherwise, can give governments a firmer basis for making decisions that are more defensible, representative, and potentially more equitable, because they arise from pluralistic, democratic processes (Habermas & McCarthy, 1985). The project also originated out of a pragmatic need for timely information about historic resources to serve the City of Austin's long range planning and regulatory functions, which includes drafting land use plans, designation of historic landmarks and historic districts, and review of demolition and remodeling permits.

This book chapter considers these three cases of city-wide survey efforts along with the idea of *citizens as sensors* (Goodchild, 2007) for capturing geographic information and local knowledge about valued places, especially those that can be described as 'historic'. Here it should be noted that when we as authors use the term "citizens," we refer generally to members of the public as distinguished from local government officials or professionals.

In this chapter, the authors offer promising developments and challenges in the incorporation of data about historic places in local government planning to aid in 'smarter' decision-making. The following section consists of a review of literature related to web-based technologies and models for online participation in planning processes. A third section delves into the use of historical surveys by local government. The next describes two city-wide efforts to survey building stock in Los Angeles, California and Detroit, Michigan. Following sections detail the development of a Wiki to support historical surveys in Austin, results of testing and launch of the Wiki, community responses, and quantitative indicators. In a concluding section, we provide insights about efforts to enhance government information systems and increase planning intelligence about building stock and historic resources using citizens as sensors and web-based technologies.

WEB-BASED TECHNOLOGY, GOVERNANCE, AND COLLABORATION

Within the US, a growing emphasis on public participation in urban planning and historic preservation is attributed to the failures of rational planning and urban renewal policies at mid-twentieth century, which neglected to include the voices of citizens in planning deliberations or account for the value of existing urban fabric (Mallach, 2011).[1] Arnstein's Ladder of Citizen Participation (Arnstein, 1969) articulated the need for citizen participation in planning processes and remains a heavily cited article in planning literature. Arnstein's conceptual ladder moves from critique of governmental processes with no participation at the bottom rungs of the ladder, to tokenism, to prescriptions for citizen power in city planning in the top rungs. Moving up the rungs of the ladder, the role of citizens in government processes becomes more direct and decision-making power is placed to a greater extent within the hands of the community. At the pinnacle of Arnstein's model is citizen control.

Today, the virtues of public participation have become more widely accepted within urban planning, historic preservation has become a common function of many planning departments (Minner & Holleran, 2016; Minner, 2016), and the technologies used to augment community participation have multiplied. Arnstein's ladder has been modified to reflect the acceptance of public participation as an essential part of planning and to incorporate new modalities of participation. For instance, Randolph adapts Arnstein's ladder for contemporary environmental land use planning (Randolph, 2012). In his conceptualization, the ladder moves from non-participation to *collaborative learning and co-management* (Randolph, 2012, 85). At this top rung, Randolph writes:

Stakeholders take part in networks and communities of place and practice to learn and develop new knowledge and build consensus for creative solutions. Beyond decisions [or decision-making], stakeholders engage in joint implementation and learn from adaptive management (Randolph, 2012, 85).

In various modified forms, Arnstein's conceptual model has been used to assess on-line tools (Senbel & Church, 2011; Hanzl, 2007). Evans-Cowley and Hollander (2010) write:

participation assessment reflects the level of control afforded participants, ranging from information-based or feedback-only options to interactive participant self-determination (Evans-Cowley & Hollander, 2010, 399).

Initiatives aimed at *open government* and e-government span this continuum; there are examples of government initiatives that range from providing better (one-way) access to government data to the use of web-based initiatives that seek to transform government decision-making processes through direct citizen involvement (Lathrop & Ruma, 2010). There are related examples of open government initiatives, especially among state and local governments in the U.S. (Lathrop & Ruma, 2010) and among some federal governments outside the U.S.[2]

Some manifestations of open government use *crowdsourcing* as a primary method of citizen participation. In fact, crowdsourcing has gained currency in planning scholarship and practice. Seltzer and Mahmoudi define crowdsourcing as:

issuing a challenge to a large and diverse group in hopes of arriving at new solutions more robust than those found inside the organization,

while Goodspeed, Spanring, & Reardon define it more instrumentally as: *dividing a large task into small pieces that can be completed by a 'crowd' of participants* (Seltzer & Mahmoudi, 2012; Goodspeed, Spanring, & Reardon 2012). Crowdsourcing is further defined as the completion of task or creation of an online body of work via contributions of distributed people (and their distributed knowledge) via the internet (Connors, Lei, & Kelly, 2012; Elwood, Goodchild, & Sui, 2012). Estelles-Arolas and Gonzalez-Ladron-de-Guevara describe "mutual benefit" as central to the concept in which:

The user will receive the satisfaction of a given type of need, be it economic, social recognition, self-esteem, or the development of individual skills, while the crowdsourcer will obtain and utilize to their advantage what the user has brought to the venture, whose form will depend on the type of activity undertaken. (Estelles-Arolas & Gonzalez-Ladron-de-Guevara, 2012)

Within the context of urban planning, crowdsourcing can range in the type and degree of information that is shared, from very basic empirical information to gathering more complex ideas from the public for municipal problem-solving.

Wikis provide one means for crowdsourcing, especially in the context of collaborative writing. The invention of the concept of the wiki is commonly attributed to Ward Cunningham and the best known example has been produced by the WikiMedia foundation called the Mediawiki (WikiMedia Foundation, 2017). Walden characterizes wikis as being associated with a *write-publish-review-edit-republish cycle rather than the traditional write-edit-review-publish sequence* (Walden, 2011, 62). Others focus less on the ongoing nature of the editing process and instead on the collaborative nature of knowledge-production. This is the case with Beth Noveck's concept of "wiki government" (Noveck, 2009; Noveck, 2010).

In an experiment of "collaborative democracy," Noveck created a custom designed wiki for citizens to directly participate in patent review for the U.S. Patent Office (Noveck, 2009; Noveck, 2010). According to Noveck, citizens who participate online may be experts in their own right, with specialized knowledge that should be directly incorporated into the process of review. The custom designed wiki that she created was focused on bringing together a small group of interested citizens to work with government to improve decision-making processes.

Central to Noveck's model is the contrast between the concepts of collaborative democracy with deliberative democracy. She describes deliberative democracy as based on the theories of Jürgen Habermas and within this model, planners strive to create the optimum conditions for *ideal speech*, which consists of egalitarian communication among participants in order to reach rational, consensual decision-making. Using collaborative democracy, Noveck proposes engaging citizens with the process of completing governmental tasks rather than focusing on forums aimed at democratic speech. This collaborative model has similarities to Randolph's ideal of collaborative learning and co-management (Randolph, 2012), while Noveck's model emphasizes an online modality for participation.

Public participatory GIS (PPGIS) constitutes another set of practices aimed at improving the interaction between citizens and government, by incorporating citizen knowledge into geographic information systems. Sieber has traced the social history of PPGIS within geography, planning, and other related disciplines describing it as *the use of geographic information systems (GIS) to broaden public involvement in policymaking* (Sieber, 2006, 491). PPGIS is an area of scholarship and practice that evolved out of critiques of the use of GIS by government agencies that are top down and that neglect local knowledge. For example, Talen (2000) has encouraged "bottom up GIS" as a means to incorporate citizen perspectives into planning processes, as has Al-Kodmany (2009), who illustrates the inadequacies of large-scale governmental efforts to create a complete GIS. Ganapati and Brabham have also pointed to the potential for the use of GIS to increase citizen participation in planning and government processes (Ganapati, S. 2010; Brabham, 2009) and Fitzjohn (2009) to the uses of GIS in layering multiple sources of data to record the meanings of heritage landscapes.

The development of social media, web-based GIS, and location-based mobile technologies have further extended PPGIS, as citizens are able to collect detailed spatial data, annotate geographic features, and share this information online (Elwood, Goodchild, & Sui, 2012; Gordon & de Souza e Silva, 2011; Tulloch, 2008). These new capabilities have enabled *volunteered geographic information* (VGI) (Goodchild, 2007; Elwood, 2008). Michael Goodchild (2007) has emphasized the importance of this shift to the practice of geography in which citizens become sensors who contribute new geographical knowledge to mapping efforts.

"Citizen science" is yet another arena in which citizens engage in the production of volunteered information, usually in empirical data collection for use in scientific studies or public policy (Bowser & Shanley, 2013; Goodchild, 2007; Newman, G., Zimmerman, D., Crall, A., Laituri, M., Graham, J., & Stapel, L., 2010; Ottinger, 2010). Data from these efforts are often related to government and nongovernmental efforts at natural resource conservation.[3] Often citizen science is seen as a means of galvanizing citizen participation and interest in science. Within the following case studies there are important parallels between efforts to conserve natural resources embedded within citizen science projects and the potential to apply this model to conservation of the built environment through historic preservation and planning.

This chapter draws from this rich context of developments in participatory planning and web-based technology to describe and analyze major efforts in three US cities with city-wide efforts to gather information about historic resources with the help of citizens and integrate it within municipal information systems. The methodology for this chapter involved participant action and observation in the development of the Austin Historical Survey Wiki, as well as qualitative case study research conducted to understand comparable municipal historical survey efforts. The next section describes the use of historical surveys within the U.S., followed by case studies.

HISTORICAL SURVEYS AS LOCAL GOVERNMENT RESOURCE

Within the United States, over 1,900 Certified Local Governments (CLGs) have agreed to survey historic resources (U.S. Department of the Interior, National Parks Service, 2017). Historical surveys are defined by the U.S. National Parks Service (NPS) as *a process of identifying and gathering data on a community's historic resources* (U.S. Department of the Interior, National Parks Service, 1985). These surveys produce lists of buildings, structures, districts, cultural landscapes, and objects (e.g. monuments and sometimes living resources such as heritage trees) deemed worthy of recognition and in some cases protection.

Municipal preservation offices typically use the information from historical surveys to nominate local and national historic districts, designate historic landmarks, and develop compatibility standards for infill within historic districts. City planners, economic development specialists and other local government officials can use this information in revitalization efforts, heritage tourism, and emergency preparedness and mitigation (Laurie, J., 2008).

Beyond conventional uses for historical surveys, information about historic resources has the potential for use in urban design and place-making initiatives. Preserving historic resources can enhance neighborhoods, retaining a sense of place in the face of urban change (Allison & Peters, 2011). Historic resources contribute to the diversity of the city as temporal collage (Lynch, 1972). Data about historic resources can also be used in sustainability initiatives aimed at conserving material resources, reducing construction and demolition waste, and retrofitting buildings to improve performance (Preservation Green Lab, 2012; Frey, 2008; Stein, 2010). In declining communities, historical surveys can be used to identify assets that should be preserved in the face of demolition initiatives aimed at addressing blight and abandonment (Bertron & Rypkema, 2012).

Given all of these uses, a historical survey is not simply a list of historic resources. The uses of historical surveys range from long-range planning to use in development review and regulatory processes. These decisions include whether buildings can be demolished, if they can be remodeled and how, as well as the kinds of new development that may be allowed within certain districts or neighborhoods.

TALES OF TWO CITIES: CITY-WIDE SURVEYS IN LOS ANGELES AND DETROIT

In Los Angeles and Detroit, two U.S. cities with vastly different economic fortunes, civic leaders, city officials, and citizen volunteers have gathered and mapped information about building stock and historic places. The initiatives in both cities have employed citizens as sensors and web-based and GIS technologies. Both of cases can be viewed as "smart city" efforts to gather information using technology to augment efforts.

In rapidly growing Los Angeles, an Office of Historic Resources was created to organize and administer a large-scale historical survey of the city's neighborhoods for historic buildings and structures that could be eligible for designation and protection. The multi-million-dollar effort called SurveyLA was first conceptualized in 2000 and over 17 years moved from idea to completion. The initiative was supported with a $2.5 million dollar matching grant from the Getty Foundation, a nongovernmental organization that

supports institutions and individuals committed to advancing the greater understanding and preservation of the visual arts in Los Angeles and throughout the world (Getty Foundation, 2017).

In 2006, after years of planning, the Office of Historic Resources matched the grant, officially launching a city-wide effort that surveyed *the entire city of Los Angeles comprised of 880,000 parcels and approximately 500 square miles* (Bernstein & Hansen, 2016, 88). The comprehensive survey effort was deemed complete in 2017.

Professional preservation consultants conducted windshield surveys, which were integral to the SurveyLA process. Consultants were aided by the development of new technological tools. This included the Field Guide Survey System (FiGSS) developed by the City of Los Angeles. FiGSS consisted of a new ESRI ArcGIS toolbox with functions that support viewing contextual information about city neighborhoods and direct entry of field observations into a GIS on a tablet or laptop. An open source web-based heritage management system called Arches, developed by the Getty Conservation Institute, was used as a platform to make a queryable online database publicly accessible (Figure 1) (City of Los Angeles, 2017; The Getty Conservation Institute, 2013). Other online technologies were used to support SurveyLA, including a blog for survey updates; the use of MindMixer, a platform for online engagement to encourage the public to get involved in SurveyLA; and integration of survey data into ZIMAS, the online land and zoning database for the City of Los Angeles. In 2011, SurveyLA won a national planning excellence award for public outreach from the American Planning Association, partly due to its *its savvy use of traditional and*

Figure 1. Screenshot of Historic Places LA, a website that displays historic places identified in SurveyLA, the Los Angeles Historic Resources Inventory. The page is displaying residential multi-family properties that are potentially eligible for designation.

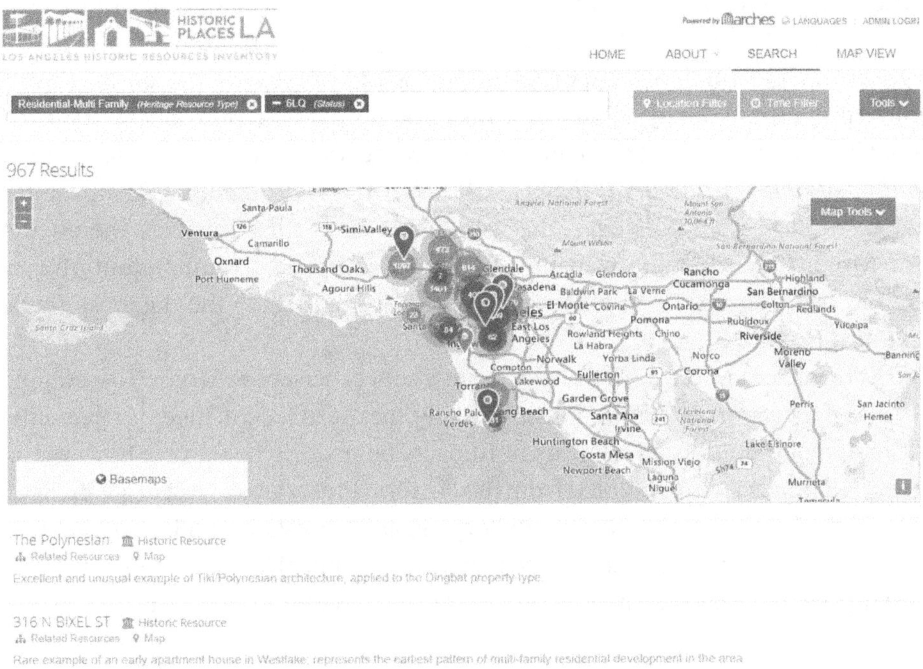

new media including the development and use of GIS and online tools, as well as multi-lingual materials and outreach to underrepresented populations, which attracted 250 volunteers (Jarmusch, 2011). In SurveyLA, volunteers were "sensors" of data about historic places; however, much of that information was provided to consultants and planners using traditional paper-based methods that had been processed for incorporating into technology later.

Meanwhile in Detroit, the Motor City Mapping project was initiated to gather information for decision-making about how best to channel funds to demolish abandoned properties. While SurveyLA was motivated to determine what not to demolish for redevelopment in a rapidly growing and expensive real estate market, Motor City Mapping and a related historic resources survey were organized in a city forced to declare bankruptcy in 2013 after years of population loss and disinvestment.

The Detroit Blight Removal Task Force (the Task Force) was organized in 2013 with three appointed chairpersons *to remove every blighted structure and clear every blighted vacant lot in the City of Detroit as quickly as possible using an environmentally-conscious approach (Detroit Blight Removal Task Force a., N.D.).*

The Task Force outlined a plan to rapidly address the demolition of properties using federal Hardest Hit Funds. By 2014, about $1.5 million in private and nongovernmental funding supported a windshield survey of the entire city for blighted properties (Davey, 2014). The Task Force's plan focused on leveraging this Motor City Mapping effort to maintain an *ongoing, open, parcel-level information portal, available on-line at all times to all residents* (Detroit Blight Removal Task Force b., N.D., 24). In this case, the web-based innovation involved the creation of the Motor City Mapping website (Figure 2), complete with the ability for any citizen to text pictures of blighted properties for addition to the database (called 'blexting') (Figure 3).

Historic preservationists in Detroit became concerned about Motor City Mapping and the prospects of mass demolitions. Normally, the use of federal funds for demolition or development requires an analysis of impacts on historic resources; however, the Hardest Hit Funds were exempted from the Historic Preservation Act of 1966 (Scola, 2014). Therefore, the nongovernmental organization Preservation Detroit launched the Detroit Historic Resources Survey, in which volunteers helped to rank properties for their potential historic value. The aim was to create a 'preservation overlay' to

Figure 2. Screenshot of Motor City Mapping website, which is now archived. Data from the website was moved to the City of Detroit's website and a national database for parcel-based data called Loveland.

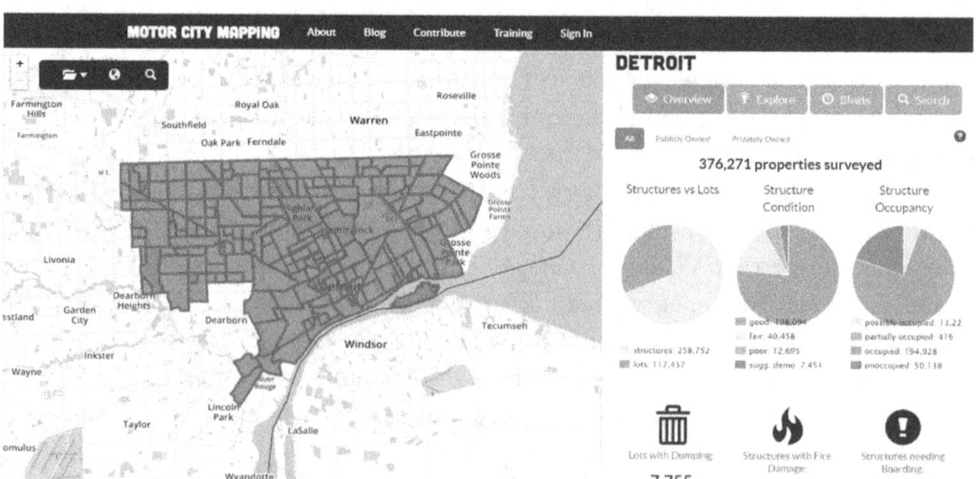

Figure 3. Queue of recent blexts with data in the Motor City Mapping website.

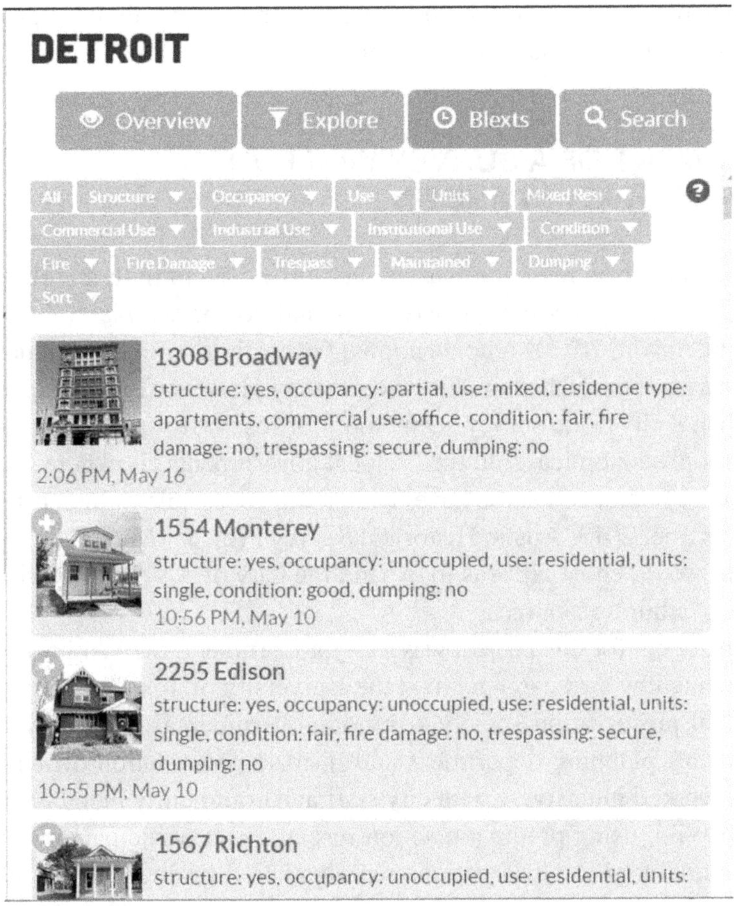

be added to the Motor City Mapping project (Ibid.) Preservation Detroit's effort also involved the use of an application for smart phones and tablets developed for volunteers to use in the field (Bragg, 2014). The survey effort was organized in partnership with the Task Force, Data Driven Detroit and Loveland Technologies, which were implementing Motor City Mapping (Evans, 2014). The results were placed in a publicly accessible location (Data Driven Detroit, 2017), but they do not appear to be integrated into either the archived Motor City Mapping portal (Loveland Technologies; Data Driven Detroit, n.d.) or a more recent version of the portal called Loveland (Loveland, 2017).[4]

Similar to the effort in Los Angeles, it appears that the use of citizens as sensors has come to a close and there are no longer opportunities for volunteers to add or correct information for historic resources. This contrasts with the following case study of the Austin Historical Survey Wiki, which was designed to accommodate cumulative and ongoing survey by both professionals and volunteers, even after a comprehensive city-wide survey is conducted.

DEVELOPMENT OF A SURVEY WIKI FOR ONGOING SURVEY IN AUSTIN

The need for a historic resources survey in Austin was initially highlighted as an action item in the city's comprehensive plan and downtown plan (City of Austin, 2011; City of Austin, 2012). A conventional yet costly solution would have been to commission a new comprehensive historical survey. However, the financial resources to accomplish a city-wide survey seemed out of reach. It was thought that a Wiki might be a more economical solution, as local governments elsewhere had gathered historical survey data, and some have made them accessible online, often through web GIS (Bertron, 2013; Austin Historical Survey Project Team, 2009). This was a solution that would enlist citizens to expand the City of Austin's knowledge of the community's cultural resources.

While the client for this project was a city government, the Wiki was developed and implemented by a research team at the University of Texas at Austin, working with the local preservation society -- a non-governmental organization, and the City of Austin's planning department and historic preservation office. Members of the team worked intensively with city staff and community members to develop and test the Wiki, using pragmatic action research and participant observation as primary methodologies. Observations were gathered through numerous research team meetings, neighborhood association and community open houses, and the use of the Wiki in graduate courses in which students worked with community members to test the tool and survey historic resources. Additionally, web analytics were used to collect quantitative data about usage of the web-based tool.

There were several key issues in designing and developing the Wiki. The first was whether the Wiki should be a highly customized website or one that was expeditiously assembled (a mashup) of existing tools. This was largely resolved with consensus that the tool needed to be custom built, which enabled new features to be incorporated that were not available among existing platforms.

A second major question was whether the tool should be primarily based on proprietary or open source software. The research team decided to use a Google Maps Application Program Interface (API), a toolkit that allows developers to build a custom application, so that the Wiki would be intuitively familiar to the public as well as easily replicable. The Wiki uses a Google Maps API to display base maps and to allow users to add historic places. It has a MySQL database and data modeling: data forms and user registration are accomplished with Drupal, an open-source PHP-based web framework.

Finally, the research team discussed at length whether the Wiki should be structured primarily by the bureaucratic needs of the historic preservation office or whether it should reflect freer forms of interaction more common to social media and digital history sites. Neighborhood groups were already using the paper survey forms for volunteer-based surveys; in effect Austin was crowdsourcing historical surveys without using any online tools. Thus, there was a research logic to simply retaining the City of Austin's paper forms as a structure and migrating to a platform that better supported crowdsourcing. Citizens already involved in historical surveys would then find the online survey questions similar to the paper form they were already using and newcomers to preservation would find sets of questions that they could respond to online.

The Wiki was designed to allow citizens to add historic places and to contribute and edit data associated with each historic place (Figure 4). The Wiki also includes data from the City of Austin and professional historical surveys. In order for the Wiki to be used in official decision-making, the sources for all contributed information needed to be known and retained as part of the City of Austin's public records.

Figure 4. Screenshot of the Austin Historical Survey Wiki.

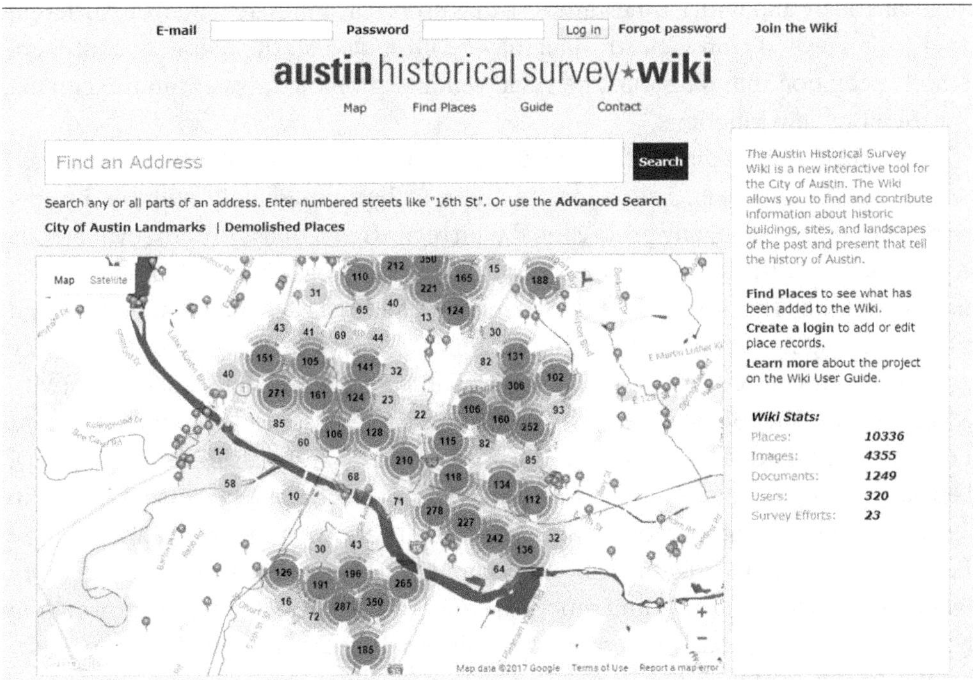

Several features of the Wiki emerged in response to these imperatives. One of the most fundamental was that information cannot be contributed anonymously; all data is displayed with the real name of the contributor and a timestamp. Additionally, each piece of data has a full revision history, in which prior contributions can be viewed along with names of the users who contributed or edited information.

Another central feature of the Wiki's design is moderation. All contributions are reviewed by a moderator before they are published. The Historic Preservation Office selects and manages the moderators. The moderator may choose to send submissions back for further editing or may publish the submission, at which point the data becomes publicly visible. Protocols were developed to prevent moderators from slowing the flow of information or overstepping their bounds. Moderators were instructed not to review information for accuracy beyond gross and obvious error (e.g. "920" instead of "1920"), but to ensure that information was neither spam nor abusive.

Central to the custom design of the wiki and its management of citizen contributions are "data review levels." A data review level is associated with each data field (see Figures 5-7), indicating whether its contents have undergone "Preservation Office Review" or "Professional Review," or if the data remains "Unreviewed." This hierarchy is rooted in the City's Development Code, because surveys reviewed by the Historic Preservation Office are used in demolition review and evaluation of historic district nominations.[5] Data at different levels of review are displayed simultaneously and with a timestamp, so that users can compare data from different levels of review. [6] "Unreviewed" data may be promoted up the hierarchy, but there is no expectation that data must be reviewed and promoted, except in the context of official review functions.

The research team uploaded information from prior professional surveys conducted over more than 30 years, allowing this data to be immediately editable by any member of the public or any professional with regard to factual accuracy, evaluations of historical significance, and updates reflecting physical change. With the Wiki, historical surveys become less of a product completed at a certain point in time, and more of an ongoing process, involving citizens, preservation professionals, and city staff in a continual process of quality control and database maintenance (Figure 8).

In addition, the tool was designed so that registered users can download data as a comma delimited file (.csv) for all historic places (or any user-defined subset). The data includes geographic coordinates, so users can display and analyze data in Excel or any geographic information system. This allows users to sift and analyze data as they wish. This feature allows community-based organizations, other city agencies, scholars, and web and mobile developers to access the city's information system for purposes that could range from original research to the development of heritage tour apps for cell phones.

Figure 5. Screenshot of historic resources page.

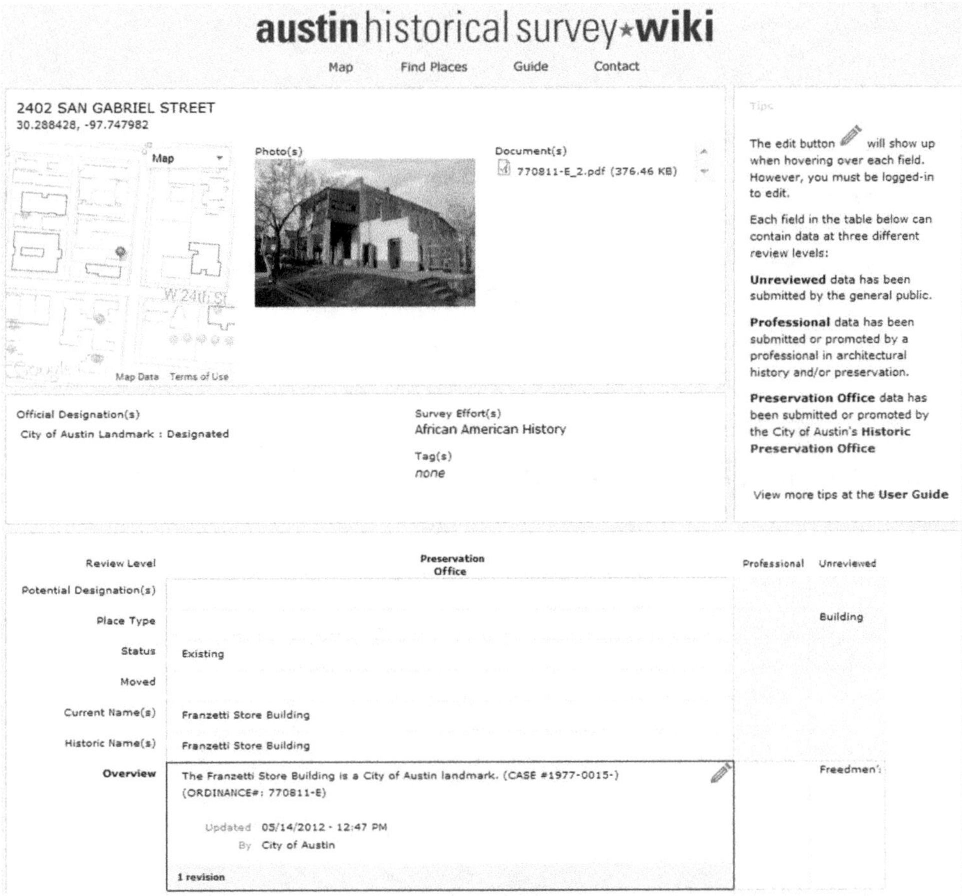

Figure 6. Close-up of historic place page showing preservation office reviewed data.

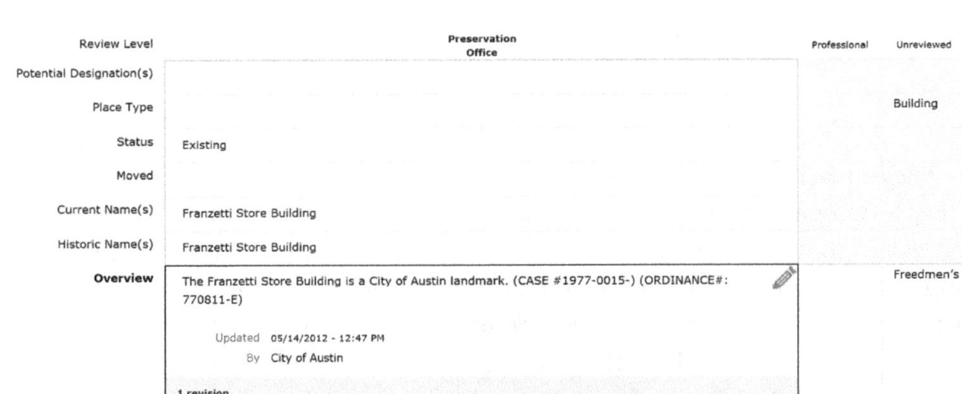

Figure 7. Close-up of historic place page showing unreviewed data.

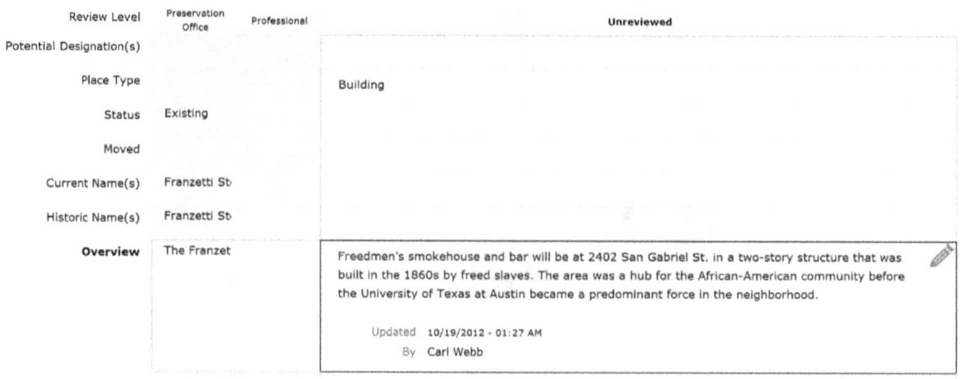

Figure 8. Comparison of traditional historical survey methods and the Wiki survey method.

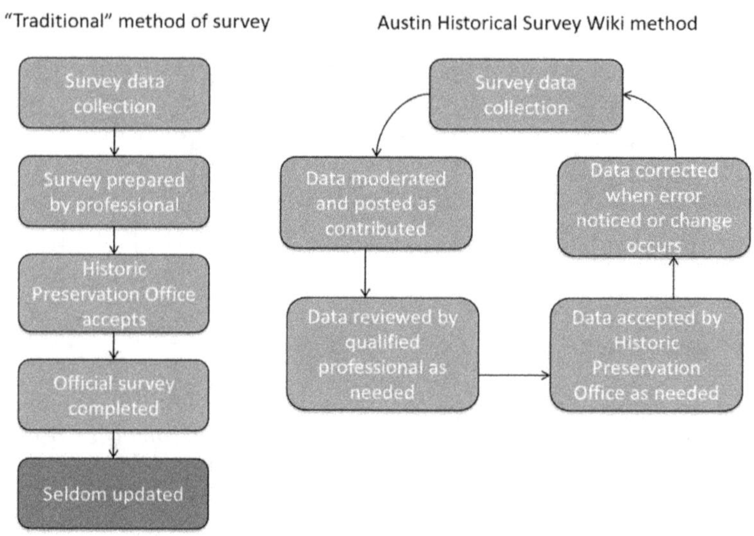

The Austin Historical Survey Wiki was designed to increase participation among a dispersed community of citizens who could contribute their time, observations, and values to widen the city's inventory of historic places. If citizens contributed information, team members reasoned that this might lead to a deeper knowledge of local heritage and enrich historic preservation and planning processes. The team hoped that the Wiki might also help the city's historic preservation office to recognize a wider range of cultural resources –a more diverse array of resources associated with ethnic and social heritage; resources considered vernacular or that are associated with cultural landscapes (Shapiro, 2007; Kaufman, 2009; Mason, 2006); and buildings and landscapes associated with the recent past. Furthermore, the team hoped that the Wiki would lead to several important outcomes including more identified and preserved historic places, more citizens engaged in historic preservation; and more transparent and participatory processes within local government planning and historic preservation.

The research team also recognized from the outset that there were going to be substantial challenges to successfully employing web-based technologies. Some of these challenges relate to the *digital divide*, or unequal access to technology according to race or ethnicity, age, and other societal divisions that may inhibit access to technological education and infrastructure (Zickuhr, 2013). Additional challenges include the differential availability of citizens to contribute time and attention; the potential for the contribution of erroneous or misleading information; difficulties in representing local or traditional knowledge using the existing web-based GIS portal; and limitations in the capacity for community-based organizations or governmental agencies to develop and maintain online initiatives (Seltzer & Mahmoudi, 2013; Sieber, 2006). Other concerns expressed in literature related to volunteered geographic information, citizen science, and wikis have to do with the validity of amateur contributions and their ability to be taken seriously by local governments, scientists and other professionals, and decision-makers (Johnson and Sieber, 2013; Ottinger, 2010; Riesch & Potter, 2014).

COMMUNITY RESPONSES TO THE WIKI

An iterative process of design, development, and community engagement was integral to the creation of the tool and ultimately to understanding barriers to participation. The Wiki was first tested in a neighborhood near the university.[7] The research team met with volunteers in their homes, introducing them to the tool. While volunteers expressed enthusiasm, some seemed discouraged with a nascent beta website that had usability challenges. Existing political conflicts in the neighborhood, including ambivalence around organizing a historic district, contributed to burnout among a few volunteers who gave up before completing a neighborhood-wide survey.

The research team continued to work with the North University neighborhood while moving on to further test the Wiki in East Austin, an area of the city with several historically African-American and Hispanic communities.[8] The research team was aware that previous historic preservation initiatives in East Austin had met with suspicion, criticism, and active resistance. In the early 2000s, gentrification became a major concern as property values rose and demographics shifted. A community-based organization dedicated to environmental justice identified historic preservation as one of the main culprits of gentrification (Chusid, 2006).[9] Mindful of this, the research team focused demonstrations and open houses on history and preservation efforts already underway, in partnership with the organizations undertaking them.

The Wiki was presented as a tool to aid community-based organizations in achieving their own goals. When presenting at an African American museum, for example, the research team explained how the new tool might help with cataloguing objects associated with particular sites and promoting exhibits. When interacting with neighborhood groups, the Wiki was presented as a low-cost approach to conducting historical surveys required for local historic district applications. Music and cultural organizations were invited to use the Wiki to upload stories about significant sites, businesses, cemeteries, or other places that could draw community users and catalyze appreciation of neglected historical figures from the East Side.[10]

Some community members associated with efforts to document the area's African American and Hispanic community histories, engaged the Wiki and the research team with skepticism.[11] Community leaders' suspicions were grounded in concerns about how data would be used. Would it be used to support gentrification or secret, long-range planning that bypassed public engagement? The Wiki's inclusion of numerous fields to catalogue architectural features may have intimidated lay and older users who felt that they needed to contribute specialized information or who only wanted to contribute stories related to inhabitants of a historic home rather than documenting its architecture. For some residents the digital divide remained a reality, especially for seniors who may be able to record oral histories or share photos, but were wary of using an online tool.

Some conversations with residents and preservation advocates led to passionate discussions about the community's changing demographics. East Austin was rapidly changing; while the Austin region is one of the fastest growing metropolitan areas in the country, its central city, including East Austin, is losing African American population. There appeared to be, at the core of many informal conversations, a desire among residents of color in East Austin to control the images, planning, and funding related to their cultural assets in the face of demographic change, growing development pressure, and concerns about the community's relationship to the City of Austin and its development-oriented initiatives. Some community leaders were more interested in having comprehensive discussions with their communities on the meaning and use of historical surveys before engaging the Wiki. The greatest concern was being able to tell stories, especially about lost places. While performing outreach with a local African American historical group, its leadership had a wealth of stories about "touchstone" sites, places that were meaningful to those who remembered vibrant businesses that served as important social spaces and community institutions (Allen, 1989). Some community members were more enthusiastic about documenting their personal and community attachments to these touchstone sites. They seemed reluctant to participate, because the Wiki was built for municipal purposes and didn't seem to encourage community remembrance of important social spaces.[12]

One neighborhood group was interested in organizing a local historic district to slow the rate of demolitions and preserve their interpretation of neighborhood character. This neighborhood group embraced the Wiki, dedicating time before one of their meetings to test the tool. However, introduction of the Wiki revealed fissures between Anglo and Latino residents, in a community undergoing demographic and socioeconomic shifts. A Latino preservation advocate pointed out a less-than-flattering photo of a property that came from a decade-old professional survey on the Wiki. He expressed concern that photos taken by strangers and posted to the Wiki would contribute to stereotypes about Latino residents. Other advocates did not see the City-sanctioned neighborhood group as legitimate representation of residents and suggested other, less Anglo-dominated groups for more "authentic" testing.

While designed for better transparency and for community-based, some viewed the Wiki with suspicion, concerned that there were ulterior motives including gentrification and the cultural appropriation African American and Hispanic heritage. The Wiki was designed for community-based organizations to use for their own purposes as well as the City of Austin's; however, its status as a municipal platform and database may have reduced participation among community-based organizations.[13]

AFTER OFFICIAL LAUNCH

The Austin Historical Survey Wiki was officially "launched" at City Hall with great fanfare on June 4, 2012. News media covered the event and general enthusiasm was expressed by participating officials and by members of the public. By 2017, 320 users have registered to edit or contribute to the Wiki, although only 117 have contributed data. As a measure of broad participation, this number is modest in a city of more than 940,000. On the other hand, the Kansas Historic Resources Inventory, another website that allows public contributions, has 456 registered users, in a state of 2.9 million (Kansas State Historic Preservation Office & Kansas Historical Society, 2014).

Table 1 shows the number of contributions as of the writing of this book chapter. The column "number of places created" represents the number of historic buildings, objects, structures, and other sites that have been added to the Wiki.

An aim of the project was to encourage community-based organizations to adopt the Wiki as a means of organizing information for their own purposes. The local heritage society continues to add content and uses the site to disseminate content, but

Table 1. Contributions to the Wiki as of July 2017

Number of Places Created	*10,336*
Images Uploaded	*4,355*
Documents Uploaded	1,249
Current Data Fields	*44,764*
Data Fields Added	*98,425**

* These numbers include data that have been archived due to the contribution of new information at the same level of review.

only on a limited basis. It remains an open question as to the number of community-based organizations that will have the interest and capacity to contribute to the City's information system in a systematic way that also benefits community-based goals. The evidence seems to indicate that adoption of the tool is limited and that major outreach efforts would be required to revive participation on the Wiki.

A test of the Wiki's capacity to support government decision-making and the use of volunteered data will come when City staff begin regularly including Wiki data in staff reports that involve deliberation on the part of the Landmarks Commission, the Planning Commission, and the City Council. Or when citizens insist on the use of Wiki data in official proceedings.[14] Neither has happened. If data from the site were incorporated into the daily practice of municipal preservation and planning, as had been intended from the start, it would more directly test the hypothesis that the public can contribute valid data to information systems used in planning and decision-making. The lack of use in public decision-making means that the connection between the Wiki and public administration is largely unrealized and the connection between the two is likely to be invisible to many potential users.

In addition, professional consultants conducting surveys for the City of Austin have not incorporated citizen outreach with the Wiki into their workflows (see Figure 9). City Preservation Office staff have neither required public engagement nor used the Wiki to do more than disseminate information. The experiment brought into high relief not only the limitations in the capacity for government agencies and community organizations to maintain online initiatives as anticipated by Seltzer & Mahmoudi (2013) and Sieber (2006), but their reluctance to apply data from the tool in decision-making processes or facilitate communities of active users willing to review and edit data already on the Wiki.

During a PowerPoint presentation before the City of Austin's Human Rights commission hearings on gentrification, Fred McGhee, an archaeologist, preservation advocate, and former city council candidate, mentioned the Wiki during a discussion of data types that should be publicly available to those wanting to understand changes to

Figure 9. This screenshot shows a survey of city-owned properties conducted by a preservation professional and disseminated on the Wiki. Volunteers were not employed in the survey process; however, citizens can still contribute new "unreviewed data" about these places.

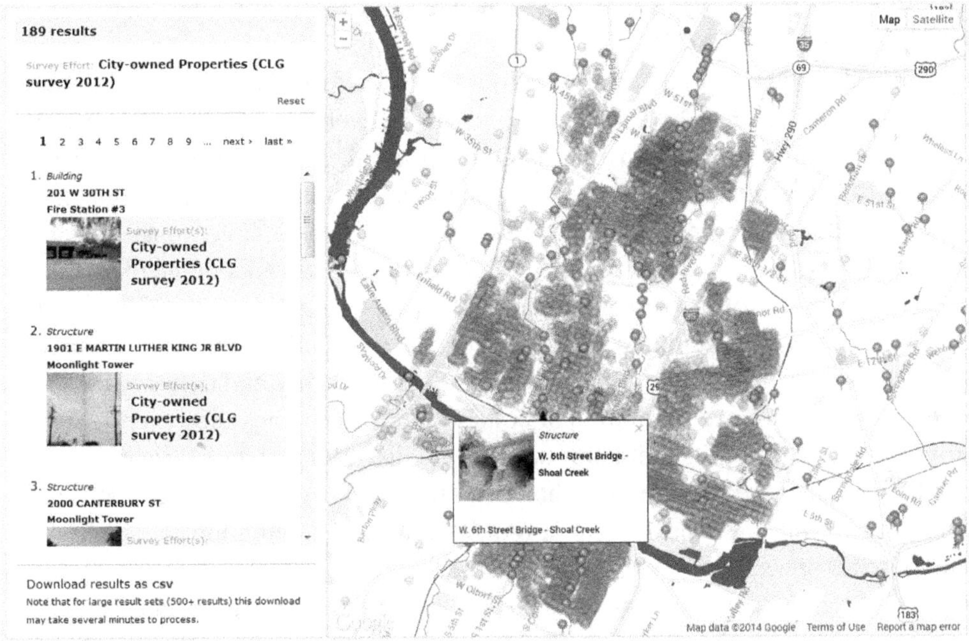

their community's built environment. He suggested greater public access to a platform that integrates data related to zoning cases, historic preservation designations, and other related data (McGhee, 2015). However, despite the opportunity to integrate the Wiki and its data into CodeNext, a major initiative of the City of Austin to reconsider its zoning and land development ordinances, there was no movement by the City to do so.

VALUE OF THE WIKI MODEL AND SUGGESTED FUTURE EXPERIMENTATION

On the Austin Historical Survey Wiki, citizens can describe a place's historic significance as well as its attributes. They have the ability to place markers anywhere on the map and record information about any place within Austin's city limits as historic resources. In this way, the Wiki can capture representations of collective places of value. This is an important modification to the current method of historical survey, which is largely driven by staff and professionals. This may have constrained participation among those who do not wish to use standardized forms or established preservation categories to share information about local history.[15] In this case, the opportunity to participate is not a free form invitation to add memories about a place; the Wiki is an invitation to share data related directly to the goals and objectives of the Historic Preservation Office.

Part of the rationale for the Wiki was to provide a platform where historic resources could be identified in advance of conflicts over development. This assumes that citizens are willing and able to articulate their attachments to historic places before they are threatened, and to share information in a way that a local government can use in its established bureaucratic processes. The Wiki does not intentionally exclude alternative narratives about place, but it does not explicitly encourage them either.

Some members of the research team were disappointed that the tool could not give citizens the opportunity to give voice to what they believe *should be done* with the data (Minner, 2015) The Wiki created a data-gathering and data-maintenance tool that relates to Noveck's notion of collaborative democracy; however, it did not provide a space for online deliberation. One can indicate that a historic resource is eligible for landmark designation; however, there is no place to share ideas as to other creative means of preservation, remembrance, or celebration.

While the inventory is crowdsourced, the solutions to *how* one might preserve a place are not. If a user has an idea for adaptively reusing an abandoned or underutilized building or for incorporating artwork that celebrates the past, there is no place to contribute these ideas. This limits the potential rewards of participation, such as giving citizens a sense of empowerment and even some limited control over how their data is used (as their data might be accompanied by their perspectives). Group preferences or consensus over the desirability to preserve a place cannot be represented on the site. This may limit the formation of communities of Wiki users who might otherwise steward data on the site.[16] The Wiki is similar in that regard to the portals created for SurveyLA, Motor City Mapping, and Preservation Detroit's historical survey.

Community participation also requires city government to have an ongoing commitment to overcoming divides that are not only digital, but also to address mistrust that has developed out of a much longer history of prior government interactions with the community. This necessity for trust cuts both ways; the Austin Historical Survey Wiki was created with the idea that preservation professionals and city staff would want to seek out public contributions and open their information to revision and additions beyond their control. [17]

Furthermore, the choice to participate on the Wiki requires a user's willingness to associate a particular piece of information with one's name. That requires trust in the intentions of local government and other users. Distrust can be based on a range of concerns, from doubt that the information will be used, to concerns that data on the Wiki will be used either to oppose or to promote preservation in ways with which the contributor might disagree.

Even with these observed issues and limitations, the authors feel strongly that there is value in the tool that was produced. The public now has access to previously inaccessible professional surveys. As the research team prepared data for upload to the Wiki, many questions were raised that required working closely with City staff. This resulted in clarifications and corrections to professional assessments as well as basic information about surveyed properties. Thus, both accessibility and the quality of information about historic places has been improved.

In addition, the system of moderation and data review levels is replicable and may be useful for other government efforts that could benefit from citizen involvement in data collection. The categories of data that has been accepted by government officials, vetted by professionals, or simply available and unreviewed is a substantial improvement over crowdsourcing sites that do not make this clear. The Wiki provides a system of accountability for data that a local government accepts as official, while allowing citizen-generated data to coexist online. Some of this crowd-sourced information will become official data; some may simply remain visible. In time, the availability of citizen contributed data may change what is considered official government data, expanding beyond current norms and standards.

Austin's Wiki should not be understood as a crowdsourcing model that simply parses out discrete and easy steps for a general public, but a tool that requires government officials and other professionals to continually interact with citizens and to facilitate the flow of information into and out of the tool. It is a tool that was built for the preservation community to share information with government officials and for both to engage in joint fact-finding. From Austin's web experiment, it is evident that the ability to realize improvements in public engagement and planning support systems for decision-making will require more than technical infrastructure.

The successful implementation of tools similar to Austin's Wiki will require continued development of both technological infrastructure and public outreach methods that can enable diverse communities of citizens to coalesce, collaborate, and deliberate online. At the fundamental level, the success of this model requires a willingness among government officials to engage with the public and a long-term commitment to both citizen participation and stewardship of online resources. Online engagement also requires willingness on the part of community-based organizations to share information, encourage their constituents to actively participate in local government data gathering, and advocate for the use of citizen-generated data in decision-making over the long term. Ultimately, this model requires government officials and professionals not only to design and maintain a usable platform, but to actively facilitate both collaboration and deliberation among citizens, community stakeholders, and decision-makers on and off-line.

CONCLUSION: CITIZENS AS SENSORS VS. CITIZEN DELIBERATION

The case studies of large-scale survey efforts in Los Angeles, Detroit, and Austin, highlight a few models for gathering and sharing information about historical resources. All three have led to technological investments in web infrastructure that range from improvements to and deployments of both open source and proprietary GIS, the innovative use of social media and texting, sophisticated moderation and tagging of data, and multi-lingual and multi-media public outreach. In all three examples, data from the historical surveys remains publicly available and is shared online.

Austin's wiki effort differs from the other two, as the city-wide survey effort has not been deemed complete. The Wiki remains open for additions and corrections, while there appears no remaining open invitation for new information or corrections to data in Los Angeles' or Detroit's historic resources surveys. The concept of the historical survey as an ongoing and cumulative process without an end date appears a significant advance in concept. After all, the state of historic resources changes with demolitions, remodeling, and reinvestment; what is valued by citizens may be ever-shifting, and the need for stewardship is continuous.

In the use of 'citizens as sensors,' all three cities had survey efforts that actively engaged volunteers in surveying for historic resources. This invitation was directly channeled into a pre-existing historical survey process and the tasks associated with gathering of empirical data that can be easily entered into a city's land use, zoning, and property records information. This is a "crowdsourcing" model in which the tasks of local government or business groups are prioritized. Perhaps the greatest limitation to making each of the cities 'smarter,' is the limited depth of holistic engagement with historic preservation. In each of the examples, citizens remain instruments of local government needs. The values or deeper intellectual engagement – the ladder of citizen engagement or collaboration -- in determining what historic resource are of value and how to respond to this information does not extend to upper rungs. While deeper engagement and thinking about historic preservation and local government control may have resulted from these efforts, it was not central to either the process or technology. To do so would have required more fully engaging citizens as the central control or 'brain' rather than sensor. This would mean extending the function of digitizing memory to direct deliberation about it. This does not mean to imply that participatory efforts or technological innovations failed – far from it. Instead, this is to point to the need to create even greater and ever present opportunities to incorporate our ability to remember the past and deliberate about its value into the creation of collective visions and actions to shape the future.

NOTE

This project was funded in part through the City of Austin, a grant from the University of Texas at Austin, a Certified Local Government Grant administered by the Texas Historical Commission, a grant from the National Center for Preservation Technology and Training, and a Preserve America grant by the Department of the Interior, National Park Service. Any opinions, findings, and conclusions or recommendations expressed in this material are those of the author(s) and do not necessarily reflect the views of the U.S. Department of the Interior, or other grantors.

REFERENCES

Afzalan, N., Sanchez, T., & Evans-Cowley, J. (2017). Creating smarter cities: Considerations for selecting online participatory. *Cities (London, England)*, *67*, 21–30. doi:10.1016/j.cities.2017.04.002

Al-Kodmany, K. (2009). Planning for the Hajj: Political Power, Pragmatism, and Participatory GIS. *Journal of Urban Technology*, *16*(1), 5–45. doi:10.1080/10630730903090289

Albino, V., Berardi, U., & Dangelico, R. M. (2015). Smart Cities: Definitions, Dimensions, Performance, and Initiatives. *Journal of Urban Technology*, *22*(1), 3–21. doi:10.1080/10630732.2014.942092

Allen, B. (1989). *Blacks in Austin*. Austin, TX: Self-published.

Allison, E. W., & Peters, L. (2011). *Historic Preservation and the Livable City*. Hoboken, NY: John Wiley and Sons.

Arieff, A. (2016, July 20). Mapping Detroit, Inch by Inch. *The New York Times*. Retrieved November 20, 2017 from https://www.nytimes.com/2016/07/21/us/mapping-detroit-inch-by-inch.html?mcubz=1

Arnstein, S. R. (1969). A Ladder of Citizen Participation. *Journal of the American Institute of Planners*, *35*(4), 216–224. doi:10.1080/01944366908977225

Austin Historical Survey Project Team. (2009). *Austin Historical Survey Wiki Phase I: Concept Background and Wiki Design*. Austin, TX: University of Texas at Austin School of Architecture. Retrieved October 9, 2014 from http://www.soa.utexas.edu/files/hp/Austin_Historical_Survey_Web_Tool_PhaseIReport.pdf

Batty, M. (2017). The Sixth Kondratieff Is the Age of the Smart City, Keynote Speech. 15th International Computers in Urban Planning and Urban Management, Adelaide, Australia.

Bernstein, K., & Hansen, J. (2016). SurveyLA: Linking Historic Resources Surveys to Local Planning. *Journal of the American Planning Association*, *82*(2), 88–91. doi:10.1080/01944363.2015.1137199

Bernstein, K., Sun, E., & Sucre, R. (2009). *Survey LA Panel Discussion*. Presented at the Association for Preservation Technology: Preservation in the City Without Limits, Los Angeles, CA.

Bertron, C. (2013). Survey Forth! Innovative Survey Methodologies. *Forum Journal*, *27*(4), 34–35.

Bertron, C., & Rypkema, D. (2012). *Historic Preservation and Rightsizing Current Practices and Resources Survey*. Washington, DC: PlaceEconomics.

Bowser, A., & Shanley, L. (2013) *New Visions in Citizen Science, Case Study Series, Vol. 3*. Washington, DC: Commons Lab, Science and Technology Innovation Program Woodrow Wilson International Center for Scholars. Retrieved November 20, 2017 from: https://www.wilsoncenter.org/sites/default/files/NewVisionsInCitizenScience.pdf

Brabham, D. C. (2009). Crowdsourcing the Public Participation Process for Planning Projects. *Planning Theory*, *8*(3), 242–263. doi:10.1177/1473095209104824

Bragg, A. E. (2014, January 8). *Preservation Advocates Launch Critical Survey of Historic Detroit Neighborhoods*. Retrieved July 1, 2017 from Preservation Detroit: http://preservationdetroit.org/call-to-action-major-preservation-survey-of-historic-detroit-neighborhoods/

Bryant, S. L., Forte, A., & Bruckman, A. (2005). Becoming Wikipedian: transformation of participation in a collaborative online encyclopedia. In *GROUP '05 Proceedings of the 2005 International ACM SIGGROUP Conference on Supporting Group Work* (pp. 1–10). New York: ACM Press.

Chusid, J. (2006). Preservation in the Progressive City: Debating History and Gentrification in Austin. *The Next American City,* (12), 23-27.

City of Austin. (2003). *Staff Task Force on Gentrification in East Austin: Findings and Recommendations*. Austin, TX: City of Austin.

City of Austin. (2011). *Downtown Austin Plan*. Austin, TX: City of Austin.

City of Austin. (2012). *Imagine Austin*. Austin, TX: City of Austin.

City of Los Angeles. (2013). *SurveyLA. City of Los Angeles Website*. Retrieved November 20, 2017 from: http://preservation.lacity.org/survey

Connors, J. P., Lei, S., & Kelly, M. (2012). Citizen Science in the Age of Neogeography: Utilizing Volunteered Geographic Information for Environmental Monitoring. *Annals of the Association of American Geographers, 102*(6), 1267–1289. doi:10.1080/00045608.2011.627058

Coppola, S. (2012, October 23). Austin's tax breaks for historic properties challenged in lawsuit. *Austin American-Statesman*.

Creative Commons. (2017). *State of the Commons 2016*. Retrieved November 20, 2017: http://creativecommons.org

Data Driven Detroit. (2017). *Historic Resource Survey*. Retrieved from Data Driven Detroit: http://portal.datadrivendetroit.org/datasets/d464738110584038bfb836f861e566a0_0

Davey, M. (2014, February 17). A Picture of Detroit Ruin, Street by Forlorn Street. *The New York Times*. Retrieved July 21, 2017, from https://www.nytimes.com/2014/02/18/us/detroit-tries-to-get-a-clear-picture-of-its-blight.html

Detroit Blight Removal Task Force. (n.d.a). *Frequently Asked Questions*. Retrieved July 21, 2017, from The Detroit Blight Removal Task Force: http://www.timetoendblight.com/faq/#1

Detroit Blight Removal Task Force. (n.d.b) *What is the context for our work?* Retrieved July 21, 2017, from Detroit Blight Removal Task Force: https://s3.amazonaws.com/detroit-blight-taskforce/CHAPTER+02.pdf

Dia, H. (2017). *When Artificial Intelligence Permeates the Modern City and Transforms Urban Life (Keynote Speech)*. 15th Annual Computers in Urban Planning and Urban Management Conference, Adelaide, Australia.

Elwood, S. (2008). Volunteered geographic information: Key questions, concepts and methods to guide emerging research and practice. *GeoJournal*, *72*(3/4), 133–135. doi:10.100710708-008-9187-z

Elwood, S., Goodchild, M. F., & Sui, D. Z. (2012). Researching volunteered geographic information: Spatial data, geographic research, and new social practice. *Annals of the Association of American Geographers*, *102*(3), 571–590. doi:10.1080/00045608.2011.595657

Estelles-Arolas, E., & Gonzalez-Ladron-de-Guevara, F. (2012). Towards an Integrated Crowdsourcing Definition. *Journal of Information Science*, *38*(2), 189–200. doi:10.1177/0165551512437638

Evans, E. (2014). Integrity, Character, and Intactness: A preservation survey model for legacy cities. In T. Schwartz (Ed.), *Historic Preservation and Urban Change*. Cleveland, OH: Cleveland Urban Design Collaborative, Kent State University.

Evans-Cowley, J., & Hollander, J. (2010). The New Generation of Public Participation: Internet-based Participation Tools. *Planning Practice and Research*, *25*(3), 397–408. doi:10.1080/02697459.2010.503432

Fietkiewicz, K. J., Mainka, A., & Stock, W. G. (2017). eGovernment in cities of the knowledge society. An empirical investigation of Smart Cities' governmental websites. *Government Information Quarterly*, *34*(1), 75–83. doi:10.1016/j.giq.2016.08.003

Fitzjohn, M. (2009). The Use of GIS in Landscape Heritage and Attitudes to Place: Digital Deep Maps. In L. S. S. Marie & J. Carman (Eds.), *Heritage Studies: Methods and Approaches* (pp. 237–252). London: Routledge.

Frey, P. (2008). *Building Reuse: Finding a place on American climate policy agendas*. Retrieved June 27, 2014, from National Trust for Historic Preservation website: http://www.preservationnation.org/information-center/sustainable-communities/additional-resources/builllding_reuse.pdf

Ganapati, S. (2010). *Using geographic information systems to increase citizen engagement* (White paper). Retrieved March 9, 2012, from IBM Center for the Business of Government website: http://www.businessofgovernment.org/sites/default/files/GanapatiReport.pdf

Getty Foundation. (2017, July 20). *Our Priorities*. Retrieved from The Getty Foundation: http://www.getty.edu/foundation/initiatives/

Glasmeier, A., & Christopherson, S. (2015). Thinking about Smart Cities. *Cambridge Journal of Regions, Economy and Society*, *8*(1), 3–12. doi:10.1093/cjres/rsu034

Goodchild, M. F. (2007). Citizens as sensors: The world of volunteered geography. *GeoJournal*, *69*(4), 211–221. doi:10.100710708-007-9111-y

Goodspeed, R., Spanring, C., & Reardon, T. (2012). *Crowdsourcing as Data Sharing: A Regional Web-based Real Estate Development Database.* Presented at the ICEGOV '12, Albany, NY. 10.1145/2463728.2463819

Gordon, E., & de Souza e Silva, A. (2011). *Net locality: Why location matters in a networked world.* New York, NY: Wiley-Blackwell.

Habermas, J., & McCarthy, T. (1985). *The Theory of Communicative Action* (Vol. 2). Boston, MA: Beacon Press.

Hanzl, M. (2007). Information technology as a tool for public participation in urban planning: A review of experiments and potentials. *Design Studies*, *28*(3), 289–307. doi:10.1016/j.destud.2007.02.003

Hu, G., Pan, W., Lu, M., & Wang, J. (2009). The widely shared definition of e-Government: An exploratory study. *The Electronic Library*, *27*(6), 968–985. doi:10.1108/02640470911004066

Jarmusch, A. (2011). 2011 National Planning Excellence Award for Public Outreach. *Planning (Chicago, Ill.)*, *77*(4), 35.

Johnson, P., & Sieber, R. (2013). Situating the Adoption of VGI by Government. In D. Sui, S. Elwood, & M. Goodchild (Eds.), *Crowdsourcing Geographic Knowledge: Volunteered Geographic Information (VGI) in Theory and Practice*. Dordrecht, The Netherlands: Springer. doi:10.1007/978-94-007-4587-2_5

Kansas State Historic Preservation Office & Kansas Historical Society. (2014). *Kansas Historic Resources Inventory*. Retrieved June 27, 2014, from http://khri.kansasgis.org/

Kaufman, N. (2009). *Place, Race and Story: Essays on the Past and Future of Historic Preservation.* New York: Routledge.

Lathrop, D., & Ruma, L. (Eds.). (2010). *Open government.* Beijing: O'Reilly.

Laurie, J. (2008). Historic preservation and cluster based economic development. *Economic Development Journal*, *7*(1), 38–46.

Long, J. (2010). *Weird city: sense of place and creative resistance in Austin, Texas* (1st ed.). Austin, TX: University of Texas Press.

Lynch, K. (1972). *What Time is This Place?* London: The MIT Press.

Mallach, A. (2011, July/August). Demolition and preservation in shrinking US industrial cities. *Building Research and Information, 39*(4), 380–394. doi:10.1080/09613218.2011.573743

Mason, R. (2006). Theoretical and Practical Arguments for Values-Centered Preservation. *CRM: The Journal of Heritage Stewardship, 3*(2), 21–47.

McGhee, F. L. (2015, October 15). *Gentrification, Historic Preservation, Public Housing and Human Rights.* Retrieved July 21, 2017, from City of Austin website: http://www.austintexas.gov/edims/document.cfm?id=242870

Minner, J. (2013) *Landscapes of Thrift and Choreographies of Change: Reinvestment and Adaptation along Austin's Commercial Strips* (Dissertation). Austin, TX: University of Texas at Austin.

Minner, J. (2015). IMHO to IOHO: A Call for Crowdsourcing Creative Possibilities for Urban Adaptation. *Forum*, 50–62.

Minner, J. (2016). Revealing Synergies, Tensions, and Silences Between Preservation and Planning. *Journal of the American Planning Association, 82*(2), 72–87. doi:10.1080/01944363.2016.1147976

Minner, J., & Holleran, M. (2016). Introduction to the Special Issue. *Journal of the American Planning Association, 82*(2), 69–17. doi:10.1080/01944363.2016.1142391

Newman, G., Zimmerman, D., Crall, A., Laituri, M., Graham, J., & Stapel, L. (2010). User-Friendly Web Mapping: Lessons from a Citizen Science Website. *International Journal of Geographical Information Science, 24*(12), 1851–1869. doi:10.1080/13658816.2010.490532

Noveck, B. S. (2009). *Wiki government: how technology can make government better, democracy stronger, and citizens more powerful.* Washington, DC: Brookings Institution Press.

Noveck, B. S. (2010). The Single Point of Failure. In D. Lathrop & L. Ruma (Eds.), *Open government: collaboration, transparency, and participation in practice* (pp. 49–69). Beijing: O'Reilly.

Ottinger, G. (2010). Buckets of Resistance: Standards and the Effectiveness of Citizen Science. *Science, Technology & Human Values, 35*(2), 244–270. doi:10.1177/0162243909337121

Preservation Green Lab. (2011). *The greenest building: Quantifying the environmental value of building reuse*. Retrieved June 27, 2014, from http://www.preservationnation.org/information-center/sustainable-communities/sustainability/green-lab/lca/The_Greenest_Building_lowres.pdf

Randolph, J. (2012). *Environmental land use planning and management* (2nd ed.). Washington, DC: Island press.

Riesch, H., & Potter, C. (2014). Citizen science as seen by scientists: Methodological, epistemological and ethical dimensions. *Public Understanding of Science (Bristol, England), 23*(1), 107–120. doi:10.1177/0963662513497324 PMID:23982281

Scola, N. (2014, February 26). *Why Preservationists Are Mapping Detroit's Historic Layer*. Retrieved from Next City: https://nextcity.org/daily/entry/why-preservationists-are-mapping-detroits-historic-layer

Seltzer, E., & Mahmoudi, D. (2013). Citizen Participation, Open Innovation, and Crowdsourcing: Challenges and Opportunities for Planning. *Journal of Planning Literature, 28*(1), 3–18. doi:10.1177/0885412212469112

Senbel, M., & Church, S. P. (2011). Design Empowerment: The Limits of Accessible Visualization Media in Neighborhood Densification. *Journal of Planning Education and Research, 31*(4), 423–437. doi:10.1177/0739456X11417830

Shapiro, K. (2007). From modernism to McDonald's: Ideology, controversy, and the movement to preserve the recent past. *Journal of Architectural Education, 61*(2), 6–14. doi:10.1111/j.1531-314X.2007.00145.x

Sieber, R. (2006). Public Participation Geographic Information Systems: A Literature Review and Framework. *Annals of the Association of American Geographers, 96*(3), 491–507. doi:10.1111/j.1467-8306.2006.00702.x

Stein, C. J. (2010). *Greening modernism: Preservation, sustainability, and the Modern Movement*. New York, NY: W.W. Norton & Company.

Talen, E. (2000). Bottom-up GIS: A new tool for individual and group expression in participatory planning. *American Planning Association. Journal of the American Planning Association, 66*(3), 279–294. doi:10.1080/01944360008976107

Technologies, L. (2017, June 21). *Detroit*. Retrieved from Loveland: https://detroit.makeloveland.com/

Technologies, L. (n.d.). Retrieved July 21, 2017, from Loveland: https://makeloveland.com/#t=search&b=none

Technologies, L., & the Data Driven Detroit. (n.d.). Retrieved July 21, 2017, from Motor City Mapping: https://www.motorcitymapping.org/#t=overview&s=detroit&f=all

Texasinvasives.org. (2011). *Texas Invasives*. Retrieved October 4, 2014, from http://www.texasinvasives.org/

The Getty Conservation Institute. (2013, January). *Arches Project*. Retrieved October 4, 2014, from http://www.getty.edu/conservation/our_projects/field_projects/arches/

Townsend, A. M. (2013). *Smart cities: Big data, civic hackers, and the quest for a new utopia* (1st ed.). New York: W.W. Norton & Company.

Tree Folks and Lady Bird Johnson Wildflower Center. (2014). *The Great Texas Tree Roundup*. Retrieved October 4, 2014, from http://treeroundup.org/

Tulloch, D. L. (2008). Is VGI participation? From vernal pools to video games. *GeoJournal*, *72*(3-4), 161–171. doi:10.100710708-008-9185-1

U.S. Department of the Interior, National Park Service. (1985). *Guidelines for local surveys: A basis for preservation planning bulletin* (Revised Version from 1985). Retrieved June 27, 2014, from http://www.nps.gov/nr/publications/bulletins/nrb24/iNDEX.htm

U.S. Department of the Interior, National Parks Service. (2017). *Become a CLG. National Parks Service website*. Retrieved June 28, 2017, from https://grantsdev.cr.nps.gov/CLG_Review/Get_All_CLG.cfm

Walden, D. (2011). Interviews: Ward Cunningham. *IEEE Annals of the History of Computing*, *33*(4), 62–67. doi:10.1109/MAHC.2011.81

WikiMedia Foundation. (2017). *MediaWiki*. Retrieved November 20, 2017, from https://www.mediawiki.org/wiki/MediaWiki

Zickuhr, K. (2013). *Who's not online and why*. Washington, DC: Pew Research Center. Retrieved June 27, 2014, from http://www.pewinternet.org/files/old-media//Files/Reports/2013/PIP_Offline%20adults_092513_PDF.pdf

ENDNOTES

1 The National Historic Preservation Act of 1966 was adopted to protect historic resources from destruction of urban fabric using federal funds. The same year, the federal Model Cities program adopted public participation as an essential feature of urban revitalization efforts.

2 "E-government" is also a common term within public administration that refers to the use of online technology to improve government "management and delivery of public services" (Hu et al., 2009).

3 For instance, scientists at the Lady Bird Johnson Wildflower Center, a research center connected to the University of Texas at Austin, have developed online platforms enabling scientists to work with groups of citizens to gather observations of invasive species around the state (Texasinvasives.org, 2011) and to map urban forests to be used in efforts to conserve and enhance municipal resource protection in Austin (Tree Folks and Lady Bird Johnson Wildflower Center, 2014). These were early prototypes for Austin's Wiki.

4 Note: The Loveland Technologies effort has now been expanded to include parcel-by-parcel mapping in cities across the U.S. (Loveland Technologies, n.d.; Arieff, 2016). At this point in time, there appears to be no inclusion of historic resource survey results in this Data portal for Detroit or any of the other communities represented on the Loveland website.

5 Austin City Code §25-11-213(B)(2); §25-2-353(C). Preservation professionals are defined by reference to federal rules: "Secretary of the Interior's professional standards for expertise in "history" or "architectural history" as described in Code of Federal Regulations Title 36, Chapter I, Part 61 (Procedures for State, Tribal, and Local Government Historic Preservation Programs)."

6 Each field's review level maintains separate revision histories. The Historic Preservation Office manages the data review process. Data fields may be promoted individually or in batches. When it becomes apparent that there are large quantities of data that should be promoted in batch, for example when a professional survey effort is completed, the preservation officer can query the Wiki for these fields and batch promote them. The overall effect is that data moves up the ladder from one level of review to another, while leaving traces of its migration behind for users to follow. Editing any data field, regardless of its review level, will submit as new non-reviewed data. Thus, users may suggest corrections without overwriting officially accepted data.

7 This phase was funded by grants from the City of Austin and a Certified Local Government grant from the Texas Historical Commission.

8 This phase of the project was funded through a National Park Service Preserve America grant.

9 In response, the City launched a staff task force that explored the potential causal link between historic preservation and gentrification. The task force concluded that there was little evidence that preservation contributed to displacement (City of Austin 2003). The task force even suggested that preservation tools had a positive effect on neighborhood stability and had the potential to mitigate the effects of gentrification. Nonetheless, there remained the potential in introducing the Wiki that prior battles could re-emerge.

10 Amateur genealogists and alumni from Huston-Tillotson, a historically black university, were two groups who expressed enthusiasm for the project after presentations. The genealogists, accustomed to utilizing public land and property records, considered the Wiki a more detailed, personalized tool with potential to make their family histories more tangible, by associating ancestors with places. Alumni saw the potential for sharing stories that they worried might soon be lost as elderly membership passed away.

11 The African American Cultural Heritage District, designated by the state in 2009, holds particular significance in the planning history of the city. Encompassing much of what was called "the Negro District" in the 1928 City Plan, the District commemorates the success of the African American community despite segregationist public policies.

12 The Wiki had been designed to allow and even encourage the documentation of places that had been demolished or otherwise no longer existed. Still, it wasn't immediately apparent to users what the City's Historic Preservation Office would do with this information.

13 In addition, language translation, a feature that supports non-Native English speakers across the city, is presently difficult to deploy on the Wiki. In an early version of the Wiki used in beta testing, a language translation feature made translation services via Google readily accessible. After transition to the City of Austin's servers, the path to activate Google translation became less apparent. This may have substantially affected the usability of the Wiki for Spanish-speaking populations or others whose first language is not English.

14 Another test of the Wiki will arise when developers and real estate agents begin to consult the Wiki for information and contribute information to it.

15 This may be particularly problematic in Austin, a city that has been noted for its geography of creative resistance, where community members fight the homogenization of place with a particularly protective stance toward beloved taco bars, music venues, and other places of the very recent past threatened with redevelopment (Long, 2010; Minner, 2013). To date, users of the Wiki have not recorded much of the more elusive attachments to place celebrated in the popular media and blog sites about Austin.

16 Unfortunately, there is now no way to know where there are conflicts over designation or over the data on the Wiki. The best one can do is to search for the records on the site to see if information has been systematically modified and then try to talk with individual users.

17 In addition, some users were unwilling to add images to the Wiki without the means of retaining greater rights to their photos (uploading images grants to the City non-exclusive rights to reproduction). Some desired the integration of Creative Commons licensing, which offers a standardized means of sharing on the web that retains certain rights, such as requiring attribution when images are used, and specifying whether images may be modified (Creative Commons 2014). The Wiki does not yet deploy the system.

Chapter 5
Citizen-to-Citizen Resource Sharing in Emergency Response

Soon Ae Chun
College of Staten Island (CUNY), USA

Vijay Atluri
Rutgers University, USA

Jaideep S. Vaidya
Rutgers University, USA

Basit Shafiq
Lahore University of Management Sciences, Pakistan

Nabil R. Adam
Rutgers University, USA

ABSTRACT

During large-scale manmade or natural disasters, such as Superstorm Sandy and Hurricanes Harvey and Irma, collaborations among government agencies, NGOs, and businesses need to be coordinated to provide necessary resources to respond to emergency events. However, resources from citizens themselves are underutilized, such as their equipment or expertise. The citizen participation via social media enhanced the situational awareness, but the response management is still mainly handled by the government or government-sanctioned partners. By harnessing the power of citizen crowdsourcing, government agencies can create enhanced disaster situation awareness and facilitate effective utilization of resources provided by citizen volunteers, resulting in more effective disaster responses. This chapter presents a public engagement in emergency response (PEER) framework that provides an online and mobile crowdsourcing platform for incident reporting and citizens' resource volunteering as well as an intelligent recommender system to match-make citizen resources with emergency tasks.

DOI: 10.4018/978-1-5225-5999-3.ch005

INTRODUCTION

Modern government has adopted the use of various technologies across its many facets of operation. In particular, emergency response stands out as a specific service provided by the government, paid for by taxpayer money that is ripe for innovation and optimization via a technological solution. The recent boom of social media services and their adoption by the public at large provide a unique opportunity to revisit the traditional problems involved in efficient and effective government emergency response and disaster management.

A frequent problem encountered during and after disasters is the lack of situational awareness of citizens' needs by government agencies. This makes it more difficult to effectively respond to the incident with appropriate resources. The communication between the responders and the victims is typically limited to one way – from the government to the citizens. The victims of the disaster are often left with few effective channels to access timely or comprehensive information regarding the assistance/cleanup/rescue services being rendered. This often fosters hard feelings between members of the government response teams attempting to do their best to manage the multitude of problems they face with limited time and resources, and the citizens who feel they deserve an immediate response, because they paid for these services with their taxes. For example, one of the most visible problems during the aftermath of superstorm Sandy (which afflicted the US East Coast in 2012) was a large response from the public (i.e., citizen volunteers) wanting to help clean up and to volunteer their time, equipment, and services to benefit Sandy victims, but were largely ignored by the government. This was due in part to the lack of a centralized way to allocate work efficiently and effectively – or to even identify areas where problems had occurred.

Some would consider such a situation a failure of the government agencies to serve its citizens, as they are chartered to be the marshals of resource planning and allocation during the response to a disaster – communicating with the public at large should be considered a top priority. Non-profits and volunteers also do themselves a disservice by not fostering a standardized method of communication amongst themselves and the official incident response teams. Squandering volunteer time or resources through poor communication of required tasks by the government and through the overloading of government employees by citizens obliging them to solve simple problems they themselves could handle with volunteer help are bad outcomes for everyone involved.

Thus, we propose a system named PEER – Public Engagement in Emergency Response to address the above-mentioned problems (Lorenzi, D., J. Vaidya, S. Chun, B. Shafiq, V. Naik, V. Atluri & N. Adam, 2013, 2015). The PEER system primarily focuses on leveraging the power of "crowdsourcing" – that is, using the citizens themselves as "sensors" in order to gather situational awareness data to aid in the recovery planning and allocation of resources during and after a disaster. PEER provides a framework for the government to allocate volunteers and their resources to solve problems in a more timely and local fashion. This is achieved through the use of social media as a platform to foster two-way communication and informational exchange between the government, citizen victims of the disaster and citizen volunteers.

We aim to create a design that implements a comprehensive, unified disaster management system focused around social media channels and smartphones for information dissemination. We use citizen-generated data using social networks to provide a means of citizen participation in disaster situation awareness and management. The government can utilize the social data for correctly pinpointing the disaster situation, hotspots for responses, and required resources. The citizens can also engage in "contributing" their skills and resources in the form of volunteerism to offset the resource scarcity of the government. All of the data generated on the public social networks can be seen by everyone – aiding transparency for both sides. Our approach also proposes the use of an incentive rewards system as policy that integrates with the technological system to achieve the desired outcome, and as such is orthogonal to the existing work.

The rest of the chapter is organized as follows. We first discuss existing work in the literature that is related to our research, and we present the overall technological and policy framework. We then present the architecture of the proposed system. An overview of a real-world prototype that has been implemented is presented with many of the major functionalities. Subsequently, we look at potential applications of the PEER framework and explore the aftermaths of Superstorm Sandy as a case study to highlight some of the relevant lessons learned. We discuss some of the challenges and risks in deploying such a system, and some of the limitations of the proposed approach and how they can be ameliorated. Finally, we conclude the chapter and discuss future work.

BACKGROUND

Prior literature exists with work done on using crowdsourcing in digital government as well as for emergency responses. Studies (Chun & Artigas, 2012; Anbalahan, 2010; Pohl, D., A. Bouchachia, & H. Hellwagner, 2012) advocate for participatory open environmental planning in a similar vein to our own proposal. They promote frameworks where environmental monitoring, environmental awareness and response planning work in tandem with high-tech sensors and crowdsourcing to shift environmental planning away from government-centric planning towards direct involvement of the citizens in the process. Various methods for gathering information on the unfolding of a disaster (Raschid, 2007; Bharosa, N., J. Lee, M. Janssen, & H. Raghav Rao, 2009; Chen, R., R. Sharman, H. R. Rao, & S. J. Upadhyaya, 2008; Therese, J., B. Fajardo & C. M. Oppus, 2010) and distributing information to individuals during the incident, as well as storing data about the disaster have been researched at length. Some prior disaster management systems have made the goal of being open source a top priority (Careem 2007; Currion, P., C. de Silva, & B. Van de Walle, 2007; Li, J.P., R. Chen, J. Lee, & H. R. Rao, 2013). Much research has been conducted on the topic of communication methods in disaster management over the years, with wireless and mobile technologies having been suggested as methods of enabling communication into and out of a disaster zone (Ray & Turuk 2011; Srinivasan 2011). The sharing of resources (e.g. equipment, hospitals, evacuation centers, etc.) and services in emergency management is emphasized using both a technological framework and a mobile component (Shafiq, B., S. Chun, V. Atluri, J. Vaidya, & G. Nabi, 2012), combining the most utilitarian aspects of prior research and case studies (Convertino, G., H. M. Mentis, A.Slavkovic, M. Rosson, & J. M. Carroll, 2011; Yates & Paquette, 2010).

Other research includes case studies, surveys, and models (Hristidis, V., S. Chen, T. Li, S. Luis, & Y. Deng, 2010, Li, J.P., R. Chen, J. Lee, and H. R. Rao., 2013, Solnit 2009, Janssen, M., J. Lee, N. Bharosa, & A. Cresswell, 2010) conducted with regards to disaster management and response systems. These studies have considered important factors such as data management and open source technologies in disaster situations as well as the resulting analysis of the gathered data to make intelligent actionable decisions. Since many different groups with overlapping responsibilities are involved in responding to an emergency/disaster, a multi-agency model (Lee, S., S. Lee, K. Kim, and J. Park, 2011) is required to control the flows of information. There is much interest in telecommunications use during disasters, and work has been done on testing the validity of models using telecommunications and Geographic Information Systems (GIS) data (Choi, Y., J. M. Blackburn, & H. Seo, 2006; Jin, G., B. Nicolai, K. Jiang, & C. Winer, 2011). One case study in particular developed an android application for use on smartphones in emergency situations (Therese, J., B. Fajardo & C. M. Oppus, 2010), providing some validity and testing of a mobile smartphone-based disaster reporting system.

Other projects (Zheng, L., C. Shen, L. Tang, T. Li, S. Luis, & S. Chen, 2011; Zheng, L., C. Shen, L. Tang, T. Li, S. Luis, S. Chen, & V. Hristidis, 2010, Wu, A., X. Yan, & X. Zhang, 2011) use data mining techniques to address disaster information management challenges on mobile devices. The research (Zheng, L., C. Shen, L. Tang, T. Li, S. Luis, & S. Chen, 2011; Zheng, L., C. Shen, L. Tang, T. Li, S. Luis, S. Chen, & V. Hristidis, 2010) addresses several critical tasks that can facilitate information sharing and collaboration between private and public sector participants for major disaster recovery planning and management through an iOS application, specifically addressing the use of mobile devices in disaster scenarios. These approaches aid in retrospective analyses of the effectiveness of public-private mobile disaster responses, such as reviewing geotagged photos of damage sites (Wu, A., X. Yan, & X. Zhang, 2011). Some of the data mining capabilities include hierarchical summarization techniques that generate brief reviews from a large collection of reports at different levels; probabilistic models that dynamically generate query forms based on users' feedback; and recommendation techniques that are adapted to help users identify potential contacts for report sharing and community organization (Zheng, L., C. Shen, L. Tang, T. Li, S. Luis, S. Chen, & V. Hristidis, 2010; Wu, A., X. Yan, & X. Zhang, 2011).

Intelligent agent AI approaches are proposed as a potential avenue for research in disaster management (Friedrich & Burghardt 2007; Ramchurn, S., A. Rogers, K. Macarthur, A. Farinelli, P. Vytelingum, I. Vetsikas, & N. R. Jennings, 2008; Cayirci & Coplu 2007) to help minimize and/or eliminate human errors. Since disaster managers are presented with complex and nuanced scenarios, artificial intelligence and/or decision logic (decision support systems) are often utilized as preliminary tools to analyze the situation, while still keeping a human in control of the final decision making. Several studies (Li, L., D. Wang, C. Shen, & T. Li, 2010; Lee, S., S. Lee, K. Kim, & J. Park, 2012, Truong, H., A. Manzoor, and S. Dustdar, 2009; Bessis, N., A. Brown, & E. Asimakopoulou, 2010) use disaster ontologies and reasoning as methods for automated decision making processes. Basit et al. (2012) have developed an ontology-based approach to enable resource sharing through the UICDS framework. This ontology allowed different agencies with different disaster/emergency management systems to share information between agencies to better utilize resources and reduce human error and delays in the responses. It utilizes the ontologies of organizational structure, workflow activities and resources, and the inference rules to discover and share resource information and interoperability from different incident management applications. This approach focuses on the information sharing among government agencies for better disaster management, while our work enables the information sharing between governments and citizens during and after disasters.

Five cases have been reported on the involvement of citizen volunteers in disaster responses by Fernandez et al. (2006), including the 1906 San Francisco earthquake, the 1985 Mexico City earthquake, and Hurricane Katrina (2005). The study suggests that large groups of citizens came out without a pre-planned role to assist those in need after disasters. The spontaneous volunteers, unlike the trained volunteers (e.g. Doctors without Borders), are "welcome" but a lack of coordination of well-intentioned volunteers often creates more chaos and overhead (Fernandez, L., J. Barbera, & J. van Dorp, 2006). In Palen et al. (2007), an online forum is proposed as a platform to share information and to coordinate citizen-led efforts but it lacks resource sharing. This chapter presents a technology-driven citizen volunteer framework that can enhance the government's awareness of disaster situations and coordinate the citizen volunteers and their resources that the government can draw on if needed.

PEER: A FRAMEWORK FOR CITIZEN-TO-CITIZEN RESOURCE SHARING

Architecture

In emergency situations, the mobile phone and social media provide the major communication platforms, disseminating situation reports (e.g. fallen trees, road blocks, etc.), making SOS help requests (e.g. required tools or vehicles to escape danger, medical needs, etc.), or simply looking for missing pets or family members. The requirements of information, services or resource have to be shared rapidly for immediate help. The transparent information sharing can be achieved by providing minute-by-minute situational awareness, prioritized tasks and resource lists, and a directory of potential providers of the tasks or resources. The PEER framework is based on achieving this information, service and resource transparency through the information shared via social media channels, and the PEER system that processes and analyzes the social media data in real time for others to easily understand the needs and to match-make the resources for particular needs or tasks. In addition, it provides an easy method of citizen volunteer registration and resource and service publication, so that other users, including government officials, businesses, or other citizens, will be able to discover the capabilities or resources needed for the urgent situations. Both the search or browsing of citizen-volunteered resources or services, and the recommender system to identify appropriate services or resources to match-make to each request are provided. These citizen-to-citizen or citizen-to-government resource sharing models will speed up some of the emergency assistance without delays introduced by an extended search for the right service providers.

More specifically, the PEER system enables communication and informational exchange between parties (citizens/volunteers/victims and the government) while providing efficiency and optimality with regard to resource identification, ranking and allocating them via connecting citizens to citizens. The PEER system provides the following functionalities:

1. **Situation Awareness:** The raw data collected from many citizens' social media postings are collected, integrated, analyzed and provide the real-time situation of the emergency event progression for citizens, government officials and any

other users. These include visualization of the situations, including geospatial clusters, and temporal progressions of the emergency situations. In addition, citizens can subscribe to the government's alerting services for up-to-date status reports of emergency situations, nearby incidents or evacuation-related information.

2. **Volunteer Registry**: Citizen volunteers can register to the system in advance of emergency events. The registry keeps the up-to-date records of the citizens' resources, capabilities, expertise and availabilities.
3. **Lookup Resource Requests**: Citizens will be able to view the prioritized list of needs and requests for help and resources. These requests may come directly from the government as well as from the citizens.
4. **Resource Recommendation**: The PEER system may analyze the resource or service requests and match-make them with available volunteer resources.
5. **Resource Sharing Registry**: Citizens will be able to sign up for resource sharing or volunteering for particular requests requiring resources or services.
6. **Monitoring and Auditing services**: The PEER system records all the resource sharing events. The government or citizen groups should be able to monitor and view the resource sharing among different parties, citizen-to-citizen, or citizen-to-government, for accountability.

Figure 1 shows the architecture of the PEER Citizen Emergency Engagement for Resource Sharing System with major components.

In this section, we provide an overview of the goals and motivations for developing the PEER framework, while detailing the various technical considerations that must be made when designing and implementing a system such as PEER. The major components include the volunteer registration and alert subscription service, the incident reporting, the visualization of situation awareness, and the matching of resource requests and volunteer resources. These functionalities are supported by the analytics and resource and incident management layers. In the analytics layer, the system contains components to assess the incidents' severity, perform risk assessment to prioritize urgent tasks, and a recommender system to match-make the resource requests with the volunteer resources or capabilities. The incentives and rewards component can be considered as a social policy tool to engage the citizens and businesses as volunteers. Both aspects of the system design are equally considered as it requires a careful alignment of technology and policy in order to function as intended. We also discuss the issues relating to two-way communications during disasters, and the utility of a hybrid platform consisting of a smartphone application/website/social media network to solve the challenges brought about in these types of conditions.

Figure 1. Architecture for the PEER Citizen Emergency Engagement for Resource Sharing

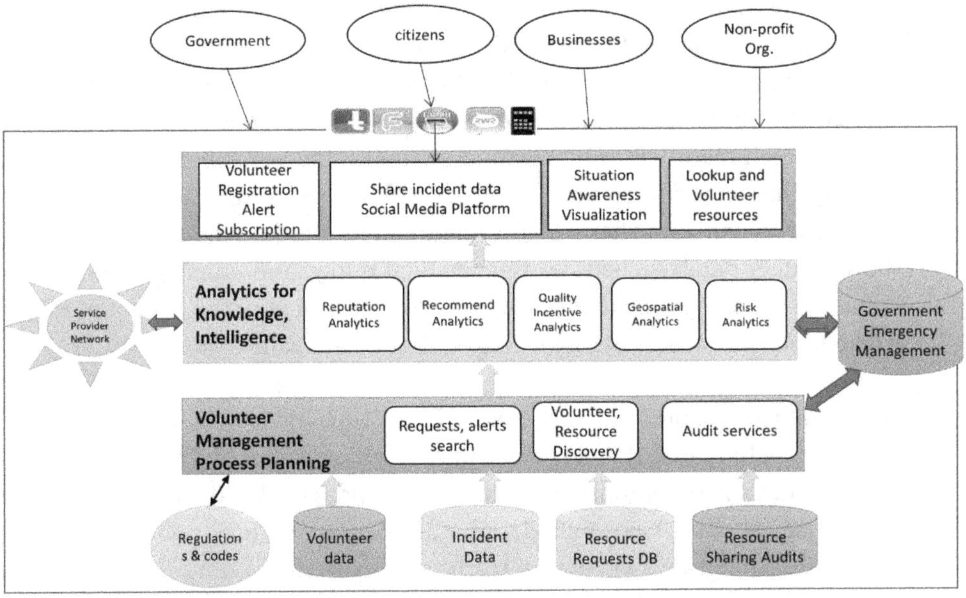

Volunteer Registry and Alert Subscription Services

The volunteer registration database is an important component of PEER, as it provides a centralized repository for information about citizens and their capabilities. The volunteer aspect, i.e., making this a subscription-based service, is important as citizens are not required by law to offer assistance during disasters – this means that individuals who sign up for listing in the directory have a desire to help other citizens and the government as they have actively chosen to register with the system. As such, the database is designed with the following features:

1. The database stores each volunteer's contact information and location (place of residence) along with their work abilities and tools, i.e., if they have a special profession that is useful to the government beyond a general worker – such as electrician, lumberjack, plumber, etc. or if they have special equipment that can be of use – large trucks for hauling, chainsaws, water pumps, etc. When a disaster strikes and an emergency management response is needed, the government can query the database and proceed to notify the volunteers nearest to each problem who have the tools and capabilities to handle the problem. This requires sharing of information between the volunteer database and the

incident reporting system mentioned earlier. Using these two systems in tandem, the government can maximize the utility of its volunteers and allocate its own resources in a more efficient manner to speed the restoration of services.

2. The database also allows for easy distribution of reward incentives after the initial disaster recovery efforts are considered satisfactorily resolved. The volunteer's contact information is stored in the database and the incident reporting system's work dispatch records can be corroborated with the volunteer's information, meaning that accurate records of which volunteers completed specific jobs can be generated from sharing information between the systems.
3. The volunteer database handles the personal information management of the volunteers for the government (and the volunteers themselves) and aids in the distribution of rewards after the disaster has been dealt with.

A well-designed volunteer database and subscription service is necessary to fully realize the ideas behind PEER and provide utility to the citizens and the government alike. We have designed appropriate online forms and a backend database system. The web interface needs to be designed for both desktop users and mobile users. The desktop version is integrated with social media and provides maximum functionality including all of the bells and whistles (and data/power intensive tasks) that can be driven by a powerful CPU.

Any comprehensive solution must also include a mobile smartphone application that can utilize the web interface to communicate with the backend database. This mobile application will perform much of the same functionality as the desktop version, but also leverages the additional capabilities of the smartphone, such as its data gathering abilities and mobile voice/data service, as these will be more tightly integrated into the functionality of the application.

Incident Reporting and Resource Requesting System

This system is designed to serve as the vehicle of communication and situational awareness between the government's emergency response team and citizens affected by the disaster. It is a two-way communication system in that the government can receive data and information from the citizens from a variety of sources and then respond to the citizens via social media and provide real time updates on the incident management web portal. The web portal component features a real-time map that shows where each problem is located, what type of problem it is, how many people have reported it and which citizens or government officials (if any) are responding to the incident. It can also provide real time GPS tracking of the government's "problem assessors" as they move around in the disaster area and manage the queue of incidents. The risk assessment tool can be used based on evidence coming from multiple sources. The incident reporting system procedures for reporting and resolving a problem can consider different strategies as follows:

1. Citizens can report a problem they have encountered to the government via their smartphones, either through a custom application running on the smartphones or through traditional means and/or social media channels. Once a predetermined threshold for the number of times that a problem has been reported by different citizens is reached, it shows up on the government's emergency blotter / live map of the disaster area. Particular types of disasters require immediate attention of the government and exceptions can be made for them, however the potential for abuse of this status by users exists and will require some form of validation.
2. The victims/citizens, using their smartphones, can send multi-modal data (text, audio, images, GPS coordinates) in support of their incident claims, which can be aggregated and analyzed by an automated system to detect the type and severity of the problem. More data gathered to support the validity and severity of the incident can give everyone involved a better idea of what to expect and how to prepare to solve the problem. This functionality is in addition to the previously described thresholding metric (Adam, N., J. Eledath, S. Mehrotra, N. Venkatasubramanian, 2012a; Adam, N., B. Shafiq, R. Staffin, 2012b), and provides additional layers of analysis to minimize the risk of the government wasting resources on problems that can be handled in other ways.
3. The use of custom designed ontologies relating to emergency management and disaster recovery can provide reasoning abilities to detect problems. For example, an incoming tweet can be analyzed for textual content – the meanings of the words can be determined by the ontology and can provide recommendations for actions that would best handle the problem in an efficient manner. In a study (Lorenzi, D., Vaidya, J., Shafiq, B., Chun, S., Vegesna, N., Alzamil, Z., Adam, N., Wainer, S., & Atluri, 2014), these kinds of functionalities have been implemented to enable progressive government responsiveness.
4. Once a problem has been identified, the government sends a certified individual to assess the problem. The job of the assessor is to determine whether the problem can be solved by citizen volunteers or needs to be handled by specialized government employees. If citizens can handle it, the incident reporting system can create a work order for the volunteers and dispatch them to the scene. After citizens report back, the system may dispatch a government official to verify that the work is completed and was done correctly.

5. The incident reporting system keeps a log of all the incidents it handles. Once these are cross referenced with the information in the volunteer database, one can easily distribute rewards when time is appropriate based on the work completed and filed with the system.

An individual incident report consists of the following information: an image of the problem (e.g. downed power line, fallen tree, etc.) GPS coordinate tags (if available) of where the problem is located – otherwise nearest crossroads are sufficient, a brief description of the problem (text), the username who reported the problem (same as stored in the volunteer database), and a timestamp (date/time). A common list of problems that occur during disasters is included so the users can select them from a dropdown box as opposed to filling in the information themselves, reducing the number of errors/inaccuracies in submissions. Unique logins enable the remainder of the information to be filled in automatically by the smartphone application or browser cookies.

The incident reporting system must be able to function with and without network connectivity. As such, it implements a queue-based system for sending and receiving incident reports. This means that if a user does not have network connectivity, it will store the incident reports on the device until it regains network access. They can also be bundled alongside the system snapshots such that anytime any users meet and share data, each of those users are carrying all of the incident reports for everyone they have encountered up to that point in time, so that the first person to gain network connectivity will be able to submit all of the incident reports to the system. Duplicates can be handled via usernames and timestamps and discarded if they are received again by the system at some later point in time. In this way, emergency response personnel can see all problems as soon as possible in case there are incidents that require higher priority attention.

Recommender System to Identify Matching Resources for Incident Reports

The incident reporting system creates an incident response request in the Resource Request Database when it is verified by the designated official or the intelligent system. It then searches the volunteer registration database for the resources that can be eligible and best qualified for the incident request – this allows for monitoring of incident-related resource request types, severity, etc. and matching them with the nearest capable volunteers. It utilizes data analytics and database queries, and uses a recommender algorithm that can optimally address the incident response requests. For example, after a problem has been entered into the incident reporting database, such as a downed tree blocking Anytown Blvd. near Main St., and entered into the resource request database, it can query the *volunteer resource database* for a lumberjack, a chainsaw owner, and/or a truck owner near those crossroads to handle the problem. These required resources may be derived by reasoning on the incident response knowledge base. All volunteers dispatched will have the incident logged in the auditing database as being responders to the problem and will be granted temporary authorization to update the status of the resource delivery. This auditing database will be used to assign to the resource provider the proper credit for solving the incident request later on (rewards from the incentives system). If it turns out the problem is too large for the volunteers to handle it, the government "problem assessor" will eventually visit the site and change the status of the problem in the system.

Another technological innovation that can provide an extra degree of automation is the use of semantic ontologies for automated processing of dispatches based on the language used in reporting the incident. By creating specific relationships in the emergency response ontology (e.g., Shafiq, B., J. Vaidya, V. Atluri, N. Adam, S. Chun & A. Lieb., 2008) that link disaster report keywords with actions executed by the volunteers in the database and incident reporting system, the system can automatically dispatch government workers and volunteers to problems in the most efficient manner. It can also help guide the government problem assessors to locations where they will most likely be urgently needed first. The emergency response ontology will work closely with the schema of the relational database system to make sure that the decisions it is supporting and executing make sense within the global framework of the efforts of disaster management. Such ontologies exist and have been used to enable resource management for emergency responses (Shafiq, B., S. Chun, V. Atluri, J. Vaidya, & G. Nabi., 2010).

Emergency Request and Response Analytics

Finally, we can utilize the hashtag feature of the Twitter platform by developing a system of codified hashtags to use for mining aggregate data about the disaster. Using the same five digit zip codes as the postal system, the name of the type of disaster, and the day/month/year we can quickly find the approximate location of the incident and when it was reported through the social network. The name of the city/town/village/etc. can be retrieved from the geotag coordinates, and the zip code serves to narrow down where in that locale the event occurred. This is the smallest level of granularity used in the system, and hierarchical structures can be used to gather and generate information for larger areas (e.g. pull the tweets for the five zip codes that comprise Anytown to see all the problems in each area within that town). The textual portion can then be analyzed by using the ontology for determining appropriate actions, or archived for later analysis for the purpose of creating metrics, e.g., response time, ontology accuracy, resolution of the problem, etc. For example, a tweet from a smartphone with geotagging enabled might be "Fallentree#sandy10/29/2012 http://twitpic.com/dSw7Fb" with the URL containing a link to the image of the downed tree and the tweet itself having the geolocational tags (GPS data) enabled. There is also the possibility of using multiple hashtags, as there is no limit to the number of hashtags in a tweet; the only limit is that the total number of characters in the message, including hashtags, cannot exceed 140. To make mining these tweets easier, we can create a special hashtag that users know will be read by PEER and either responded to by the government or used as an aggregate data source for generating performance metrics from the system.

Incentive Reward System

Average people are likely to participate in the volunteer system only if there is a reason beyond simply being a good citizen who has a strong desire to help the community, since this does not typically motivate them sufficiently. Some motivations for online crowdsourced work, e.g., contributing to Wikipedia, include academic interest, charity, money, fun, community participation, forced participation, self-benefit from the product, and interest in the content (Organisciak, 2008). In crowdsourced tasks such as Mechanical Turks, payment is used as an incentive to ensure the quality of the tasks completed by crowds (Kamar & Horvitz 2012) or to balance the trade-off between price and desired completion time (Faradani, S., B. Hartmann, & P. Ipeirotis, 2011).

To overcome this common apathy, a system of rewards based on results garnered from the incident reporting system – that are focused on the community level can be developed to encourage people to sign up with PEER and report problems. These rewards should be designed in such a way that they are beneficial to both the citizens affected by the disaster and the government. Aligning the incentives and rewards in such a way that they serve everyone's interests requires an understanding of the circumstances of the participants in the system, a high degree of transparency and shared goals. Let us consider a possible scenario when the PEER becomes active.

Incentives will be allocated based on the number of individuals who sign up with the volunteer database and provide assistance by solving problems during the duration of the disaster. Using the postal zip code system, areas can be logically defined making distribution and management of people and money easier for the government. A separate database is not required to store the incentives as they could be generated by joining data from the VD/SS and the IRS systems. Since each reported instance is unique within the system, its unique identifier can serve as the primary key for joins when derivative computations need to be made.

Some potential options to reward volunteers are small property tax reductions for the tax year in which the disaster occurred, or if it happened after tax season, the following year. Property tax breaks are a double-edged sword for the government, however, as this is the primary means for raising money for local projects. This fact forces governments to be aggressive in saving for disasters in the immediate future, and offer smaller property tax discounts, until the disaster reserve grows. Areas that fall under the PEER management system could potentially have lower insurance bills, since citizens are looking after one another and are providing services that would normally cost money that the insurance company would have to pay if a claim were filed. These rewards would be distributed by referencing the number of volunteers in a predefined area that worked on problems reported by the incident management system. This way even if an area was not heavily affected by the disaster, its volunteers still have a reason to go out and help in other areas.

There is some concern that this model will be difficult to administer and fund, considering that towns do not have administrative jurisdiction to provide tax breaks to individuals in other towns – or simply may choose not to do so. Also volunteers can and should be able to help out wherever they are needed, and should not be restricted by the reward system. A potential solution is to have the federal government allot the tax breaks based on areas that "qualify" for breaks based on the destruction in that location and breaks for those in nearby areas – making it known that volunteers beyond this location will not receive any rewards from the government. Notably, social networks themselves can also be used to publicize the contributions of the volunteers and indeed such postings can be considered one of the rewards (having the ability, for example, to put a super volunteer flag on your Facebook wall, etc.).

Preventing abuse of this system is important when cost-based incentives are to be distributed amongst the populace. Beyond the government assessors' role as the monitor/verification method, citizens themselves can be advocates for prevention, detection and correction of fraud and abuse within the system. By asking citizens to review and evaluate work in addition to government oversight, more work can be completed and verified than would otherwise be possible with the government workers alone. For example, a "threat tier" system could be implemented to categorize the danger of each incident on the map. Low ranking incidents can be solved by citizens and verified by citizens, saving the government assessors extra work. This system can be imagined in a similar vein to the Facebook "like" system, where people "appreciate" work. After an incident is solved and moved to the completed queue, citizens can verify that the work was actually done and then "appreciate" the success accordingly.

Methods to detect and prevent misuse can be implemented through a series of system controls and time metrics, along with a weighting algorithm that uses time of incident reported along with the threat tier rating to determine a validity ranking. The most pressing issues will be dealt with first, and priority will be given to the incidents that require government help. Most severe issues will be reported in the initial time period after a disaster has been declared. Severe issues declared at a later date are more suspect. For example, an incident report of a downed tree that comes in two weeks after the initial disaster is reported would be suspicious. The downed tree might have fallen without a causal connection to the initial disaster and might have been reported because the owner wanted to get volunteers' aid or the government to pay for its removal from the property. Based on the system controls, this would be flagged as a suspicious incident and investigated further. If it is determined that this individual acted in a deceitful manner, their PEER system privileges will be revoked and they can be fined by the government.

Methods of Communication

During disasters, it is important to note that there may exist situations where communication channels are either over carrier capacity, crippled in some way that makes communication inefficient or ineffective, or simply down altogether. We must consider the following scenarios in detail to make sure that communication options exist under each of the following scenarios: full electrical power and cellular/data connectivity, partial electrical power and cellular/data connectivity, and no electrical power and cellular/data connectivity. In order for citizens to offer their services to the government in times of crisis that require emergency responses to extraordinary disruptions of services, a robust number of options for communication needs to be made available to ensure that information can be disseminated (both to and from the government) throughout the population. This is where the utility of the smartphone's multitude of communication options and portability become invaluable for spreading information amongst a population.

When there is a total breakdown in the delivery of electrical power and telecommunications a few options still remain. Thankfully most modern smartphones have multiple of communication options beyond the carrier network they are registered on, such as Wi-Fi and Bluetooth, and possibly even cellular radios. While most of these are useless when telecommunications cease to function, the more localized vectors such as Bluetooth become attractive options. The smartphone application can use the hardware's built in ad-hoc Wi-Fi & Bluetooth sharing capabilities to provide the framework for what can be essentially a modern, mobile hybrid implementation of a mesh network and a sneakernet. For example, the smartphone application can store localized copies of itself via the Bluetooth 4.0 wireless interface that can be provided to other individuals who perhaps did not download and install the application before the disaster struck. When electricity or power are non-existent, it gives the owner of the smartphone some time (the length of time they have remaining battery power and connectivity) to make decisions that can impact their survival chances or help restore services in their area more quickly.

System snapshots are another potential solution to the problem of information dissemination. For example, the states of the incident reporting system can be encapsulated in a data package, time stamped and stored on both the government's servers and locally within the application on the smartphone. This way, if communications are lost, the user has a point of reference and some information to operate on. When two users of the application enter into each other's ad-hoc Wi-Fi area or Bluetooth area, the application can compare the timestamps of the snapshots and push the newest snapshot to users it encounters. Thus, if someone enters an area of zero carrier connectivity, people can still get information updates. This also provides options for incrementalism in these backups, decreasing the size/volume of data that needs to be transmitted, as only the data after the timestamp needs to be sent. Figure 4 gives an overview of the communication structure that would be utilized by a system such as PEER.

PEER PROTOTYPE SYSTEM

In this section we provide a prototype system of the PEER framework, called "New Jersey Meadowlands Flood Warning Community System (NJMFWC)". It is aimed at sharing flood warning alerts and citizen-based flood situation reports through various communication and social media tools such as Twitter messages, discussion forums, emails etc. Citizens do not have to make an effort to find such information from a specific source; rather the information comes to them automatically using programmed feeds and alerts. In addition, the system provides a web platform wherein registered citizens can share their voluntary services and equipment that they can offer during flood emergencies. It makes this information available via a database. During emergencies, this information can be very useful. Citizens do not have to solely depend on governmental agencies for flood relief; often there may be people around who can provide help faster, and this platform helps to leverage this fact to the benefit of all. This is another distinctive feature of the developed system. Government agencies and citizens of the region can equally benefit from such a framework.

The Flood NJ Twitter account automatically tweets the latest flood warning messages generated by the New Jersey Meadowlands Commission (NJMC) Flood Warning System (Artigas, F., S. Chun & Y. Sookhu, 2009). The Flood NJ Online

Forum engages users to post and participate in discussion topics on the disaster. The report and incident feature allows users to post any incidents and incident-related information (e.g., flood damage pictures) that they have witnessed or have knowledge of. The resource sharing feature enables citizens to see voluntary services and equipment that are available near them. These resources are location-aware, so the GIS-based resource sharing enhances the situational awareness and location-based sharing.

Citizen Resource Sharing for Flood Incidents: NJMFWC

The NJMFWC is a PEER web portal (Figure 2) that is dedicated to a community of citizens in and around the New Jersey Meadowlands region, who can contribute towards better emergency preparedness and management before and during flooding.

The following list provides some of the key features of NJMFWC:

1. Web portal to manage flood-related information and volunteer resource sharing portal.
2. Share flood warnings and report flood incidents to enhance situation awareness.

Figure 2. A PEER prototype system to share Flood Warnings, Incident Reports and Citizen Resources

3. Citizen volunteer registration.
4. Volunteer services request and resource sharing request.
5. Search and notify of available services and resources.
6. Forum to discuss and share flood related information.

The NJMFWC is developed using Joomla, a content management systems (CMS) to report and share flood incidents, volunteer management, volunteer service registration and service requests. NJMFWC provides the infrastructure for community-based resource sharing during and after flooding incidents.

Report and Share Flood Incidents

The NJMFWC system allows citizens to report and share any incident that they have knowledge of. The report and share component is based on an open source platform called USHAHIDI - that enables information collection, visualization and interactive mapping.

An incident report can be done in two different ways. A citizen can fill out the online incident report form or use social media, such as Twitter. Figure 3 shows the incident report by the system's online form, where the end users can specify the location, time, incident category/type, title and brief description, and also has the capability of uploading images of incidents. In this case, a downed tree photo is uploaded.

The incident report form shows a map displaying all the events reported, which are depicted by the red dots on the map. Users can click on those to see the topic of the report. Each incident report specifies a location and an incident type. The location of the incident can be explicitly defined via a form-based interface or implicitly via selecting a location on the map.

The report module provides a mechanism for citizens to submit reports using their mobile phones or the internet, while simultaneously creating a temporal and geospatial archive of events. The reporting component allows anyone to gather distributed data via Short Messaging Service (SMS) messages sent from cell phones, email or web and visualize it on a map or timeline. Our goal is to create the simplest possible way of aggregating information gathered from the public for use in crisis responses.

Additionally, the system could use Twitter to tweet about the incidents along with the images related to the incident. Figure 4 shows the Twitter-based incident report.

Once a report and its related data are captured in the incident report database and is verified by an administrator or by the system (e.g. multiple reports of the same incident or realistic images or videos), it shows up in the incident reports map. Users can click on the report to see its detailed information, along with any multimedia

Figure 3. Incident report Form

Figure 4. Incident Report, using Twitter

content associated with it. The interactive map allows users to zoom in or zoom out of the map to see reported incidents. The incident report map also has a list of incident categories that classify the types of incidents reported, such as Power Outage, People Injured, etc. Users have the ability to see the reports belonging to a particular incident category (Figure 5A).

In addition, the reported incidents are awaiting verification for authenticity as shown in Figure 5B, and the reported incident in Figure 5A can be zoomed in for a detailed information window that displays the list of incidents reported in that location and the images (Figure 5C). The report can contain images, associated videos and a news source of the incident referring to the location where it occurred via a Google map. The users can also search for the exact place by using the 'Find Location' search bar below the map.

Volunteers and Resource Registration System

Resource Sharing is one of the key features of the PEER system. It allows citizens to come forward and share information about services or equipment that they

Figure 5. (A) The interactive incident reports map; (B) The incident report verification; (C) Verified Incident Details

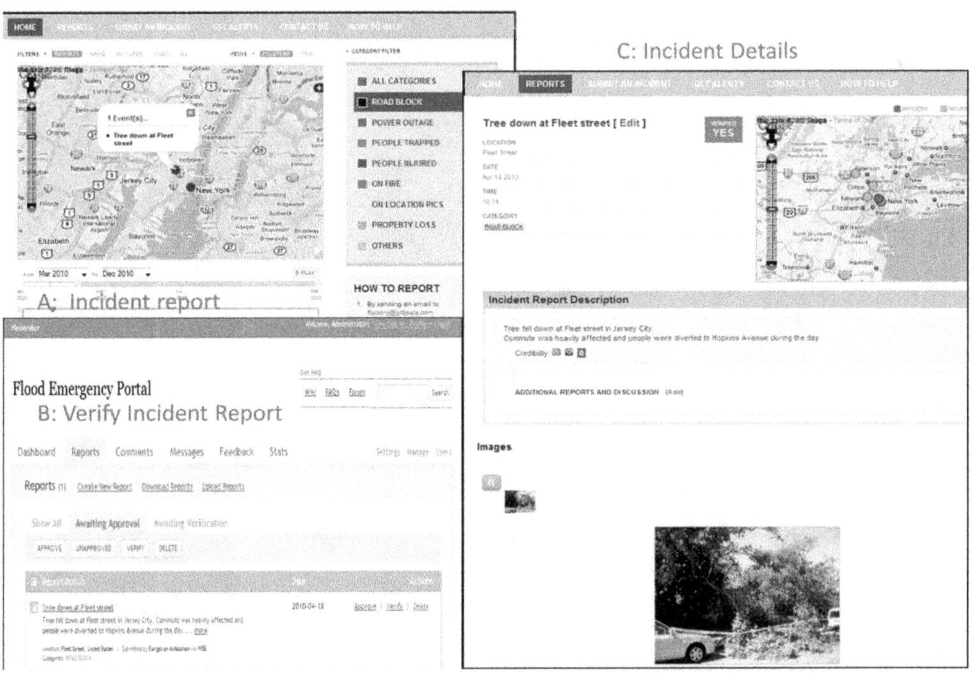

wish to volunteer at the time of an emergency. Local authorities and people can equally benefit from this feature by better preparing and combating any emergency situations and by raising awareness of flooding problems. This is aimed at making citizens better equipped to face unexpected situations. The Volunteer Registration form in Figure 6 lets citizens register with the website and become members of the NJMFWC. Citizens can opt to select any voluntary services and equipment they wish to share at a time of emergency. A volunteer's profile records not only the contact information but also the volunteer services or resources. The types of volunteer services include firefighter, doctor, electrician, plumber, etc. Equipment resources include chainsaw, bulldozer, ambulance, pump, generator, etc. Citizens can add new services and equipment if they like. The site administrator will check and add the fields to the existing ones at their discretion.

Resource and Service Requests

The registered volunteers, their services and their resources can be searched for with keywords by citizens and government officials. All the services and equipment registered with the NJMFWC can be seen on the Resource Map Page (Figure 7). It shows a map with locations where a particular service or piece of equipment is

Figure 6. The Volunteer Registration Form

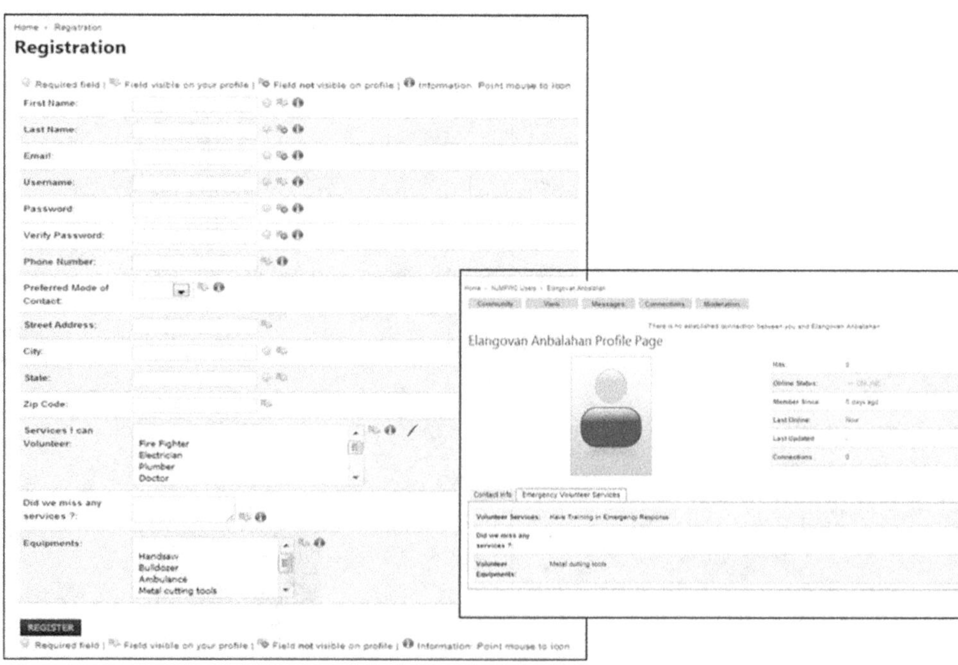

Figure 7. Available Resource Locations on a map and the detail information can be opened

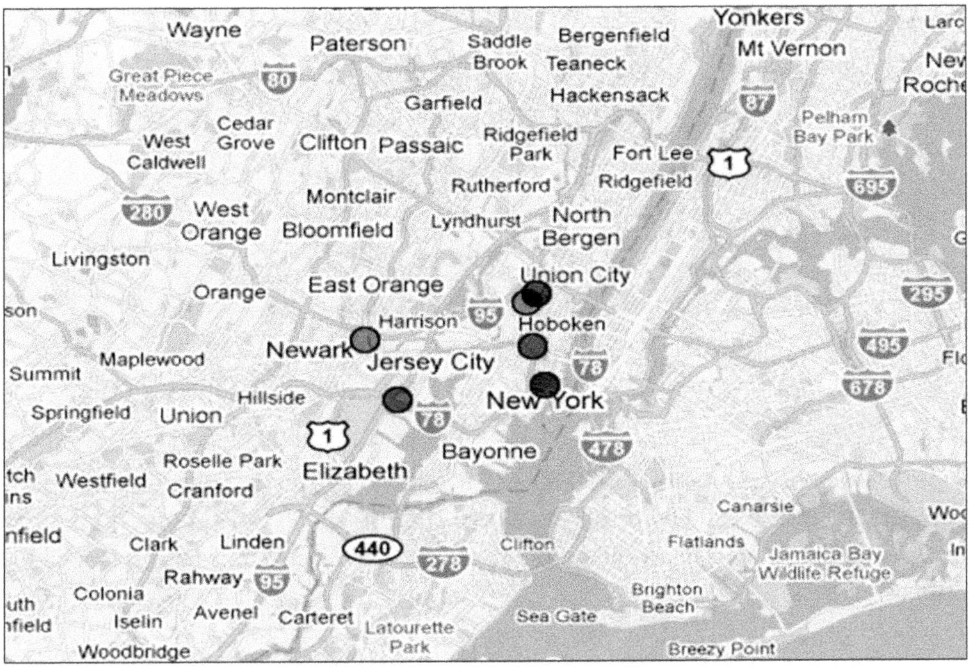

available. This provides a general idea of how near or far a volunteered service or equipment is from a disaster location.

Citizens can use the Resource/Service Request form (Figure 8) to search for available services or other resources near their area. The search result does not disclose any information about the member who has volunteered to offer the service or share the resource. This is done to keep the anonymity of the NJMFWC members intact. The requesting citizens can opt to receive email alerts when a particular service becomes available, by clicking on the 'Alert Me' button, which leads to a form to enter his/her contact details and message methods (SMS, email, etc.)

Alternatively, the incident reports and available resources can be reasoned about and analyzed to recommend the matching services or resources for particular incidents, based on the proximity (accessibility of resources), type, and severity of the incident.

Discussion Forum to Promote Citizen-to-Citizen Interactions

The Flood NJ forum is an online bulletin board, also referred to as online discussion site. The online forum lets members post messages and comment on messages of others. Users have to register to become members, and only members are allowed

Figure 8. Service/Resource Request form and alert notification option

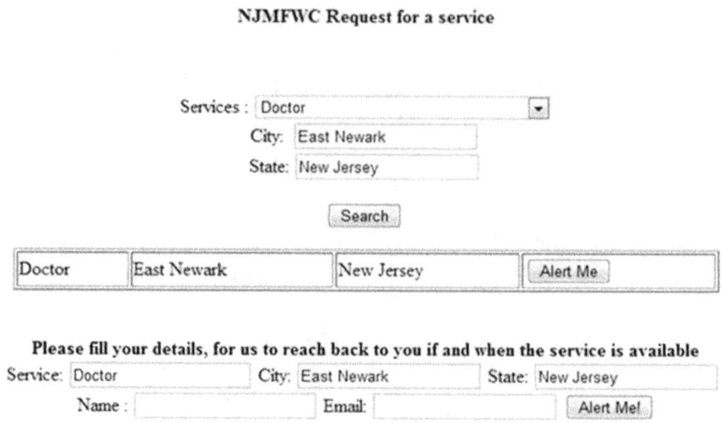

to post topics or comments. Specifically, it allows to view recent posts made by members, by threads and by discussion topics. It also shows "my Discussions" that displays the latest entries posted by the member. The discussions are organized by topic categories.

BENEFITS OF PEER FRAMEWORK

The primary benefits of the proposed citizen-to-citizen resource sharing system include better management of unaffiliated, often spontaneous, volunteers in times of disasters and emergency. A review of the news in the aftermath of superstorm Sandy and recent hurricanes can provide us with some perspective of real life events that transpired, which can provide motivations for design criteria for augmenting PEER. We begin a review of an altruistic private company providing a service they were in no way required to provide to the public. We then follow up with a review of some examples of unsavory situations that can occur in the aftermath of a disaster and how systems like PEER can help avoid these types of problems.

Benefits for Citizen Volunteer Management

Faced with disaster – natural or man-made, governments exercise emergency response plan and most of registered nonprofit organizations automatically respond according to a pre-established plan. However, spontaneous and unaffiliated volunteers such as neighbors and ordinary citizens pose management challenges for the professional responders and governments (FEMA 2002, 2005). They may not be registered properly and their capacities and resources are not recognized and often underutilized. The limited resources available at the federal, state, and local governments can be alleviated with citizen involvement and their resource sharing during all phases of the emergency management, i.e. to prepare for, respond to, recover from, and mitigate the effects of disasters. National and local governments or volunteer centers can adopt and implement the PEER system and have the volunteer resource registration database in advance of any disaters. At the time of a particular resource needs, the government can leverage these unaffiliated volunteers and their resources. This volunteer database can in turn be used for reaching out for training them with the emergency response plans and processes, and for informing them of limits of liability protection as unaffiliated volunteers.

Potential Benefits for Private Companies

One interesting news story covered private telecommunications firms like Verizon and AT&T sending out WECC (Wireless Emergency Communications Centers) (FEMA, 2002), to provide people without cellular voice/data service and power the opportunity to recharge their phones and contact their loved ones or conduct business to help bring order to their affairs thrown into turmoil by the disaster. These units are highly mobile and can be deployed as long as there is the infrastructure to support their movement, namely clear roads. Factoring WECC's into modern emergency response plans is a necessary part of the realities of today's world.

The story notes that, "the four major U.S. carriers - AT&T, Verizon, Sprint, and T-Mobile -- have issued statements about their continued efforts to restore full service on the East Coast". It is easy to see why these companies would want to support a system like PEER so they can see field reports from their customers about the extent of damage that their networks have sustained. By using PEER, they can also be aware of other hazards or situations that may impede their crews from restoring service. Finally, it was noted that AT&T and T-Mobile announced that they were allowing for their networks to interoperate. If AT&T service is down in one area, their customers will be able to use T-Mobile's service and vice versa. This type of sharing is important in times of disaster, to provide people with a means of communication, and such network sharing could easily be coordinated through a system like PEER. This assumes that the cellular networks and the handsets they support are compatible (AT&T and Verizon use different technologies). Another interesting point of note is that AT&T made possible recovery donations via text messages – this let people who want to contribute financially to disaster relief efforts do so by incurring charges on their phone bills.

Potential Benefits for Fraud Prevention

One important aspect to consider is that there are predatory individuals and businesses looking to exploit a disaster and its victims for quick personal gains. One should consider the types of behaviors and practices that are used, and attempt to offer official channels or procedures for handling these types of situations in a disaster response system to minimize the options available to scammers and to provide some degree of government protection.

One such publicized example occurred in New Jersey during the aftermath of Sandy. The state Division of Consumer Affairs conducted an investigation of APK Auto Repair and Towing, a Toms River towing company accused of price gouging Seaside Heights residents (Associated Press 2012). After Central and South New Jersey were crippled by storm damage, Seaside Heights residents complained that APK Auto Repair and Towing charged exorbitant prices for vehicles removed from town. Under the settlement, the company agreed to not offer for sale or sell merchandise at an excessive price increase during a state of emergency, which was declared before Sandy struck, or 30 days after it ends. It also agreed to adhere to the state's consumer fraud and predatory towing prevention laws.

This entire scenario could be prevented by a system like PEER. For example, if towing of disabled vehicles is needed, the government could set a price per vehicle towed per incident and make the rate publically known. This can serve as a price gauge for people to use when dealing with private towing companies. It could also specify a list of approved vendors of towing services, or allow volunteers to tow cars to a government approved lot, where owners can reclaim them within a specified time period before they are sold at auctions, with the funds raised to be used towards future disaster recovery efforts. Again, it is worth emphasizing the idea of transparency, so that this type of behavior can be brought to light and stopped as quickly as possible.

CHALLENGES AND RISKS

As with all proposed frameworks, their success or failure is based on the prototype systems that implement their methodology and attempt to validate the claims made by the framework. While we are certainly not the first to suggest technology-based systems for use in disaster management, we want to raise attention to the idea of utilizing existing/popular channels of interpersonal communication to help with disaster/emergency reporting and responses. While our framework is designed to be generalizable, our examples and prototype focus primarily on the flood warning system implementation; it is serving to demonstrate an effective case for a single type of disaster and that other types of emergencies/disasters can be added in due time with adequate resources. Research (Li, J.P., R. Chen, J. Lee, & H. R. Rao., 2013; Lee, J., N. Bharosa, J. Yang, M. Janssen, & H. R. Rao, 2011) supports our suggested methods in the proposed framework and the goals achieved through the implementation and testing of the prototype.

However, this system relies heavily on volunteer participation, which is outside of the realm of technology. We make a number of assumptions relating to human behavior and motivation to provide a justification for a rewards system. The intention behind the incentive system is to motivate those who would otherwise not participate out of a sense of shared civic duty or personal values toward helping a fellow citizen during a time of need. There is research that supports the fact that improper incentivizing can cause more problems than it solves, for example a popular behavioral psychology study involving the Israel daycare by Dan Ariely. In this case, the daycare's incentive to get parents to pick up their kids earlier had the reverse effect of making them stay later, because the parents could pay a fine instead of arriving on time, providing a justification for their tardiness. In our example, we attempt to use a verification system to prevent fraud and abuse. We acknowledge that the proposed system framework can end up causing a slower response time due to registering, reporting, etc., but this problem will affect any system that attempts to record and reward actions on behalf of the public trust. One must weigh the immediacy of each case, and determine whether helping a neighbor is easier and more beneficial to both parties by not reporting it and just getting the work done.

While we focus on the technological aspects of leveraging social media and crowdsourcing, we cannot neglect the numerous potential legal risks and issues that will be faced by a system similar to our framework. Several examples were suggested to us during the review process of the PEER framework. For example, a citizen task assigned and accepted via the PEER system may violate trespassing laws, cause damage of property or citizen injury, or may destroy evidence of a crime. The reporting system is not immune to legal problems either, for example, reports of a disaster may be construed as slander/libel, or it may cause citizen information to be inappropriately shared or lost, which can lead to crimes such as robbery of identity theft. There even exists the possibility that work done in good faith violates legal contracts (e.g., a sole-source towing franchise exists, but a citizen tows cars out of the way) or law (only a qualified person allowed to do code-related work). When private companies disclose information in a public fashion, it can be used by competitors to poach customers or use other underhanded business practices to undermine one another. As such, we mean to sketch out the broad strokes of policy and provide a framework for consideration. Legal work and policy work are best left to the experts; as such we reserve this for future study.

From a technological perspective, PEER is a system that relies on three core technologies: smartphones and their associated sensors, data exchange (cell service and associated mobile data), and electric power. In a disaster scenario these are all resources that can be in short supply. Much in the same way that 911 calling rules on mobile phones are regulated by the government, we advocate for a similar policy for emergency data exchange. Strict and explicit rules can be created to ensure that the minimum amount of bandwidth used to get the required information to all parties is available in a similar fashion to the 911 call routing. This way the citizens get the information, the government gets a disaster management data policy in place, and the telecommunication companies get minimal impact to their systems. The intention is for PEER to be another channel of communication between citizens and government during disasters/emergencies in addition to the existing channels. Social media is a natural outlet for people to share information with each other, with its main power coming from the scope of reports (public postings) and the speed at which they can be reported (new information appears constantly.) Another important aspect worthy of mentioning is that social media access is provided by private companies for public consumption. Since they control the flow of information, they perhaps have more power than the citizens or government would like them to have. If their networks fail or are overloaded, PEER fails. While social media by its very nature is designed for high volume, high number of user traffic and is very robust, outages are not unheard of. Also since their APIs are closed source, any changes they make can potentially break applications that use them to interact with their services. This contrasts starkly with the open source model we advocate, yet no satisfactory alternative exists in the marketplace at this time.

Balancing the need to communicate relevant information with personal privacy preservation for citizens creates a constant risk for citizens, the government, and private companies. This is especially true with regards to the citizen database, which contains information about tools, manpower etc. Disclosing too much information can lead to some of the aforementioned legal problems (ID theft, robbery, loss of competitive advantage) and too little information can lead to inefficient decisions, underutilization etc. A nuanced privacy model much such as those used in social media, e.g., Facebook, can provide relevant information to the desired parties who need it to make decisions, while minimizing unwanted data exposure. This is not a trivial task, and will require careful consideration and cooperation between the government and citizens/private parties to protect everyone's interests. Another challenge stemming from this gathering and centralization of data is the "freshness" of the data. Information can go out of date quickly, as people and situations change over time. Data that is not current can lead to problems with forecasting accurate responses (over/underestimations of resources). An update policy will need to be in place to ensure that the government can make accurate decisions with the information it has on hand.

LIMITATIONS

As with any new centralized system, PEER is not without its limitations. In its current implementation, it has three core needs that must be met in order to function properly. Without these needs, its ability to function in a timely manner is diminished; however it can still provide some of its services, albeit in a limited capacity.

First, it requires a population that is heavily invested in smartphone technology and its related telecom infrastructure. While problem reports can still be filed with PEER through other means, the ability to have a large segment of the population capture and report clear, accurate data from the field on a portable device relating to problems from the disaster is essential to the assumption of efficiency.

Secondly, PEER relies on the popular social networks as the primary vehicles for information dissemination functioning during times of crisis. Generally these networks are large enough and designed at a scale to handle high volumes of users and traffic - their infrastructure can be considered “global” so that even multiple disasters in geographically distributed locations will not cripple the networks’ capability to function. Also more traditional means of communication are made available in the event social media are down or not functioning properly.

Finally, this system is built around the idea of volunteerism. Without citizens volunteering and registering with the system, we cannot achieve the level of engagement or efficiency possible that a focused effort on the part of the community can achieve. While the system can function without volunteers, as the VD/SS, and can simply be populated with government employees while the incident reporting system can still be made accessible to the citizens – this would imply that all of the work falls on the shoulders of the government, and response times to problems will increase considerably as a consequence.

FUTURE RESEARCH DIRECTIONS

In this chapter, we have proposed a PEER technology framework that enables citizen-to-citizen based emergency responses and demonstrated a prototype system. As the system stands now – its focus is mainly on generating and capturing field data relating to disaster incidents and to providing incident responses. While it does have some limited data analysis capabilities, in a future revision of the system it will have the ability to retroactively mine the data it gathered to make suggestions about improvements to managing disasters based on the types of problems it encountered and the metrics associated with their resolution. The emphasis is on using these performance metrics, over time, to lower the amount of money taxpayers must spend on disaster management and relief. The improved efficiency of response/resolution and distribution of resources, coupled with volunteer work will help to offset some of the costs of disaster management. PEER can also aid in improving government transparency in budgeting procedures. By making it clear to the populace what amount of funds is set aside for disaster management and relief – and making sure unused funds are saved in the account and not spent elsewhere, it can lower the overall costs to the government itself as well as to taxpayers.

Liability Issues

From a legal perspective also come into play when developing a system such as this. Unless there is an absolutely clear outlining of responsibility and who is liable if problems arise on either government or private property, using this system would become difficult and would have to be restricted to private parties receiving help from other private parties or private parties receiving help from the government.

The important take away from this chapter is to consider the power and reach of social networks, and the speed and ease with which information can be shared amongst a population through them. While existing disaster management and emergency reporting methods have been tested and proven to work, we want to explore the use of this new channel of communication to aid in emergency/disaster management and recovery. The first step is to develop a strong framework that addresses the complex issues that government workers and citizens face when beginning to respond to the aftermath of a disaster. While we advocate for a technology-based solution, much effort is required on developing good public policy that aligns with the goals of both the government and the citizens alike. Also, creative use of a ubiquitous technology such as smartphones with their extraordinary capabilities, allows to organize them a "network of sensors," much in the same way a traditional smart city has sensors to monitor situations. The citizens will generate the data for free as long as the government responds to it in a timely manner and in an appropriate fashion. Future research is required to fully satisfy the public policy and legal requirements of putting a system such as this into place.

CONCLUSION

In conclusion, we proposed a citizen-to-citizen resource sharing framework called PEER. This system intends to motivate citizens to volunteer resources related to responding to emergency incidents, such as equipment, tools, vehicles, services or expertise, that are often on demand to provide emergency responses. The PEER system provides the resource sharing capabilities through the creation and development of a centralized database for incident reports, volunteers and their resources. In addition, capturing the incident reports from social media as well as the PEER system is important, because many citizens use mobile phone-based social media as main communication channels in emergency situations. The collected incident reports are organized for further verifications by officials, and indexed. Citizens gain situation awareness through various analytics supported by the PEER system, and the citizen volunteers and resources are also visible to others to request assistance. The citizen-to-citizen resource sharing mechanism can utilize analytics and reasoning engines to recommend needed resources for a particular incident. The utilization of nearby resources is possible due to the location visibility of available resources, speeding up the incident responses. By aligning citizen and government incentives and by encouraging and rewarding altruistic behavior, costs of recovery can be lowered and time to recovery can be accelerated. While we discussed some of the technical and sociological issues, there are many more complex human factors and privacy issues that need to be addressed. In the future, we plan to further explore these issues and identify ways to take these into account. We will also address the validity of the proposed PEER framework in future publications, after testing of our prototype system have been conducted.

ACKNOWLEDGMENT

This research was partially supported by the New Jersey Meadowlands Commission; Department of Homeland Security, and partially by grant# CNS-1624503 from the National Science Foundation. The prototype system was developed by E. Anbalahan as part of his Master's Project at New Jersey Institute of Technology. This chapter is an extended version of the articles (Lorenzi, D., J. Vaidya, S. Chun, B. Shafiq, V. Naik, V. Atluri & N. Adam, 2013, 2015).

REFERENCES

Adam, N., J. Eledath, S. Mehrotra, N. Venkatasubramanian. (2012a). Social media alert and response to threats to citizens (SMART-C). *Proceedings of CollaborateCom*, 181-189.

Adam, N., Shafiq, B., & Staffin, R. (2012b). Spatial computing and social media in the context of disaster management. *IEEE Intelligent Systems, Volume*, *27*(6), 90–96. doi:10.1109/MIS.2012.113

Anbalahan, E. (2010). *Community Based Flood Management System Using Web 2.0, Master's Project*. NJIT.

Artigas, F., Chun, S., & Sookhu, Y. (2009) Real-time Ocean Surge Warning System, Meadowlands District of New Jersey. *Proceedings of the 10th International Conference on Digital Government Research (dg.o 2009)*, 216-222.

Associated Press. (2012). *Vehicles towed in Seaside Heights by company accused of price gouging to be released*. Retrieved from http://www.nj.com/news/index.ssf/2012/11/vehicles_in_seaside_heights_to.html

Bessis, N., Brown, A., & Asimakopoulou, E. (2010, July). A Mathematical Analysis of a Disaster Management Data-Grid Push Service. *International Journal of Distributed Systems and Technologies*, *1*(3), 56–70. doi:10.4018/jdst.2010070104

Bharosa, N., Lee, J., Janssen, M., & Raghav Rao, H. (2009) A case study of information flows in multi-agency emergency response exercises. *Proceedings of the 10th Annual International Conference on Digital Government Research: Social Networks*, 277-282.

Careem, M., De Silva, C., De Silva, R., Raschid, L., & Weerawarana, S. (2007) Demonstration of Sahana: free and open source disaster management. *Proceedings of the 8th annual international conference on Digital government research: bridging disciplines \& domains*, 266-267.

Cayirci, E., & Coplu, T. (2007, June). SENDROM: Sensor networks for disaster relief operations management. *Wireless Networks*, *13*(3), 409–423. doi:10.100711276-006-5684-5

Chen, R., Sharman, R., Rao, H. R., & Upadhyaya, S. J. (2008, May). Coordination in emergency response management. *Communications of the ACM*, *51*(5), 66–73. doi:10.1145/1342327.1342340

Choi, Y., Blackburn, J. M., & Seo, H. (2006, November). A connection oriented telecommunications disaster management model using mobile GIS. *International Journal of Mobile Communications*, *4*(1), 63–81. doi:10.1504/IJMC.2006.008198

Chun, S., & Artigas, F. (2012). Sensors and Crowdsourcing for Environmental Awareness and Emergency Planning. *International Journal of E-Planning Research*, *1*(1), 56–74. doi:10.4018/ijepr.2012010106

Convertino, G., Mentis, H. M., Slavkovic, A., Rosson, M., & Carroll, J. M. (2011, December). Supporting common ground and awareness in emergency management planning: A design research project. *ACM Transactions on Computer-Human Interaction, 18*(4), 22. doi:10.1145/2063231.2063236

Currion, P., de Silva, C., & Van de Walle, B. (2007, March). Open source software for disaster management. *Communications of the ACM, 50*(3), 61–65. doi:10.1145/1226736.1226768

Faradani, S., Hartmann, B., & Ipeirotis, P. (2011) What's the Right Price? Pricing Tasks for Finishing on Time. *Human Computation: Papers from the 2011 AAAI Workshop*, 26-31.

FEMA. (2002). *Preventing a disaster within the disaster: The effective use and management of unaffiliated volunteers*. UPS Foundation. Retrieved from https://www.ncjrs.gov/pdffiles1/Archive/202852NCJRS.pdf

FEMA. (2005). *Managing Spontaneous Volunteers in Times of Disaster: The Synergy of Structure and Good Intentions*. UPS Foundation. Retrieved from https://www.fema.gov/pdf/donations/ManagingSpontaneousVolunteers.pdf

Fernandez, L., Barbera, J., & van Dorp, J. (2006). Spontaneous volunteer response 36 to disasters: The benefits and consequences of good intentions. *Journal of Emergency Management (Weston, Mass.), 4*(5), 57–68.

Friedrich, F., & Burghardt, P. (2007, March). Agent-based systems for disaster management. *Communications of the ACM, 50*(3), 41–42. doi:10.1145/1226736.1226763

Hristidis, V., Chen, S., Li, T., Luis, S., & Deng, Y. (2010, October). Survey of data management and analysis in disaster situations. *Journal of Systems and Software, 83*(10), 1701–1714. doi:10.1016/j.jss.2010.04.065

Janssen, M., Lee, J., Bharosa, N., & Cresswell, A. (2010, March). Advances in multi-agency disaster management: Key elements in disaster research. *Information Systems Frontiers, 12*(1), 1–7. doi:10.100710796-009-9176-x

Jin, G., Nicolai, B., Jiang, K., & Winer, C. (2011) Distributed image processing and classification for GIS based disaster management and communication system. In *Proceedings of the 2nd International Conference on Computing for Geospatial Research & Applications*. ACM. 10.1145/1999320.1999347

Kamar, E., & Horvitz, E. (2012) Incentives for truthful reporting in crowdsourcing. *Proceedings of the 11th International Conference on Autonomous Agents and Multiagent Systems*, 3, 1329-1330.

Lee, J., Bharosa, N., Yang, J., Janssen, M., & Rao, H. R. (2011, January). Group value and intention to use - A study of multi-agency disaster management information systems for public safety. *Decision Support Systems*, *50*(2), 404–414. doi:10.1016/j.dss.2010.10.002

Lee, S., Lee, S., Kim, K., & Park, J. (2012) Bursty event detection from text streams for disaster management. *Proceedings of the 21st international conference companion on World Wide Web*, 679-682. 10.1145/2187980.2188179

Li, J. P., Chen, R., Lee, J., & Rao, H. R. (2013, April). A case study of private-public collaboration for humanitarian free and open source disaster management software deployment. *Decision Support Systems*, *55*(1), 1–11. doi:10.1016/j.dss.2012.10.030

Li, L., Wang, D., Shen, C., & Li, T. (2010) Ontology-enriched multi-document summarization in disaster management. *Proceedings of the 33rd international ACM SIGIR conference on Research and development in information retrieval*, 819-820.

Lorenzi, D., Chun, S., Vaidya, J., Shafiq, B., Atluri, V., & Adam, N. (2015). PEER: A Framework for Public Engagement in Emergency Response. *International Journal of E-Planning Research*, *4*(3), 30–47. doi:10.4018/IJEPR.2015070102

Lorenzi, D., Vaidya, J., Chun, S., Shafiq, B., Naik, V., Atluri, V., & Adam, N. (2013). Community Based Emergency Response. *Proceedings of The 14th International Conference of Digital Government Research*, 82-91. 10.1145/2479724.2479739

Lorenzi, D., Vaidya, J., Shafiq, B., Chun, S., Vegesna, N., Alzamil, Z., ... Atluri, V. (2014) Utilizing social media to improve local government responsiveness. *Proceedings of the 15th Annual International Conference on Digital Government Research*, 236-244. 10.1145/2612733.2612773

Nicolai, B., Jin, G., Jiang, K., & Winer, C. (2010) Disaster assessment with parallel image processing for GIS based local area disaster decision support system. In *Proceedings of the 1st International Conference and Exhibition on Computing for Geospatial Research; Application*. ACM. 10.1145/1823854.1823901

Organisciak, P. (2008). *Motivation of Crowds: The Incentives That Make Crowdsourcing Work*. Retrieved at http://crowdstorming.wordpress.com/2008/01/31/motivation-of-crowds-the-incentives-that-make-crowdsourcing-work/#comments

Palen, L., Hiltz, S., & Liu, S. B. (2007). Online forums supporting grassroots participation in emergency preparedness and response. *Communications of the ACM*, *50*(3), 54–58. doi:10.1145/1226736.1226766

Pohl, D., Bouchachia, A., & Hellwagner, H. (2012). Automatic sub-event detection in emergency management using social media. In *Proceedings of the 21st international conference companion on World Wide Web*, 683-686. 10.1145/2187980.2188180

Ramchurn, S., Rogers, A., Macarthur, K., Farinelli, A., Vytelingum, P., Vetsikas, I., & Jennings, N. R. (2008). Agent-based coordination technologies in disaster management. *Proceedings of the 7th international joint conference on Autonomous agents and multiagent systems: demo papers*, 1651-1652.

Raschid, L. (2007). Information integration and dissemination for disaster data management. *Proceedings of the 8th annual international conference on Digital government research: bridging disciplines \& domains*, 333-335.

Ray, N. K., & Turuk, A. (2011). A framework for disaster management using wireless ad hoc networks. *Proceedings of the 2011 International Conference on Communication, Computing & Security*, 138-141. 10.1145/1947940.1947970

Shafiq, B., Chun, S., Atluri, V., Vaidya, J., & Nabi, G. (2012). Resource Sharing using UICDS Framework for Incident Management *Transforming Government: People Process and Policy*, *6*(1), 41–61.

Shafiq, B., Vaidya, J., Atluri, V., Adam, N., Chun, S., & Lieb, A. (2008). Ontology Driven Resource Management for Emergency Response. *Workshop on Secure Knowledge Management (SKM 2008).*

Shafiq, B., Vaidya, J., Atluri, V., & Chun, S. (2010). UICDS compliant resource management system for emergency response. *Proceedings of the 11th Annual International Digital Government Research Conference on Public Administration Online: Challenges and Opportunities (DGO '10)*, 23-31.

Solnit, R. (2009). *A Paradise Built in Hell*. New York: Penguin Books.

Srinivasan, P. K. (2011). An emergency GSM satellite portable terminal for disaster management. *Proceedings of the 1st International Conference on Wireless Technologies for Humanitarian Relief*, 455-460. 10.1145/2185216.2185333

Therese, J., Fajardo, B., & Oppus, C. M. (2010, June). A mobile disaster management system using the android technology. *WTOC*, *9*(6), 343–353.

Truong, H., Manzoor, A., & Dustdar, S. (2009). On modeling, collecting and utilizing context information for disaster responses in pervasive environments. *Proceedings of the first international workshop on Context-aware software technology and applications*, 25-28. 10.1145/1595768.1595779

Wu, A., Yan, X., & Zhang, X. (2011). Geo-tagged mobile photo sharing in collaborative emergency management. *Proceedings of the 2011 Visual Information Communication - International Symposium.* 10.1145/2016656.2016663

Yates, D., & Paquette, S. (2010). Emergency knowledge management and social media technologies: a case study of the 2010 Haitian earthquake. *Proceedings of the 73rd ASIS&T Annual Meeting on Navigating Streams in an Information Ecosystem, 47.*

Zheng, L., Shen, C., Tang, L., Li, T., Luis, S., & Chen, S. (2011). Applying data mining techniques to address disaster information management challenges on mobile devices. *Proceedings of the 17th ACM SIGKDD international conference on Knowledge discovery and data mining*, 283-291. 10.1145/2020408.2020457

Zheng, L., Shen, C., Tang, L., Li, T., Luis, S., Chen, S., & Hristidis, V. (2010). Using data mining techniques to address critical information exchange needs in disaster affected public-private networks. *Proceedings of the 16th ACM SIGKDD international conference on Knowledge discovery and data mining*, 125-134. 10.1145/1835804.1835823

Chapter 6
Sharing Semantic Territorial Interpretation:
The Contribution of OnToMap Project to Spatial Planning

Angioletta Voghera
Politecnico di Torino, Italy

Luigi La Riccia
Politecnico di Torino, Italy

ABSTRACT

This chapter explores the theme of collaborative construction of territorial knowledge through the use of ICTs, proposing a new approach to spatial representation based on semantic ontologies. The theoretical perspective is applied in a recent experimentation conducted in collaboration between the Politecnico di Torino and the Università degli Studi di Torino, called OnToMap – Community Maps 3.0. The chapter is divided into two parts. The first part aims to define the theoretical framework of collaborative mapping practices and VGI, social semantic mapping, GIScience, and Web 3.0 applications, specifying which approach to participation is assumed as conceptual background. The second part focuses on the web-based application OnToMap, which relies on a territorial ontology for encourage construction of collective and shared knowledge of places. The chapter ends with a reflection on semantic and cartographic representation of urban space and its potential in terms of citizen empowerment.

DOI: 10.4018/978-1-5225-5999-3.ch006

INTRODUCTION[1]

In 1968, long before our digitally increased cities were made available by companies like Google, the French philosopher Henri Lefebvre described what he named *the right to the city*. He argued that the great potential of urban life should be available to all, not just the most powerful elites that own and control most of our cities. The goals of his battle for more just and inclusive cities were those who traditionally play mediation roles in the development of urban policies that produce inequalities: property owners, the state and the police, etc. The *right to the city* has never been thought of as a set of codified rules, but has become part of the claims and slogans against exclusion that have brought different types of benefits to the inhabitants of the cities.

Territorial planning has always been involved in the representation of space since the 1970s (Dennis, 1970), when it recognized as essential the support of social actors in the decision-making process. The participatory construction of spatial representations evolved ever since: starting from the paper representation (low usability and scalability) it moved to the use of GIS and PGIS systems (Steiniger & Hunter, 2013).

Recent innovations in ICT are significantly changing the technology-society relationship, with implication on spatial planning. At the same time, the procedures used to conduct scientific research are being revolutionized by the exponential increase in the amount of available data. This condition is having a important influence on methodologies which are used to analyse, study and develop scientific research related to urban planning and primarily concern territorial data usage and, for this reason, are independent from the method used for data processing and mapping. As an example, big data analysis is directly connected with the possibility to include in the used databases data and information that previously could not be treated as data. This information, as well as providing the necessary databases for the development of statistical high correspondence analyses, also interacts with new kind of data basins, thus increasing the range of possible applications: it is possible, for example, to build and obtain geo-information regarding flows (people, things and information), perceptions and personal evaluations of specific places and landscapes.

The European Smart City model, meaning the set of strategic actions to make cities intelligent, digital and inclusive, is closely connected to the development of infrastructures dedicated to communication and social participation. The smart city concept has been quite fashionable in the policy arena in recent years: compared with the concept of digital city or intelligent city, the main focus is not limited to the role of ICT infrastructure but is mainly concentrated on the role of social, economic and environmental issues. There are various different forms of expression of the smart city in practice: "e-government", or "e-gov" (AOEMA, 2005) identifies the possibility to improve the quality, rapidity and reliability of services provided to citizens by public administrations, thanks to the digitalisation of administrative apparatus (Clift, 2004). However, this model also refers to "learning communities", meaning communities focused upon sharing knowledge (European Commission, 2015).

In this area, attention is paid to the use of ICT for sharing geographical knowledge in support of spatial design and urban planning. In line with institutional practices of transferring geographical knowledge on digital media through the construction of Geographic Information Systems (GIS), within web 2.0, new possibilities were configured for citizens to construct voluntary and spontaneously geographic knowledge, using different digital media (Goodchild, 2007, 2009).

This chapter constitutes a reflection about the theme of collaborative production of geographical knowledge by constructing a technical framework at international level, showing a recent experimentation conducted in Politecnico di Torino with the students of Urban Planning Course (Architecture Course). In particular, the focus of the experimentation was oriented towards the implementation of a synergic exchange with OnToMap – Mappe di Comunità 3.0[2], which is based on an ontology in the terms described by Gruber (1995) as *an explicit specification of a conceptualization.*

Constructing a territorial ontology involves choosing concepts and relationships between them, through which the representation of the territory can be expressed. In general, the application domain is that of participatory decision-making processes, aimed at favouring the reflection on territorial identity, territorial needs and on inclusion of citizens in the territorial transformations.

BACKGROUND: THE PATHWAY TOWARDS E-PARTICIPATION

The issue of the World Wide Web as a space for new opportunities and spaces of democracy does not arise with Internet, but at least ten years before. This is not the place to remember important researchers as Terry Winograd (Stanford University) or Douglas Schuler (promoter and theorist of the community networks), and others, but that path opened new frontiers for the evolution of computing and new research areas as the participatory design and participatory GIS, which are now well known.

Since 1970, public participation in decision-making processes has been considered as a useful and effective practice for encouraging dialogue between administrators and citizens, essential for building consensus around policy choices (Dennis, 1970). In particular, as the field of territorial planning is interested to build open and agreed processes, traditional methods of consultation and information do not appear sufficient to ensure that all public application will be heard (Turkucu & Roche, 2008) and numerous research focuses on defining new theoretical approaches capable of operationally integrating grassroots applications.

Arnstein (1969) defines the participation of citizens as,

the redistribution of power that enables the have-not citizens, presently excluded from the political and economic processes, to be deliberately included in the future.

The central principle of this interpretation model develops around the use of participation as a tool for increasing the empowerment level of citizens. The eight proposed stages (Figure 1) of participation match to different purposes, ranging from the manipulation of the public to control by citizens of decision-making processes. The concept of "control by citizens" is the highest level of participation, in which all citizens, even those usually excluded, take part in the process of planning and devising public policies and controlling their implementation. The lowest step of the ladder consists of "manipulation", which sees the purpose of the participation process as maintaining consolidated power by building public consensus, through the education of citizens. Those stages of participation are grouped into three subsections, which represent general degrees of participation, from the level of "non-participation" to "tokenism" to "citizen power". Arnstein also considers the levels that cannot really be said to be participation – non-participation – but which conversely imply efforts by the decision-making politicians to prevent contrasts and conflicts on decisions already processed (manipulation and therapy). Arnstein even considers two different levels (placation and delegated power), which take account of the different nuances of the degree of power attributed to citizens.

Figure 1. Ladder of citizen participation. Interpretation of the scale proposed by Arnstein (1969). Source: Authors' Reworking

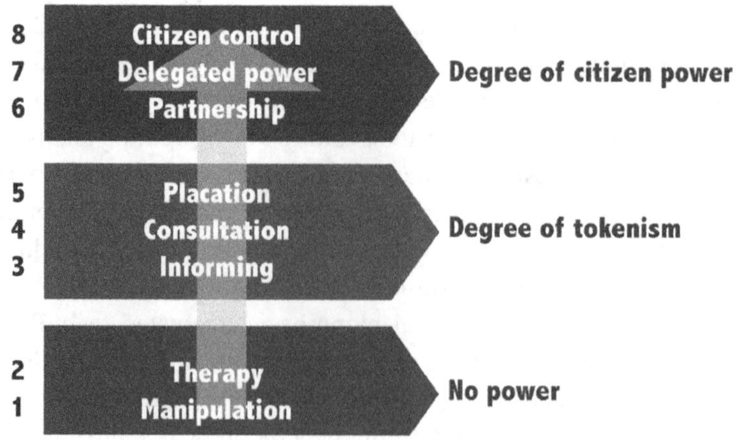

The use of ICT in participation and decision-making processes is relatively recent; different forms of technologies of participation have been developed since the end of twentieth century (Cunha, E., Allegretti, G., & Matias, M., 2011). The concepts of government, democracy and participation applied to the virtual context are accompanied by the prefix "e-" (electronic), thereby acquiring new meaning. In particular, the distinction between participation and e-participation is particularly clear in the different methods of constructing participated processes for an "e-decision-making" or "e-planning".

The digitalization developments allowed bringing democratic processes closer to citizens and the first ideas on ICT-based participation were introduced in the 1980s. In this regard, Kersten (2003) compared the forms of e-voting, e-consultation and e-participation for involving citizens in societal decision-making processes through the use of ICT.

Since e-participation today receives attention from different disciplines and scientific backgrounds it becomes increasingly perceived as an autonomous research area. Despite this, the scientific literature about this research area is diversified: this is seen as an impeding element concerning the consolidation of the topic and the development of theory building (Susha & Grönlund, 2012).

For this reason there is not consensual definition of e-participation. An interesting definition of OECD (2003) describes e-participation as active participation via ICT in the form of *a relationship based on the partnership with government in which citizens actively engage in defining the process and content of policy-making*. Macintosh (2006) defined e-participation as,

the use of information and communication technologies to broaden and deepen political participation by enabling citizens to connect with one another and with their elected representatives.

According to the description of the United Nations (2014), e-participation includes the three dimensions of e-information, e-consultation and e-decision-making. Thus, e-participation enables *participation by providing citizens with public information and access to information without or upon demand*, engages *citizens in contributions to and deliberation on public policies and services*, and empowers *citizens through co-design of policy options and co-production of service components and delivery modalities*.

Although these definitions show differences, there are substantial overlaps: a common element is that e-participation uses ICT interconnections between the government and stakeholders. Thanks to the closer connection, the stakeholders practice an empowerment, allowing them to have influence on policies and planning processes, and thus supporting their active engagement in policy-making.

Software such as Ideascale[3], LiquidFeedback[4] and OnToMap have been developed to allow the collection of people' opinions and could be considered as more complex tools, models and processes than those of direct democracy (*one head, one vote*). As Tim Berners-Lee et al. (Hendler, J., Shadbolt, N., Hall, W., Berners-Lee, T., & Weitzner, D., 2008) have noted in the article *Web Science: An Interdisciplinary Approach to Understanding the Web*, the web can promote greater public engagement in the political sphere, but there is a need to proceed to trial, use and refinement of this tool. In the life cycle they suggest and that we have modified (Figure 2), experiencing within the "macro-scale" representation serves to verify what in the real-life ("micro-scale") works or not: a software application is designed based on an appropriate technology (such as algorithm and design) and with an envisioned "social" construct. The system is generally tested in a small group or deployed on a limited basis; the system's "micro" properties are thus tested. In some cases, when more and more people accept the micro system, accelerating "viral" scaling occurs.

The e-participation tool can be used in different ways: to involve citizens in defining public policies, to monitor the impacts of large works on social and environmental contexts and/or to develop territorial analyses with the involvement of local entities. There is no standardised e-participation application model; instead, it is necessary, by assessing the objectives of the research and the relevant social actors, to devise the method suited to the participatory model to be applied each time (Maier & Reimer, 2010).

Figure 2. The web presents new challenges to solve territorial issues through the application development (Hendler, J., Shadbolt, N., Hall, W., Berners-Lee, T., & Weitzner, D., 2008). Source: Authors' Reworking.

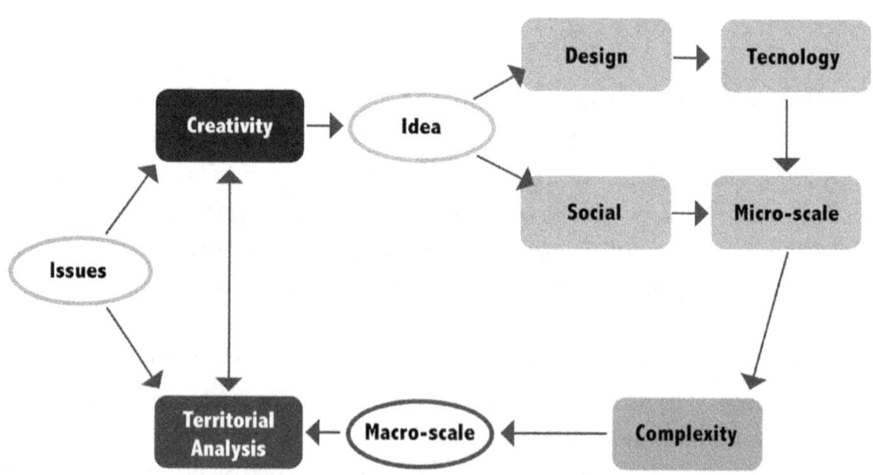

Finally, e-participation cannot be considered an increasingly positive tool, able to ensure outright transparent and inclusive processes: each case must be evaluated individually. Different experiences show, in practice, the criticality of the instrument. For a discussion on the critical issues of e-participation see Prieto-Martin et al. (2012); in many cases the result of e-participation needs to be developed by a traditional process (Medaglia, 2007).

While the concepts of participation and e-participation have various different conceptual nuances, linked mainly to scope and methodology, it is, however, possible to attribute the two cases to a theoretical paradigm of common reference, connected to the objectives and entities involved in the projects.

FROM GIS TO DIGITAL ONTOLOGIES

In the mid-1980s, GIS tools, which enable the acquisition, transformation, organization and visualization of spatial data *from the real world* (Burrough, 1986) were mainly used in companies, due to the high costs and the need for specialized technicians for their use (Obermeyer, 1998). GIS is a field that in the past has been widely explored by planners and geographers since the late 1960s and later integrated by other disciplines specialists around a core of experts in management of spatial databases.

Today, although GIS applications are still too similar to those of 30 years ago, effectively described by Samet (1990) and Laurini and Thompson (1992), these tools have been optimized to become an essential component for all public and private organizations dealing with territorial planning. In parallel to the increasingly widespread use of these technologies, research on GIS has added to the more purely technical issues, questions about the "multidisciplinary potential" of the tool, with special reference to applications in the social sphere (Craig, W. J., Harris, T. M., & Weiner, D., 2003; English & Feaster, 2003). In fact, new collective web-based geographic production practices include references to territorial governance concepts and participative democracy applied to the virtual context, pointing to a bottom up approach, meaning greater relevance placed on the role of citizens "from the bottom", favouring the construction of a network of relationships between social and political actors. Since the mid-1990s, questions related to the social aspects of GIS applications have become an issue of interest for the participation of participatory processes using the so-called PGIS (Participatory GIS) (Schroeder, 1996), thanks to the highly representative visual character (Al-Kodmany, 2001).

The potentials of Geography 2.0 and the diffusion of geospatial technologies (e.g. Google Earth, Maps API) propagate the principles of PGIS, influencing how GIS technologies empower or marginalize individuals and communities in terms of their social, political, economical, and technical conditions (Sieber, 2006). In this regard, Kingston (2002) argued that web-based GIS application could engage public participation in environmental and/or territorial decision-making process by a progressive public participation ladder through their rights to:

1. Know;
2. Be informed;
3. Be objective;
4. Define interests, actors and determine agenda;
5. Assess risk and recommend solutions;
6. Make a final shared decision.

Helped by the improved web infrastructure, general public is more aware of, more attentive, and able to use innovative tools to interact with the social web and physical environment more than ever before. The enhanced interconnectivity and interactivity extend the research agenda about PGIS and open up new frontiers in the so-called neo-geography and GIScience.

The realization of new cartography, using ICT technologies from non-expert users, takes the name of "neogeography" (Eisner, 2006) to differentiate it from geography linked to traditional cartography and is based on the overlapping or combination of two or more sources of geographic information. The voluntary dimension of these practices is at the heart of the volunteered geographic information (VGI) approach that sees citizens "as sensors" (Goodchild, 2007) involved in gathering information using mobile devices equipped with GPS sensors and web access.

Whether or not the true meaning of such practices differs from the objectives of conventional geography is still open for debate: neogeography implies extending the power of mapping and spatial exploration to non-geographers. The construction of social networks within the web domain has produced what is called "crowdsourcing" (Howe, 2006), one of the most significant developments in Web 2.0 and neogeography. Crowdsourcing applied to neogeography produces the "crowdmapping", that is a collection of information that does not come from professionals alone, but from all those who have access to the Internet and want to share data, representations, prospects with other people.

Today, crowdmapping sees the most disparate uses, from environmental complaint to signalling car parks available in cities, to emergency management and crisis situations (an example is the "Ushahidi" application[5], which can be translated as "testimony" in Swahili, born in Nairobi and developed to map violence reports in Kenya after post-election turmoil in 2008. Since then, thousands of people use this crowdmapping application to *make their voices heard*).

But VGI is not the only innovation within the set of tools dedicated to spatial representation. A second innovation is marked by the so-called "ontological turn"

(Jogan, 2002), operated by the late 1990s, with which GIS is integrated with theories of language. Ontologies represent an *explicit specification of a conceptualization* (Guarino, N., Oberle, D., & Staab, S., 2009) and allow to describe objects and relationships of an application domain. If until now the link between GIS and ontologies has appeared more implicit, mainly through the spatial translation of "relational" (first) and "object" (then), now affiliation to the themes of representation of knowledge is quite clear. Referring to the concept of ontology in informatics, it becomes a way of formally representing the structure of a system, that is the relevant entities and relationships that emerge from its observation, and which are useful for our purposes. A domain conceptualization based on ontologies allows to model a subset of reality that surrounds us by making understandable concepts and relationships relevant to the software application within a common language and a universal vocabulary.

It therefore radically changes the point of observation with respect to spatial thinking: spatial relations are no longer the privileged object of the scientific investigation, the priority problem becomes the conceptualization of the territorial system, that is, its translation into corresponding object systems (lexicon), defined in the unambiguously and universally intelligible semantic profile.

Ontologies are today an innovative element of programmes and projects found within the smart city approach. Developing an urban ontology means, at first instance, constructing technical-scientific consensus on a glossary that includes conceptual definitions (Laurini 2007, 2016). The use of ontologies for representing the territory places particular attention on semantic aspects of territorial representation, which seems useful in this sense as it leads to a more dynamic pattern of representation. Space constantly changes over time, based upon the construction of adequately defined elements and their intrinsic relationships. An ontology allows for a fixed but flexible structure to be constructed, enabling the creation of different representations of changing data (Las Casas & Scorza, 2013).

In the case of OnToMap, as explained later, the structure of the ontology was organised based upon a specific research topic: to build a useful tool for technicians and for the population, capable of putting citizens and institutions into contact by promoting language uniformity. The construction of ontological representation, particularly attentive to processes of construction of identity and the perceptive dynamics of the locations, guides the interpretation of the territory, which is assessed in accordance with three constituent dynamics: natural, artificial and regulatory. The natural dimension refers to all physical elements not altered by man; the artificial dimension includes elements produced by man; and the regulatory dimension considers the technical and design aspects relating to the sphere of responsibility of the public administration. The three proposed levels of observation consider issues relating to different domains of representation, linked to tangible aspects, represented with

GIS techniques, socio-economic aspects, highlighted by semantic relationships, and perceptive elements, linked to the voluntary construction of knowledge.

In the ontological representation we have defined different types of relations, based upon the guidelines on the subject (Noy & McGuinness, 2001). The interrelationships between the concepts are mainly based upon inclusion relationships (is part of, is composed by) and upon regulatory constraints (norm, has focus, connects) identified from the main planning instruments.

Based upon this initial subdivision into categories, the subsequent phase of identifying the concepts and attributes which specify the identified dimensions is developed; each concept and attribute is assigned a summary definition, as formal and agreed as possible, thereby developing a dictionary of ontology. Each definition was developed in a general perspective, processed by consulting dictionaries and encyclopaedia, and a technical perspective, processed on the basis of the relevant regulations.

The decision to structure a double glossary was made in order to guarantee the objectives of language uniformity; however, it was prepared within a research programme attentive to the themes of urban planning and territorial design. Any changes to the ontological structure may be proposed by scholars of different and related disciplines, such as geography, planning, architecture, sociology or economics, etc. However, it is useful as a starting point, albeit limited, for initiating a discussion on the issue, still scantly investigated both in the national and international context.

The OnToMap project is mainly focused upon social and planning issues; the former identify an emotional approach to cartographic representation, oriented towards the development of social cohesion and to stimulating local planning; the latter take account of the legislative implications, expressed as institutional constraints and planning. In the structure of the ontological representation, the interrelations between the concepts are based upon regulatory constraints identified on the basis of the main urban instruments, at local and supra-local level. OnToMap is designed as a useful tool for public administrations, which could benefit from its use both as a channel of disclosure but, above all, as a tool of empowerment, fostering the social discussion and individual identification of criticalities and values.

The purpose of the project was to create a description of data that would be as complete as possible.

SOME UPDATE RESULTS FROM THE ONTOMAP PROJECT

Harper (1988) defines visual sociology as recording, analysing and communicating social life through photography, movies and videos. We can consider in this interpretation four fundamental aspects of visual sociology:

1. The purpose of the analysis and the knowledge the researcher seeks to bring to light;
2. The importance of support from registration;
3. The use of images to represent a culture;
4. The analysis of images produced by a culture to understand it.

The OnToMap project has been pursued with this aim: applying the methodological perspective of visual sociology to the collective production of virtual maps (or "community maps") through web-based tools, which allow easy dissemination of information over the Internet.

Traditionally, community maps are made up of cartographic representations developed by the community through a participatory approach (Parker, 2006). OnToMap allows for online consultation of spatial data and the creation of interactive virtual maps that reflect individual information needs and point out to the public administration and planners the territorial critical aspects and project proposals, noting geographic areas and individual elements displayed in the maps.

OnToMap is based on an ontological site structure that provides a formal model for man-machine interaction management in search, display, and sharing of spatial information. Unlike traditional GIS that generally use digital ontologies to integrate heterogeneous sources of information (Fonseca, F., Egenhofer, M., Davis, C., & Borges, K., 2000; Buccella, A., Cechich, A., Gendarmi, D., Lanubile, F., Semeraro, G., & Colagrossi A., 2011), OnToMap uses ontologies to mediate the specification of information elements with different perspectives on the data that users can adopt:

- As regards the information representation, ontology supports the management of data such as *linked data*[6], that is, data linked to each other through hyperlinks in the web. The links between the data result from the semantic relationships that associate them, allowing to describe the relationships between information layers to express spatial or thematic associations, as well as the different levels of abstraction in their description (providing a taxonomy that binds general concepts to the specific ones);
- From the point of view of information search, ontology offers two main features:
    - The first provides the user with a graphical tool to explore the information space navigating the concepts defined therein and the semantic relationships between them. The selection of concepts on the graph allows you to populate the map with individuals who are their instances. Starting from ontology, graphic exploration is performed by navigating thematic views, each containing a subset of relevant

concepts from a specific point of view ("semantic around", for example, the natural or the normative);

- ◦ The second feature is support for search query information based on text queries, typical of search engines, but based on semantic interpretation rather than on a simple match of keywords. This function is based on the association of encyclopaedic knowledge with the concepts of ontology, to describe its meaning, and to enrich the concepts with linguistic terms that allow flexible recognition from user-written phrases (e.g. using synonyms of the concepts, and synonyms of the terms that they need in their encyclopaedic description, to recognize the concepts to which the user refers). Matching textual queries and concepts of interest is established using natural language analysis techniques that allow you to abstract from the user's specific vocabulary (Ardissono, L., Lucenteforte, M., Mauro, N., Savoca, A., Voghera, A., & La Riccia, L., 2016).

- Finally, as far as crowdsourcing is concerned, ontology supports the description and classification of new information elements that can be integrated into the information space in a structured way, making it easier to retrieve when searching for data.

The general aim of the OnToMap project is to create a tool capable of responding to requirements of an informative nature and encouraging inclusive urban processes. It is clear that applications of different types must be used to create a unique tool that contains within it the potential of each of them. It is precisely a case of creating a hybridisation between different cognitive systems, implementing what is defined here as a semantic approach to cartographic production, "social semantic mapping". The social dimension is understood in this sense to be a generator of the mapping process, built, therefore, on the basis of the social actors. This element must not, however, be confused with the purposes of mapping, which do not concern the strictly social dimension but, more generally, the territorial aspect, meaning the set of social, spatial and regulatory elements.

The process of social semantic mapping is divided into four phases. In the first instance, the social dimension refers to the collective construction of territorial knowledge, as a process of issuing interpretations, perceptions, tales of urban space and landscape, by the community. The territorial visions elaborated by the comparison between the actors are articulated in unitary representations of knowledge obtained thanks to digital media and, specifically, thanks to the territorial ontology (semantic representation of knowledge, see Figure 3). The knowledge representation takes place on a map (spatial representation). The knowledge gained in the process, and its formal representation, can then be used to supplement the institutional knowledge and to support policy decisions.

Figure 3. Social semantic mapping, a definition. Source: Authors' elaboration from Voghera and Crivello (2015)

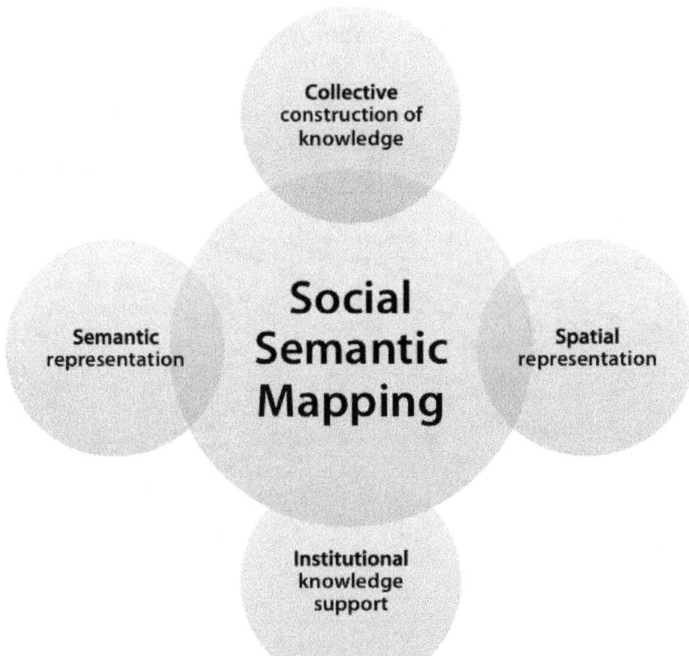

The OnToMap homepage shows a map (the platform uses OpenStreetMap cartography), on which the results of the semantic queries made through the ontology graph are displayed. Despite the page having characteristics similar to those of traditional PPGIS in terms of display of categories, by selecting a specific type of data of interest, a profoundly different action is performed.

Figure 4 shows a portion of the basic ontology of the OnToMap system. The representation guides the interpretation of the territory, which is read in the project according to three constituent dimensions: natural (natural dimension), artificial (cultural dimension) and normative (norms). The natural dimension refers to the physical elements considered unaltered by the hand of man; The cultural one includes all the anthropic elements; Finally, the regulatory dimension considers all the technical and project aspects related to the sphere of competence of the public administration, spatializing the rules and rules of local government, expressed as institutional constraints or planning.

These three dimensions take into account problems related to different domains representing the material aspects represented by GIS; socio economic aspects, highlighted by semantic relations; and perceptual aspects, typical of the voluntary construction of territorial knowledge. Therefore, ontology is the means used to

Figure 4. A conceptualisation of ontology in OnToMap: thin arrows represent the relations of specialization (from general concepts to the more specific ones); heavy arrows point to the normative and the dashed ones represent the conceptual relationship between concepts. Source: Authors' elaboration

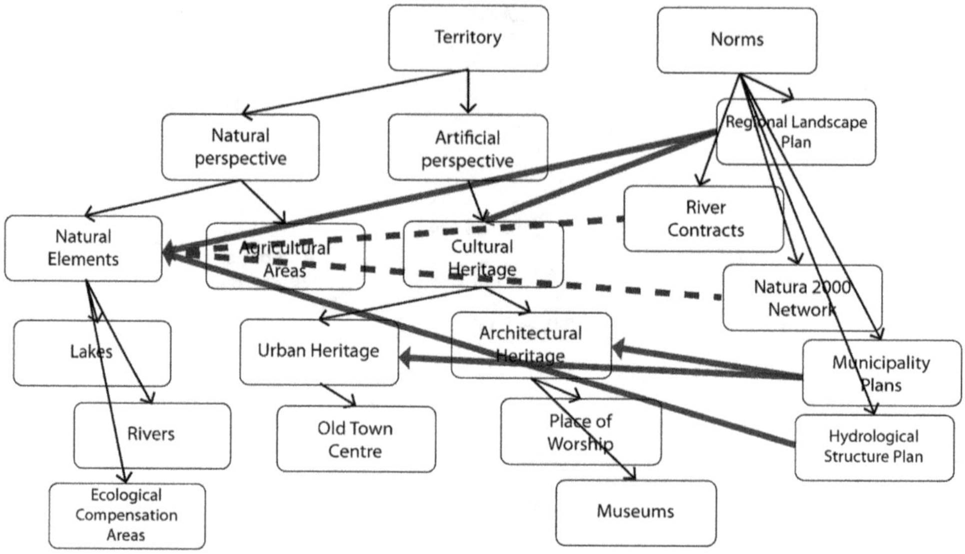

provide a unified description of the territory. In addition, it also plays the role of the "common language" among the different subjects involved in content analysis from different points of view and the retrieval of information at various levels of detail: for example, a geographic element could represent at the same time a service for the citizen and an element of the cultural heritage, and so on.

At first instance, it is necessary to define whether to query the entire ontology or just a portion of it. By selecting a point of view, it is, in fact, possible to use predetermined points of view, to speed up the data search operation. Currently, six separate points of view are set, excluding the complete one which returns the ontology in its entirely. The points of view are designed to meet the information requirements closest to citizens: the tourist point of view selects all elements of the territory that may be of interest to those visiting the city; the childhood point of view is focused on places of leisure and education. The young point of view cites transportation services, spaces of sharing and sports areas; the older point of view gathers together health services, trade and transport. Finally, the technical point of view cites all regulatory instruments underlying the government of the territory. Those points of view can be integrated and customised by users and will constitute an initial useful filter on which to articulate the analyses of the data collected; in addition, they can be redefined and re-aggregated during the course of some experimentations on the application, with the aim of articulating plural approaches to ontology, as close as possible to the participatory requirements.

At this point, when selecting a precise item within the ontology, the items with which it is related can be displayed. The relationships that connect the elements are essentially conceptual and regulatory relationships. For example, by selecting the theme of connecting infrastructures, the networks of vehicle roadways, public transport systems, cycle paths, pedestrian walkways, tourist routes and natural paths will be displayed. In addition to this, the Urban Mobility Plan, which regulates the mobility planning, will also appears as connecting element.

Each item that makes up the ontology includes a spatial representation, a shapefile that is superimposed on the map, and a set of metadata displayed in a separate window. That information, taken from the vector data, is more or less rich, depending on the selected concepts. Each regulatory instrument is linked to the web page through which it can be read and downloaded.

User reports follow the same mechanism, allowing information to be added to an element by selecting it by type and geographical location (Figure 5). The importance of this element is clear as regards the informative purpose; in fact, all data is re-processed so that it can be queried in a semantically relevant manner. Users can create customised private or public maps, selected areas and concepts. All the maps are, on the other hand, visible to the facilitator, which performs the role of filter between the different representations of the territory.

Figure 5. Example of a query within a selected point of view on ontology graph. Source: Screenshot of the OnToMap web interface.[7]

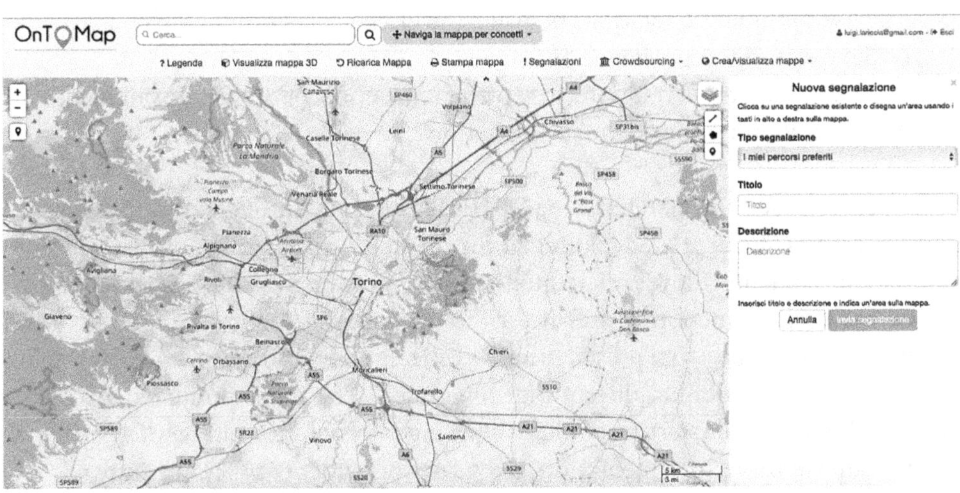

The territorial views gathered using the maps produced by users are summarised in a single integrated map created by the research team. The OnToMap programme does not provide, in its current version, a specific function for combining or overlapping maps. It is therefore the researcher's task to gather the maps produced and to analyse them, defining homogeneous interpretation criteria and producing a summary view. It is worth noting that the territorial ontology facilitates the task in this sense, providing uniformity of the concepts inserted by users. The latter must in fact choose from the concepts present in the ontology which ones to combine with the design on the map. In the working phase on the maps, it is therefore possible, using OnToMap, to assess which concepts have been used, with what frequency and in what territorial context. In future, OnToMap will be equipped with a special function for automatically overlapping the maps and selecting certain concepts of the ontology, to facilitate the comparison between representations and to simplify the summary work. The figure of the researcher is, however, essential and irreplaceable as a mediator and interpreter of the data gathered.

OnToMap has been tested in several areas. A first experimentation was carried out within the "urban planning project" atelier (2016/2017) of the Politecnico di Torino[8]: students (39, divided into 10 groups) used the application in a laboratory equipped with computers and internet access, under the supervision of the research team that accompanied the creation of maps for Venaria Reale (TO) and the Regional Park of La Mandria.

The experimentation consisted in using the application to represent the themes of interest of its territorial project in an innovative way: 10 analytical maps were created. These products have been verified and analysed according to four reading keys:

- How the instrument is used to read the territory;
- The use of semantic relations between the elements useful for the construction of the territorial representation;
- The use of semantic relations to represent the concept of "territorial identity";
- The production of possible design actions.

The representations generated using OnToMap have been articulated on two distinct levels: a "spatial" level related to direct drawing on the virtual map (crowdsourcing, Figure 6); a "semantic" level related to the use of concepts defined in the ontology (Figure 7). This distinction is useful for the purposes of the research as it allows understanding how the two representations are in relationship to each other, producing visions that may be similar, different or integrated. The maps produced were also considered as a possible starting point for building a "perception index" of a particular social group (in this experimentation, the group of students of architecture).

Figure 6. Classification of new design actions and proposals created in the 10 project maps (crowdsourcing). Source: Authors' elaboration

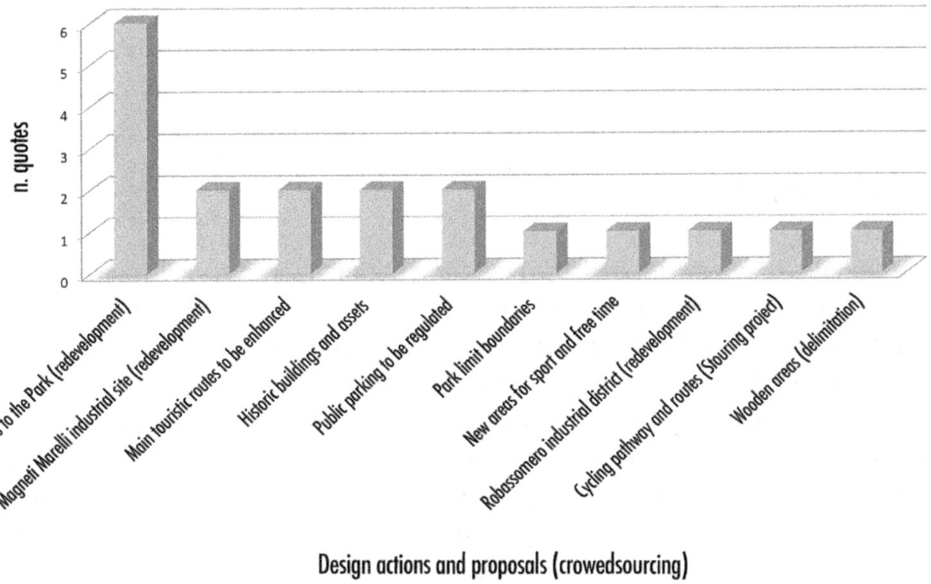

Figure 7. Classification of the use of concepts selected from ontology in the 10 project maps. Source: Authors' elaboration

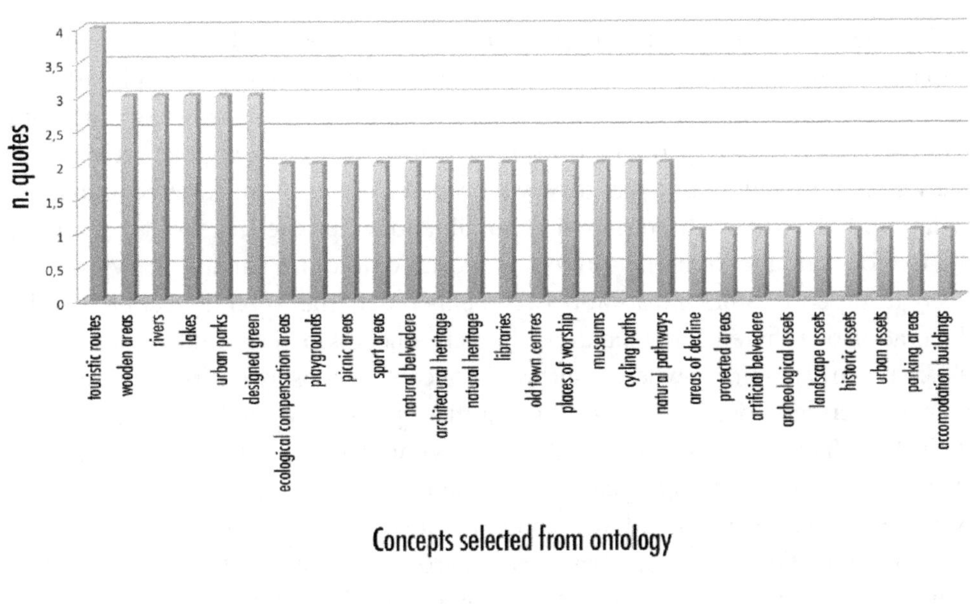

The conducted experiment has shown that some territorial elements, appropriately selected, displayed and modified by students on OnToMap, are recurring in most of the maps created. We can therefore deduce that the highlighted territorial elements reveal the expectations and the vision of a given social group with its own specific cultural background.

CONCLUSION

The digital is no longer, if ever it has been, an additional separate or virtual dimension of life: it is part of everybody's everyday life and the way we live in the city. Digital data, the ones we produce with our smartphones and the administrative ones, can be considered the starting point for controlling the urban and territorial transformations.

The urban planning field is concerned with ontologies: OnToMap project has been designed with the aim of encouraging new territorial interpretations, bringing to light new points of views of reality in an analytical and design perspectives. One of the aims of our approach is to provide the user with an integrated view of an urban project. As we have seen above, the information about an urban project is represented in different media (databases, documents, 3D city models, etc.) and at different scales (from regional plans to local plans). The ontology at the basis of the application supports an articulated spatial representation, positively contributing to urban and territorial planning that even more requires a comprehensive and inclusive territorial reading.

Our view is different from the traditional semantic web applications: the use of ontologies is a distinctive way to look at community maps that have always played a key role in urban planning processes. Traditionally, in the drafting phase of a community map, the mediator's ability to adequately represent the view of the population referring to a vision of the future of a territory was crucial in decision-making negotiations. Today, with the use of OnToMap, it is possible to widen the stakeholder audience and the focus of attention on a project proposal: ontologies can also be viewed as a way to support communication between experts and between lay people via the elaboration of a shared understanding of given concepts. This is quite a challenging task in the urban domain where a series of concepts is loosely defined and remains open to interpretation. Studies concerning the types of geographic concepts, their distinctive characteristics, and their differences compared to other types of concepts are also highly relevant to increasing the usability of geographic information. For example, geographically based socioeconomic concepts, such as administrative units (e.g. municipalities, counties, urban areas, metropolitan areas and neighbourhoods), exhibit great heterogeneity in their definitions among different countries or societies. Furthermore, they are often designated and analysed according to different spatial, environmental, and socio-economic criteria complicating the efforts for information exchange and reuse. The distinction between experts and non-experts also relates to the this consideration: the current wide access to geographic information by various groups of non-experts raises the question of proper information representation and organization in order to facilitate its understanding and support its use. Non-expert users may easily misunderstand the meaning of GIS if it is not properly represented and without a mediator. Studies of ontological conceptualizations of geographic entities and their corresponding representation within GIS are more crucial than ever.

In conclusion, it is worth citing the innovative aspects that have been included in OnToMap project and that have been emphasized the contribution for the research in the urban studies field. There are essentially two aspects to which attention should be paid: the implications of a semantic analysis for spatial representation (including the 3D representation) and the relationship between these and the processes of territorial governance.

Reasoning on the semantic dimension means making a conceptualisation, thus a structured interpretation that can be used to think and to communicate. The cartographic representation of the space is therefore connected to a mental representation, related to the possibilities of meaning implicit in the space. The representation of space is traditionally entrusted to the geographic map, a selected representation of places. Its aim is to reproduce, as effectively as possible, the characteristics existing in the real world, through a symbolic representation on the level of the map. The semantic representation, on the other hand, implies a scientific study of meaning, oriented in particular towards defining concepts and the type of information that they record. These are therefore two methods of representation that bring to the field intrinsically different types of knowledge. Connecting the two perspectives allows us to produce a complete level of knowledge of visual dimensions and meaning. As regards the territorial dimension, the subject of this study, the ontological perspective is particularly interesting as it allows us to break down all the conceptual levels that traditionally make up the definition of the territory, expressing their relationships. The semantic perspective is then linked to the cartographic representation thanks to the digital platform, which connects to each concept (meaning) a depiction (sign) on the map. Signs and meanings are then expressed through the cartographic representation. This step is an important element which has direct implications in the participation processes.

However, the new experimentation conducted in the last year has identified some missing concepts that individual users could not add directly as this requires specific technical skills of the system. This limitation, however, is intrinsic to knowledge-based systems, can be addressed and overcome by supporting a participatory approach to the specification of semantic knowledge of data. A specific work can be directed directly at the construction of a shared and participatory space ontology, with the aim of representing not only the territory but also the community that perceives it, inhabits it and foresees future transformations by expressing its own opinions and ideas. If, as hypothesized by visual sociology, visualization is a "power activity" (Faccioli & Losacco, 2010) that allows the definition of social and cultural differences by explaining them visually, working on maps and visualization, it allows to break down the different layers of meaning of produced territorial images. This operation is strongly related to the construction of a conceptual model of knowledge that is also visual, such as that of computational ontologies.

In a learning process, the geographic map is not only a tool for displaying locations and territorial phenomena. Based upon the idea that the map is an extension of the process of representation of meanings, this becomes capable of producing semantic functions and, thanks to these, of producing specifications of mapped places. In this semantic approach to cartography, the representation becomes the expression of cognitive and symbolic control (sign and meaning) of the (territorial) reality.

This aspect is worthy of particular attention, precisely because it could contribute to guaranteeing a level of empowerment of citizens (as defined based upon Arnstein's participation approach) which not only results in the stimulation of local planning, but is expressed, more profoundly, in the cultural training of citizens.

The experience of viewing the graph of territorial ontology gives users a complete but schematic conceptualisation of the relationships between the economic, social and regulatory components of urban space. The visual expression (through the graph) of the concepts and their relationships becomes important for a more effective understanding of the system itself. The presence of a technical definition and a general definition, for each aspect, further clarifies the more specialist aspects. The very possibility of viewing all the regulatory instruments that regulate the government of the territory, the cartographic space on which they act and the concepts directly linked to the individual official plans, texts and tables, makes a substantial contribution to the public participation and to some planning scenarios.

A second level is, on the other hand, directly related to the implementation of the new awareness of citizens, which can be translated into planning actions of the territory and the landscape, part of a unique process for the generation of cultural and territorial identity.

The ontology is then the mean used to provide a unified description of the territory, which abstracts from specific data representation formats and supports data integration. However, it also plays the role of an "inter-lingua" between human stakeholders, who might be interested in analyzing content under different points of view and in retrieving information at different levels of detail: e.g., a geographical item could represent, at the same time, a service for the citizen or a cultural heritage element, and it might be described taking different information needs into account.

The basic question underlying the future research activity revolves around the possibility of using the semantic approach and text queries to cartographic representation as a tool with the value of knowledge and education. The response achieved by the research is positive, considering that the semantic representation attributes meanings, expressing relationships and the causal link between the things represented, while the cartographic representation acts by codifying the concepts into visible signs.

Our future work includes various aspects, among which the validation of our query interpretation model in larger datasets and the analysis of the usefulness of different properties of concepts from the viewpoint of information filtering. At the current stage, the system analyses all the properties of items for answering a search query; however, some of them are less useful than others, and could probably be ignored. Our future work also includes the development of personal ontologies, inferred by analysing users' search behaviour and the acquisition of user models rejecting individual information preferences, given the user's interaction with a map. Both aspects are aimed at further improving the system's support to data retrieval, filtering and visualization, in order to reduce the information overload on users.

REFERENCES

Al-Kodmany, K. (2001). Bridging the gap between technical and local knowledge: Tools for promoting community-based planning and design. *Journal of Architectural and Planning Research, 18*(2), 110–130.

AOEMA. (2005). *E-government from a user's perspective*. Retrieved July 15, 2017, from http://www.aoema.org/EGovernment/Stages-Phases_of_e-government.htm

Ardissono, L., Lucenteforte, M., Mauro, N., Savoca, A., Voghera, A., & La Riccia, L. (2016). Exploration of Cultural Heritage Information via Textual Search Queries. *Proceedings. of the 18th International Conference on Human-Computer Interaction with Mobile Devices and Services Adjunct*, 992-1001. 10.1145/2957265.2962648

Arnstein, S. (1969). A Ladder of Citizen Participation. *Journal of the American Institute of Planners, 35*(4), 216–224. doi:10.1080/01944366908977225

Bobrow, D. G., & Winograd, T. (1976). *An Overview of KRL, a Knowledge Representation Language*. Report n. CS-TR-76-581. Stanford University Press.

Buccella, A., Cechich, A., Gendarmi, D., Lanubile, F., Semeraro, G., & Colagrossi, A. (2011). Building a global normalized ontology for integrating data sources. *Computers & Geosciences, 37*(7), 833–916. doi:10.1016/j.cageo.2011.02.022

Burrough, P. A. (1986). *Principles of geographical information systems for land resource assessment*. Oxford, UK: Clarendon Press.

Clift, S. (2004). *E-Government and Democracy. Representation and citizen engagement in the information age.* Retrieved July 15, 2017, from http://www.publicus.net/e-government/

Craig, W. J., Harris, T. M., & Weiner, D. (2003). *Community participation and Geographic Information Systems*. London: Taylor and Francis.

Cunha, E., Allegretti, G., & Matias, M. (2011). Participatory Budgeting and the Use of Information and Communication Technologies: A Virtuous Cycle? *RCCS Annual Review, 3*. Retrieved July 15, 2017, from http://rccsar.revues.org/316

Dennis, N. (1970). *People and Planning. Faber*. London: DOE.

Eisner, D. (2006). *What is neogeography anyway?* Retrieved July 15, 2017, from http://platial.typepad.com/news/2006/05/what_is_neogeog.html

English, K., & Feaster, L. S. (2003). *Community geography: GIS in action*. Redlands, CA: ESRI Press.

European Commission. (2015). *Digital Single Market strategy*. COM(2015) 192 final. Retrieved July 15, 2017, from https://ec.europa.eu/digital-single-market/en/news/digital-single-market-strategy-europe-com2015-192-final

Faccioli, P., & Losacco, G. (2010). *Nuovo manuale di sociologia visuale: dall'analogico al digitale*. Milan: Franco Angeli.

Fonseca, F., Egenhofer, M., Davis, C. Jr, & Borges, K. (2000). Ontologies and knowledge sharing in urban GIS computer. *Environment and Urban Systems*, *24*(3), 232–251. doi:10.1016/S0198-9715(00)00004-1

Goodchild, M. (2007). Citizens as sensors: The world of volunteered geography. *GeoJournal*, *69*(4), 211–221. doi:10.100710708-007-9111-y

Goodchild, M. (2009). NeoGeography and the nature of geographic expertise. *Journal of Location Based Services*, *3*(2), 82–96. doi:10.1080/17489720902950374

Gruber, T. R. (1995). Towards Principles for the Design of Ontologies Used for Knowledge Sharing. *International Journal of Human-Computer Studies*, *43*(5/6), 907–928. doi:10.1006/ijhc.1995.1081

Guarino, N., Oberle, D., & Staab, S. (2009). *What is an ontology? In Handbook on ontologies*. Dordrecht: Springer.

Harper, D. (1988). Visual sociology: Expanding sociological vision. *The American Sociologist, 19*(1), 54–70. doi:10.1007/BF02692374

Hendler, J., Shadbolt, N., Hall, W., Berners-Lee, T., & Weitzner, D. (2008). Web science: An interdisciplinary approach to understanding the web. *Communications of the ACM, 51*(7), 60–69. doi:10.1145/1364782.1364798

Howe, J. (2007). *Crowdsourcing: a definition. Wired Blog Network: Crowdsourcing*. Retrieved July 15, 2017, from http://crowdsourcing.typepad.com/cs/2006/06/crowdsourcing_a.html

Jogan, I. (2002). Alcune considerazioni a margine della 'svolta ontologica' della GIS Science. *Archivio di Studi Urbani e Regionali, 74*, 183–196.

Kersten, G. E. (2003). E-democracy and participatory decision processes: Lessons from eNegotiation experiments. *Journal of Multi-Criteria Decision Analysis, 12*(2-3), 127–143. doi:10.1002/mcda.352

Kingston, J. (2002). Merging top level ontologies for scientific knowledge management. *Proceedings of the AAAI workshop on ontologies and the semantic web.*

Las Casas, G., & Scorza, F. (2013). Interoperabilità e condivisione della conoscenza supporto della sviluppo regionale: nuove istanze e prospettive. Proceedings of Conferenza Italiana di Scienze Regionali XXXIV.

Laurini, R. (2007). Pre-consensus ontologies in urban databases. In J. Teller, R. Lee, & C. Roussey (Eds.), *Ontolgies for urban development*. Berlin: Springer Verlag. doi:10.1007/978-3-540-71976-2_3

Laurini, R. (2016). The geometric and semantic interoperability of the territorial databases. *Spatial data interoperability for a geographic intelligence, Seminar at Politecnico di Torino.*

Laurini, R., & Thompson, D. (1992). *Fundamentals of spatial information systems*. London: Academic Press.

Lefebvre, H. (1968). *Le Droit à la ville. Suivi de 'Espace et politique*. Paris: Anthropos.

Macintosh, A. (2006). eParticipation in Policy-making: The research and the challenges. In Exploiting the knowledge economy: Issues, applications and case studies (pp. 364–369). Amsterdam, The Netherlands: IOS Press.

Maier, E., & Reimer, U. (2010). Process Support for Increasing Participation in eParticipation. *eJournal of eDemocracy and Open Government, 2*(1), 46-55. Retrieved July 15, 2017, from http://www.jedem.org/article/view/16/53

Medaglia, R. (2007). The diffusion of eParticipation in English and Italian local government. In Understanding eParticipation: Contemporary PhD eParticipation Research in Europe (pp. 131–151). Örebro: University Library.

Noy, N. F., & McGuinness, D. L. (2001). *Ontology Development 101: A Guide to Creating Your First Ontology*. Retrieved July 15, 2017, from http://www.ksl.stanford.edu/people/dlm/papers/ontology-tutorial-noy-mcguinness.pdf

Obermeyer, N. J. (1998). The evolution of public participation GIS. *Cartography and Geographic Information Systems*, *25*(2), 65–66. doi:10.1559/152304098782594599

OECD. (2003). *Promises and problems of E-Democracy: Challenges of online citizen engagement*. Retrieved July 15, 2017, from http://www.oecd.org/gov/digital-government/35176328.pdf

Parker, B. (2006). Constructing Community Through Maps? Power and Praxis in Community Mapping. *The Professional Geographer*, *58*(4), 470–484. doi:10.1111/j.1467-9272.2006.00583.x

Prieto-Martín P., de Marcos L., & Martínez J.J. (2012). The e-(R)evolution will not be funded: A transdisciplinary and critical analysis of the developments and troubles of EU-funded eParticipation. *European Journal of ePractice*, *15*, 62-89.

Samet, H. (1990). *The design and analysis of spatial data structures*. Boston: Addison- Wesley.

Schroeder, P. (1996). Report on Public Participation GIS Workshop. In *GIS and Society: The Social Implications of How People, Space and Environment are Represented in GIS.* NCGIA Technical Report n. 96-97, Scientific Report for Initiative 19 Specialist Meeting, South Haven, MN.

Schuler, D. A., Rehbein, K., & Green, C. D. (2016, November). Is Corporate Political Activity a Field? *Business & Society*.

Sieber, R. (2006). Public Participation Geographic Information Systems: A Literature Review and Framework. *Annals of the Association of American Geographers*, *96*(3), 491–507. doi:10.1111/j.1467-8306.2006.00702.x

Steiniger, S., & Hunter, A. J. S. (2013). The 2012 free and open source GIS software map – A guide to facilitate research, development, and adoption. *Computers, Environment and Urban Systems*, *39*, 136–150. doi:10.1016/j.compenvurbsys.2012.10.003

Susha, I., & Grönlund, A. (2012). eParticipation research: Systematizing the field. *Government Information Quarterly*, *29*(3), 373–382. doi:10.1016/j.giq.2011.11.005

Turkucu, A., & Roche, S. (2008). Classification fonctionnelle des Public Participation GIS. *Revue Internationale de Géomatique*, *18*(4), 429–442. doi:10.3166/geo.18.429-442

United Nations. (2014). *E-Government Survey 2014: E-government for the future we want*. Retrieved July 15, 2017, from https://publicadministration.un.org/egovkb/portals/egovkb/documents/un/2014-survey/e-gov_complete_survey-2014.pdf

Voghera, A., & Crivello, R. (2015). Semantic Spatial Representation, an Experimental Proposal in the Framework of eParticipation. *International Journal of E-Planning Research*, *4*(4), 18–35. doi:10.4018/ijepr.2015100102

ENDNOTES

1 The paragraphs "Introduction" and "Background: the Pathway Towards e-Participation" have been written by Angioletta Voghera. The paragraphs "From GIS to Digital Ontologies" and "Some Updates Results from the OnToMap Project" have been written by Luigi La Riccia. The "Conclusion" paragraph is the results of a conjoint work of the two Authors.

2 https://ontomap.ontomap.eu/. The OnToMap project was coordinated by Angioletta Voghera (Inter-University Department of Sciences, Planning and Territorial Policies of the Polytechnic of Turin) and Liliana Ardissono (Department of Computer Science of the Università degli Studi di Torino), with support from Mauro Velluto (Csi Piemonte). Maurizio Lucenteforte (Department of Computer Science of the Università degli Studi di Torino), Adriano Savoca (Department of Computer Science of the Università degli Studi di Torino) Rossella Crivello and Luigi La Riccia (Inter-University Department of Sciences, Planning and Territorial Policies of the Polytechnic of Turin) also collaborated in the project.

3 https://ideascale.com/

4 http://liquidfeedback.org/

5 https://www.ushahidi.com/

6 As reported in http://linkeddata.org/, the so-called "linked data" allows the web to correlate previously unrelated data over the web, or to use the web to break down the obstacles associated with linked data between them with other methods.

7 To interact with the ontology, it is necessary to use the top left key "Selection of data by type", which allows you to select individual concepts and the complete view of the ontology. This can be viewed also by parts, thanks to

the dropdown menu "Points of view" which allows you to select one of the views pre-set in the system. The example in the figure shows the Point of view of childhood (Children) and, more specifically, it shows three main concepts, Private Services, Public Services, Equipped Greenery, to which, respectively, Childhood centres and Educational Farms (Private Services), Nurseries and Schools (Public Services), Play Areas, Urban Parks, Picnic Areas (Equipped Greenery) are connected. At the bottom, three keys allow you to move from the view on ontology to the view of the territorial data on the map (View results) or to cancel the search (Remove all). Each individual concept presents a description (Show description) and the concepts linked to it by relations (View related). The Show Reports key allows you to see the comments sent by users and made public.

8 Atelier "Urban Design Project" 2016/2017 "Il parco fuori le mura: percezione e innovazione del ruolo del Parco Naturale Regionale della Mandria nel territorio di contesto/Park outside the boundary: Mandria Natural protected area and its context" (Prof. Angioletta Voghera), Politecnico di Torino.

Section 2

Creativity and Innovation in Data Collection

Chapter 7
Enhancing Creativity in E-Planning:
Recommendations From a Collaborative Creativity Perspective

Paul B. Paulus
University of Texas at Arlington, USA

Jonali Baruah
Tarleton State University, USA

ABSTRACT

Sharing ideas efficiently and effectively in groups is a challenge groups and teams face on a daily basis. In typical face-to-face meetings, many factors can serve to inhibit a full sharing of ideas and thus the development of effective decisions and plans. To overcome the limitations of face-to-face meetings, computer-based group decision support systems have been developed to facilitate both idea exchange and evaluation. Evidence suggests that such systems can lead to beneficial outcomes. However, unless they are utilized effectively, even electronic meetings may not effectively tap the intellectual and creative potential of groups. The authors summarize some of the major findings of collaborative creativity and their implications for effective e-planning.

INTRODUCTION

Good planning requires good information and good decision-making. Most planning is probably done by groups during meetings or over a series of meetings. Although the diversity of perspectives available in group settings are potentially valuable for planning efforts, much research on group dynamics suggests that groups often have

DOI: 10.4018/978-1-5225-5999-3.ch007

trouble living up to their potential (Nijstad, 2009; Paulus & Nijstad, 2003). Relevant to our discussion are the facts that groups often focus on common information rather than the unique information available in group members (Stasser, Abele, & Parsons, 2012), they may have reduced motivation to perform at a high level (social loafing, Karau & Williams, 1993), they engage in social matching of low performance (Paulus & Dzindolet, 1993), and they may be hesitant to express ideas that are different from the group norm or majority perspective (Nemeth & Nemeth-Brown, 2003). Concern with potential negative evaluation of novel ideas may further limit the number and novelty of the ideas shared (Camacho & Paulus, 1995). Furthermore, during face-to-face meetings only one person can talk at a time limiting the opportunity for full sharing of perspectives (production blocking, Diehl & Stroebe, 1987). This is particularly evident in group brainstorming. When face-to-face groups share ideas, they typically generate fewer ideas than a similar number of individuals sharing ideas alone (nominals, Diehl & Stroebe, 1987; Paulus & Coskun, 2012). This production loss in face-to-face groups appears to be due in part to production blocking—the inability to express ideas as they occur because of competition for speaking turns (Diehl & Stroebe, 1987; Nijstad & Stroebe, 2006). To overcome some of these problems in groups, alternative means have been suggested for conducting the idea exchange process. These have been examined in the literature on brainstorming in which the focus is on generating a large number of ideas in groups in order to increase the chances of good ideas (Paulus, Kohn, & Arditti, 2011). The good ideas can then be selected for further review and elaboration. One major focus has been to use interaction modalities that reduce production blocking.

In this chapter we will provide a theoretical perspective of group creativity and some of relevant findings. We will discuss the use of brainwriting and electronic brainstorming (EBS) to enhance group creativity, as well as a combination of solitary and group brainstorming. Finally, the implications of the group creativity literature for e-planning will be highlighted.

A COGNITIVE-MOTIVATIONAL PERSPECTIVE ON BRAINSTORMING

Paulus and his colleagues (Paulus & Brown, 2003, 2007) and Nijstad and colleagues (Nijstad & Stroebe, 2006; Nijstad, Stroebe, & Lodewijkx, 2002) have provided cognitive theoretical perspectives that are the basis for predicting both positive and negative effects of collaborative ideation. The model suggested by Paulus and colleagues involves both cognitive and motivational factors (Paulus, Dugosh, Dzindolet, Coskun, & Putman, 2002; Paulus & Dzindolet, 2008), and we will primarily focus on that perspective. Whenever people share ideas in a group context, they can be influenced both by the content and the rate of ideation. So if some group members are highly productive, others may be motivated to increase their performance (Coskun, 2000; Munkes & Diehl, 2003; Paulus & Dzindolet, 1993). Of course, a social loafing perspective might suggest that when others are highly productive, one has a chance to reduce one's efforts since one may feel that the efforts of others are sufficient (Karau & Williams, 1993). However, in either implicit or explicitly competitive situations, upward comparison motives may be dominant (Paulus, et al., 2002). That is, participants may be more concerned about meeting or exceeding the performance levels of others in the group. To motivate a high level of performance in collaborative groups, it is important that they set high goals. Studies indicate that groups that either set more challenging goals or are provided challenging goals perform at a higher level than other groups (Larey & Paulus, 1995; Paulus & Dzindolet, 1993; Shepherd, Briggs, Reinig, Yen & Nunamaker, 1995). Individuals or groups that feel they are in competition with other individuals or groups also generate more ideas (Paulus & Dzindolet, 1993; Ziegler, Diehl & Zijlstra, 2000) than those who are not in competition. Providing group members feedback about their performance during the session can similarly enhance EBS performance (Roy, Gauvin, & Limayem, 1996).

In addition to the potential motivating effect of exposure to a high number of ideas from others, these ideas may be cognitively stimulating. That is they may lead people to think of other ideas or other categories of ideas (Nijstad & Stroebe, 2006; Paulus & Brown, 2007). It is often difficult to differentiate the motivational and cognitive stimulation effects of exposure to the ideas from others (Dugosh & Paulus, 2005; Dugosh, Paulus, Roland, & Yang, 2000), but various studies have shown that both factors are important.

Idea generation is basically a cognitive process. It involves searching one's own memory for relevant categories and ideas within those categories (Nijstad & Stroebe, 2006; Paulus & Brown, 2003; 2007). Initially one may search the most accessible categories and ideas. As one has difficulty coming up with more ideas in one category or area, one will tend to switch to another category. The key to generating a large number of ideas is to search deeply within each category to find unique ideas (Rietzschel, Nijstad & Stroebe, 2007) and then to keep going from one category to another repeating that process.

One problem with both individual and collaborative brainstorming is that people typically tap only a small percentage of their relevant memory store. Studies have demonstrated that participants are likely to remain in the same category at any point in time rather than shifting from one category to another during the brainstorming session (Brown, Tumeo, Larey, & Paulus, 1998; Nijstad & Stroebe, 2006; Larey

& Paulus 1999). In our experience and that of others, most groups feel that they have exhausted their ideas after a relatively short period of time. However, when we restart groups and ask them to continue, they are able to generate many more ideas. This is critical since the most original ideas tend to come later in the process as people tap more unique categories or dig more deeply into categories (Baruah & Paulus, 2016; Paulus, Kohn, Arditti, & Korde, 2013; Rietzschel, E. F., Nijstad, B. A., & Stroebe, W., 2007). There are a number of ways that groups can be encouraged to more effectively tap the broad range of ideas and categories in their repertoire. One useful approach is to present the groups with one category of the problem at a time and ask them to focus on each category in turn (Coskun, Paulus Brown, & Sherwood, 2000; Dennis, Valacich, Connolly, & Wynne, 1996). This approach has been found to increase the number of ideas generated both in electronic and face-to-face contexts. One can also ask participants to generate their own categories prior to brainstorming. This will insure that they at least consider a broad range of ideas and do not stop brainstorming prematurely (Deuja, Kohn, Paulus & Korde, 2014).

Another approach is to simply provide brief breaks during the brainstorming process. These breaks allow not only cognitive rest but also provide an opportunity for reflection, incubation, and overcoming of idea fixation (Paulus, Nakui, Putman, & Brown, 2006; Smith, 2003). At this point it is not clear how long the breaks should be or how many breaks are optimal in a particular period of time. Also, although Smith suggests that a change in focus or activity during the break may be helpful in overcoming fixation, research thus far is not clear about what kinds of breaks or activities are most helpful.

BRAINWRITING

There are several paradigms that allow for a reduction in process losses and that facilitate the cognitive stimulation potential of groups. One of these is brainwriting which involves the sharing of ideas on slips of paper (Heslin, 2009). This technique has been used for a number of years in various forms (Geschka, 1993; Goodman, 1995; VanGundy, 1981). In one version of this approach, participants wrote ideas on slips of paper and passed them on to the person next to them (Paulus & Yang, 2000). Participants in groups of four were asked to read slips of paper as they were shared, add their ideas and pass them on until all had read them. Then the slips were placed in the center of the table. Each person wrote in a different color pen so their performance could be identified relative to that of the others. This was designed to reduce potential social loafing (Karau & Williams, 1993). Also, the fact that ideas were shared on pieces of paper probably reduced the degree of evaluation apprehension since ideas could be expressed without concern for evaluative feedback. It was found that the groups generated more ideas than the nominals. Furthermore, this difference persisted in a subsequent session where the group participants continued to generate ideas individually. Their performance in this session was even more elevated compared to the nominals than in the first session. So apparently the benefits of shared ideation carried over into the second session. This study thus clearly demonstrates that group idea sharing can be beneficial when the problems of production blocking, social loafing, and evaluation apprehension are reduced.

An important aspect of the collaborative ideation process is the opportunity to build on the ideas of others, but there have been few studies that have actually examined this process. In one study, Kohn, Paulus and Choi (2011) presented groups of three individuals with ideas that had already been generated in prior brainstorming sessions. In some conditions participants were exposed to a set of very common ideas on that topic and in another condition they were exposed to ideas that were mentioned only by one person in prior brainstorming sessions. Using a brainwriting procedure, it was found that individuals (nominal groups) generated more combinations than groups, but groups generated more novel and feasible ideas in the unique idea condition. So it appears that group collaboration can be helpful in processing and building on novel ideas.

Although brainwriting appears to be a useful method for sharing ideas in groups, only a small number of studies have examined this technique and the ways to optimize it (Heslin, 2009). Group brainwriting may not always be superior to individual brainwriting (Goldenberg, Larson & Wiley, 2013). It may depend on a number of procedural and motivational factors. The sharing process needs to be efficient. When a person reads a slip of paper and passes it on, it may stimulate a new idea but can also distract from the flow of thought. Furthermore, the time taken to read others' ideas reduces the amount of time to generate one's own. This may not be a problem if the session is fairly long or there are multiple sessions. A more optimal system may be one in which participants can tap ideas from the pile of shared ideas when they have a temporary "lull" in their idea flow (pooling, Michinov, 2012). However, there has to be some system in which ideas already read are not mixed back into the pile. By using different colors for each of the participants, participants can easily avoid picking up slips of ideas they have generated. Michinov (2012) reported that because brainwriting required extra effort to pick the slip and read other's ideas, the participants did not always make that effort and hence there was less cognitive stimulation in such groups. Conversely, computer mediated communication gave participants easy access to the ideas produced by others on a computer screen. Hence the attention and cognitive stimulation was likely to be higher in such paradigm.

CREATIVITY THROUGH COMPUTER-MEDIATED COMMUNICATION

A more efficient way of sharing ideas may involve the use of electronic systems (Dzindolet, Paulus, & Glazer, 2012). With the advent of computer systems for group decision-making, there have been a large number of studies that have used digital means of idea sharing. Many studies have used group decision support systems (GDSS, Limayem, Banerjee, & Ma, 2006; Nunamaker, Dennis, Valacich, Vogel, & George, 1991). In the context of e-planning, GDSS can be considered as a technology that supports person-to-person collaboration, structuring and aligning the information exchange among individuals and groups (Laurini, 2001). These types of synchronous systems enable large groups of participants to share ideas electronically without having to wait their turn. Ideas shared by others are typically available on one part of the screen while the ideas the participant generates are on another part of the screen. With the technology used in many studies, the ideas are sent to different folders. When a participant submits an idea, the program typically shows the ideas in different folder. Of course, at first there are very few ideas in the folders. But as the session continues, the number of ideas in each folder increases. There are other approaches that do not place ideas in folders. It appears that using a folder approach rather than providing one file with all of the ideas can be more effective in generating more ideas, possibly because it avoids overloading participants with ideas as they are trying to generate their own (Aiken, Vanjani, & Paolillo, 1996). However, another study found that the use of folders was not critical for generating more ideas (Paulus, P. B., Kohn, N. W., Arditti, L. & Korde, R., 2013).

Advancements in information technology and globalization have enabled interaction in virtual groups in which members are often distributed geographically and cannot contribute at the same time. Hence asynchronous electronic brainstorming is a common scenario in the current business world (Chen, Liu, & Wang, 2007). Asynchronous communication technology provides participants with an opportunity to work together without the constraints of time as the group members can generate ideas at their own pace. A variety of tasks such as planning, voting, negotiation, decision-making, problem-solving, and idea generation can be performed using this method (Michinov & Primois, 2005). Research on the use of asynchronous brainstorming is limited (Michinov & Primois, 2005), but the existing literature suggests that asynchronous electronic brainstorming and traditional electronic brainstorming involve some of the same social psychological processes. For example, upon seeing the asynchronous contributions of the other group members, participants adjusted their level of performance (upward social comparison) during the idea-generation task, increasing the overall group performance especially when the participants were informed about each other's performance (Michinov & Primois, 2005). Additionally, groups generating ideas in an asynchronous paradigm (serial mode) created ideas with significantly higher elaboration compared to the groups generating ideas in synchronous mode (parallel groups) (de Vreede, Briggs & Reiter-Palmon, 2010). In contrast, synchronous processing was found to result in information overload (McLeod & Liker, 1992).

A review of literature reveals that both the synchronous and asynchronous methods of brainstorming have unique benefits and drawbacks. In organizational context, one study found that teams in the asynchronous paradigm share less information, are less comprehensive in decision-making tasks and are less effective than face-to-face teams in problem-solving tasks (Shen, Lyytinen, & Yoo, 2015). In terms of the benefits of synchronous communication, such a communication was found to be effective since it freed the participants from turn-taking (Nunamaker, Jr., J. F., Dennis, A. R., Valacich, J. S., Vogel, D. R. &George, J. F., 1991). The use of feedback and leadership were found to be effective in the synchronous paradigm (Hiltz, Johnson & Turoff, 1991). The presence of social and verbal cues explained superior performance in idea generation in synchronous communication mode (Shirani, Tafti & Affisco, 1999. However such a communication also reportedly caused information overload (McLoed & Liker, 1992). In educational contexts, asynchronous discussion was reported to be useful in encouraging in-depth, richer and inclusive interchange and better recall of concepts (Branon & Essex, 2001; Johnson, Howell & Code, 2005). However the drawbacks included the: "lack of immediate feedback, students not checking in often enough, length of time necessary for discussion to mature, and students feeling a sense of social disconnection" (Branon & Essex, 2001, p. 36).

Synchronous chat was reportedly effective in holding virtual office hours, brainstorming on various topics and in building community. However the limitation of synchronous mode of communication included difficulty in having students log in online at the same time, not enough time for reflection (Branon & Essex, 2001) and multiple errors in discussions (MacDonald, Garcia-Carbonell, & Carot-Sierra, 2013). Finally, some studies revealed that a combined sequence of asynchronous and synchronous method was the best as it maximized personal engagement in learning among college students (c.f., Johnson, 2006). More research is obviously needed to determine the best electronic idea exchange paradigm.

Electronic Brainstorming

There has been little if any research on the use of GDSS systems for electronic brainstorming (EBS) in e-planning (cf., Suchocki, 1995). Thus we will provide a summary of the general EBS literature. This summary can suggest some general guidelines for those who wish to use EBS systems in the e-planning process.

The EBS system is a method of exchanging text messages electronically in a synchronous or asynchronous environment (Dennis, Minus & Williams, in press). Today EBS can be used in a variety of forms such as instant messages (WhatsApp, MSN messenger, Facebook chat, Twitter) or web-conferences (Google hangouts, Skype, Zoom). All these tools have unique features that help make communication between groups and individuals easy and convenient.

The EBS paradigm has a number of advantages. It allows participants to share ideas without experiencing production blocking (Gallupe, Bastianutti & Cooper, 1991). The ideas can be shared anonymously so that evaluation apprehension can be minimized. Since the ideas are in an electronic form, they can be easily downloaded for further evaluation or data analyses. Many systems also allow for voting or decision making options so that groups can select the top ideas. We know of only one study that has compared brainwriting and EBS. This study found that when the process makes similar demands on the participants, a similar number of ideas are generated by brainwriting and EBS (Michinov, 2012). However, the EBS paradigm was found to yield higher satisfaction compared to brainwriting (Michinov, 2012). Additionally, EBS has the advantage of digital recording of ideas and a potentially more efficient process for voting on the pool of ideas generated. In an e-planning context, collaborative planning using EBS that includes involvement of the ideas, preferences, feedback and voting from all the participants should result in a more efficient and satisfying urban planning.

Research on EBS has shown that participants will generate and recall more ideas (DeRosa, Smith, & Hantula, 2007; Hinds & Payne, 2016) with this procedure than in face-to-face groups. Furthermore, EBS groups can also outperform similar size nominal groups. However, this benefit seems to depend in part on the size of the group. Reviews of the literature (Dennis & Williams, 2003; Derosa, Det al., 2007) have shown that for groups of 9 or larger, group sharing of ideas increases the number of ideas shared relative to a similar number of individual brainstormers. This process gain or synergy may be due to the increase in cognitive stimulation and task motivation in large groups (Dugosh & Paulus, 2005; Paulus, P. B., Kohn, N. W., Arditti, L. & Korde, R., 2013). Thus, as the group size increases, there is an increased stimulation of semantic networks resulting in a higher number of shared ideas (Brown & Paulus, 2002; Nijstad & Stroebe, 2006; Paulus & Brown, 2007) and higher motivation as the members use other's ideas as reference points to generate more ideas (Dugosh & Paulus, 2005). However, a large number of ideas in EBS context may also cause cognitive overload (Grise & Gallupe, 2000; Nagasundaram & Dennis, 1993; Santanen, Briggs, & de Vreede, 2004) and a high rate of conformity (Paulus, P. B., Kohn, N. W., Arditti, L. & Korde, R., 2013). Nevertheless, conformity is not necessarily detrimental to groups because it indicates that participants paid attention (Paulus, P. B., Kohn, N. W., Arditti, L. & Korde, R., 2013) and generated ideas that are closely linked to the ideas of other in the group, a phenomenon that actually results in higher productivity (Baruah & Paulus, 2011).

Even small EBS groups may outperform small nominal groups (Dugosh & Paulus, 2005). A key factor may be the extent to which participants achieve an appropriate balance between attention to the shared ideas and generation of their own ideas (Brown & Paulus, 2002: Paulus & Brown, 2007). Thus instructing participants that it is important to attend to the shared ideas may increase cognitive stimulation (Paulus, P.B., & Yang, H.C., 2000). If participants feel that the other group members are similar in ability, this may also increase their motivation to attend to the shared ideas and their impact (Dugosh & Paulus, 2005).

The potential for anonymity has been emphasized as an important advantage for EBS, and indeed there is some evidence that anonymity can increase the number of controversial and nonredundant ideas (Cooper, Gallupe, Pollard, & Cadsby, 1998). Some studies found that anonymity resulted in critical and probing comments (Kiesler, Siegel, & McGuire, 1984; Siegel, Dubrovsky, Kiesler & McGuire, 1986) and embellished ideas (Jessup, Connolly, & Galegher, 1990). However, a meta-analytic review found no conclusive evidence for the influence of anonymity in EBS (Derosa, et al., 2007). One reason for this lack of effect is that the potential benefit of anonymity in reducing evaluation apprehension may be counteracted by the tendency to loaf when one is not identifiable (Karau & Williams, 1993).

The benefit of electronic brainstorming paradigm also depends on temporal effects. In general, in short term (15 to 20 minutes) brainstorming sessions, the productivity is often found to be high in the first few minutes of the brainstorming session after which the number of ideas diminishes drastically over time (Kohn & Smith, 2011). Although the productivity slows down over time, the originality of ideas increases significantly (Baruah & Paulus, 2016; Kohn & Smith, 2011). Deuja et al. (2014) further reported that the originality of ideas increased over time when members were assigned with specific topics over time. Some studies reported that the increase in originality is attributed to enhanced focus on a narrow set of ideas resulting in a deeper level exploration of categories over time (Baruah & Paulus, 2016; Rietzschel, E. F., Nijstad, B. A., & Stroebe, W., 2007). A laboratory study revealed that prompting groups every two minutes in a forty-minutes brainstorming task enhanced the quantity and creativity of solutions generated (Santanen, E. L., Briggs, R. O., & de Vreede, G., 2004).

Studies that investigated creative performance over a long duration were able to investigate several social and cognitive factors that play role in creativity. Kanawattanachai and Yoo (2007) studied virtual teams over a period of eight weeks and reported that in the early stage of virtual team communication, frequency of task related communication played the key role through which the members developed expertise and trust. However, towards the end collaborations based on task knowledge emerged as the key factor that influenced team performance. In another study, geographically dispersed virtual teams working together for a period of seven weeks exhibited difficulty in communication and understanding of salience of the issues, which hampered the mutual knowledge sharing among the group members (Cramton, 2001). Some studies investigated the effect of time constraints (deadlines) and reported that groups brainstorming under time constraint produced ideas at a higher rate than those that did not have the constraint (Dennis, Valacich, Connolly & Wynne, 1996).

More studies need to be done to develop a better understanding of the effect of time on electronic brainstorming. Shen et al. (2015) reported that although some research has been done in the area of temporal effects, the research is fragmented with weak theoretical foundations. In most studies the variable of time has been oversimplified because time is often considered as a "clock view". For example, they stated that time has often been studied in the form of duration or deadlines. However the variable of time is a complex factor that incorporates not only the time that ticks away, but it can be classified in various ways based on the interpretations of the team members. For example, time can be classified in many forms such as time as a cycle, frequency, sequence, pace or duration (Ancona, Okhuysen & Perlow, 2001). The dynamics in teams will vary based on their interpretation of the time along with social and cognitive factors associated it. For example in an urban planning scenario, introduction of a new technology may be viewed as beneficial by a team with long-term vision even though there is a possibility of temporary disruption of workflow. On the other hand, another team with short-term planning horizon may be more concerned about temporary disruption of the workflow that the change may cause. Thus, more systematic empirical research is required where the temporal lens can be defined in terms of various independent and dependent variables.

Hybrid Brainstorming

Although there has been a focus on comparing the creative performance of interactive groups with that of nominal groups, it has been suggested that alternation of alone and group brainstorming or hybrid brainstorming may be optimal (Brown & Paulus, 2002; Osborn, 1953; Paulus & Brown, 2007). At the beginning of a brainstorming session, participants may have little trouble coming up with highly accessible ideas. However, at some point they will have more difficulty coming up with new ideas and exposure to ideas from others may remind them of other ideas or categories of ideas. Some studies have indeed found that an alone to group order of brainstorming can lead to generation of more ideas than a group to alone order (Baruah & Paulus, 2009). However, there may also be a benefit of generating ideas alone after group brainstorming since the solitary session allows the participants to build on the associations stimulated by the group session (Paulus, Korde, Dickson, Carmeli & Cohen-Meitar, 2015). Research by Korde and Paulus (2016) using hybrid brainstorming with a brainwriting task found that alternating individual ideation sessions with group sessions was more beneficial than all alone or all group sessions. In particular, elevations in performance were noted in the alone sessions that followed the group sessions. In group sessions the exposure to shared ideas can be very stimulating. However, the joint demands of reading ideas from others and generating one's own can limit the ability to fully tap the benefits of the cognitive stimulation. The alone session allows participants to tap more fully the benefit of the prior group stimulation. In an e-planning context, the use of EBS to generate novel ideas in the form of a mixture of group and individual sessions might be helpful.

The hybrid procedure has not been tried with electronic systems. However, theoretically the same type of benefit of alternating individual and group ideation should occur. We do not know how the sessions should be distributed in time in synchronous EBS. We have found alternating relatively short sessions to be beneficial (Korde & Paulus, 2017). Short group sessions may build up sufficient stimulation to tap during a subsequent solitary session. In asynchronous EBS, this process may be difficult to orchestrate. However, participants could be asked to take time after their sharing session to reflect on the shared ideas and to generate some additional ones while the cognitions are "hot". Thus in an e-planning context, the use of EBS to generate novel ideas in the form of a mixture of groups and individual session may facilitate the development of more novel ideas.

Diversity and Creativity

It is often assumed the diversity in background enhances creativity. However, the literature on team performance and diversity has generally found mixed results. Most reviews indicate that demographic diversity has either a slightly negative effect or no overall effect on team performance (e.g., Bell, Villado, Lukasik, Belau, & Briggs, 2011; Reiter-Palmon, Wigert, & de Vreede, 2012). Functional diversity as reflected in diversity in skills or experience may have some benefit (Hulsheger, Anderson, & Salgado, 2009). However, van Dijk, van Engen, and van Knippenberg (2012) found that functional diversity is positively related to team performance only for complex tasks. In a review of the subsets of studies that examine diversity and creativity in groups and teams, Paulus and van der Zee (2015) found that most relevant studies have found that demographic diversity can have positive effects on creativity. Possibly, creativity tasks may be more sensitive to variations in diversity than other tasks. This appears to be the case as well for EBS. A study by Daily, Watley, Ash and Steiner (1996) compared homogeneous and heterogeneous ethnic groups who were asked to develop ideas for three campus issues using a GDSS system. The heterogeneous groups generated more ideas and more realistic ideas than the homogeneous groups. Martins and Shalley (2011) examined the role of demographic and skill diversity on the functioning of dispersed virtual teams and found that nationality, age and technical skill all influenced creativity. However, gender and race composition did not influence creativity. Given the general inconsistency in findings on diversity and the paucity of research on this factor in electronic groups, strong predictions for its effect on e-planning are not warranted.

However, we would expect that a diversity dimension that is relevant to the planning task is likely to have an impact, especially if the participants are highly engaged in elaboration of each other's contributions (van Knippenberg, De Dreu, & Homan, 2004. For example, in a recent study we examined the role of various dimensions of diversity on the generation of ideas for improving the healthcare system using an electronic discussion board (Coursey, Kenworthy, Paulus & Doboli, 2017). Groups of five interacted asynchronously over a period of four weeks. It was found that age and gender diversity were related to generation of lower novelty ideas, but political diversity was related to increased novelty. The key factor was the extent to which participants replied to the shared ideas. Age and gender diversity were related to lower levels of replying and political diversity to higher levels. Thus if a topic is highly relevant to some diversity dimension and motivates responsivity to shared ideas, this diversity is likely to enhance the creative outcomes of the planning process.

Research in team innovation, in which the measures are typically verbal reports of innovation, has found that job relevant diversity is related to higher levels of reported innovation. However, thus far, no experimental studies have examined the effects of functional diversity on creativity. It seems likely that such diversity would also increase creativity since groups that have a variety of experiences and domains of knowledge should be more likely to generate novel ideas or connections. Paulus and van der Zee (2015) suggest that it is important that the creativity task be relevant to the type of diversity represented in the group. Thus for an urban planning group trying to come with novel ideas related to urban transportation issues, it obviously would be most beneficial if the group contained members with different areas of expertise related to transportation.

Creativity Training and Facilitation

As we have discussed, brainstorming in groups is a challenging task, no matter the interaction format. Some training on how to work effectively as a group may be helpful in increasing the effectiveness of the process. Research on creativity has demonstrated that training can be effective for increasing individual creativity (Scott, Leritz, & Mumford, 2004). Research on teams has also demonstrated the benefits of training for effective teamwork (Salas, E., DiazGranados, D., Klein,

C., Burke, C. S., Stagl, K. C., Goodwin, G. F., & Halpin, S. M., 2008). So it is not surprising that creativity training or the use of facilitators in group brainstorming can be effective. Several studies have shown that facilitators who insure that the group is following the brainstorming directions and encourage involvement of all members can significantly enhance the performance of face to face brainstorming groups (Kramer, Fleming, & Mannis, 2001; Offner, Kramer, & Winter, 1996; Oxley, Dzindolet, & Paulus, 1996). Facilitators tend to be used when groups use complex GDSS such as Group Systems (e.g., Adla, Zarate, & Soubie, 2011; Kay, 1995), but we know of no published studies that have examined the impact of facilitators on the number and quality of ideas generated in EBS. It is also likely that experience or training with EBS would enhance the ideation and decision-making process. In fact, some degree of training is required for participants to effectively use GDSS systems (Adla, A. Zarate, P. Soubie, J.L., 2011; Dickson, Lee-Partridge, Limayem, & Desanctis, 1996).

A study with face-to-face brainstorming groups demonstrated that about an hour of training on effective brainstorming can increase both the number and quality of ideas in face-to-face and e-nominal (group members brainstormed on the same problem using computers but in isolation without sharing ideas with each other) brainstorming groups (Baruah & Paulus, 2008). Presumably similar benefits would accrue for EBS groups, but systematic studies on this issue are not evident in the literature.

Creativity and Decision Making

Once a group has generated a large number of ideas, there is a need to sift through the ideas to determine the best ones if there is a focus on potential application. Although there has been much research on the generation of ideas, there has been much less research on the selection of ideas. Several studies have examined the process of selecting the top ideas after brainstorming. In general it appears that groups have a difficult time picking the most original ideas (Putman & Paulus, 2009; Rietzschel, Nijstad, & Stroebe, 2006). The average originality of the top ideas picked is not much different than the average originality of the total set of ideas (as judged by trained raters). There seems to be a tendency to pick the more feasible ideas instead. That is not necessarily inappropriate, but presumably the most creative ideas are both novel and useful (Amabile, 1996). It appears that it is important to explicitly instruct participants to select original ideas (Rietzschel, Nijstad & Stroebe, 2010). It is also possible that participants may experience some degree of cognitive overload when faced with selecting the best ideas (Grise, & Gallupe, 2000; Santanen, E. L., Briggs, R. O., & de Vreede, G., 2004). Furthermore, doing so in a collaborative fashion makes it even more difficult. Thus group members may satisfice rather than exert a high level of effort to agree on the best ideas. One solution may be to have people outside the group judge subsets of ideas independently and then average the scores from multiple sources. However, Faure (2004) found that those who generated the ideas were better judges of the quality of the ideas than outsiders. Another approach would be to have participants evaluate the ideas at various points during the ideation process. This would avoid the overload problem. However, it may interrupt the flow of ideas and participants may become prematurely satisfied with the ideas generated. Osborn (1953) emphasized the importance of deferring judgment of ideas during the idea generation phase. This seems to be reasonable advice, but there is not much research that has evaluated this procedure. Nemeth, Personnaz, Personnaz, and Goncalo (2004) found that groups that were encouraged to criticize and debate performed better overall than those that were told not to criticize. Rietzschel et al. (2006) found no difference in the number of ideas generated or their quality between groups who had separate idea generation and evaluation phases versus one that allowed participants to begin the evaluation process whenever they wanted to do so. So the separation of idea generation and evaluation may not be as critical as has been generally assumed. However, both of these studies were done with face-to-face groups and the issue has not been examined with EBS.

Although EBS seems to be an effective means of generating ideas, it appears less well suited for making decisions (Dennis, 1996; Kerr & Murthy, 2004). Decision making processes in which individuals sort through ideas and come to a consensus on the best ideas may be better done in a face to-face-setting. In that setting there is more chance of immediate and richer feedback than in EBS systems. This type of feedback may be important for good decision making. Decision making is typically considered a convergent process since it requires coming to some sort of agreement or consensus. In contrast, idea generation is considered to be a divergent process since it requires the generation of a broad range of ideas that differ from one another and from other participants. In one study of divergent and convergent process in idea generation teams, it was found that teams that communicated with a chat system generated more suggestions than face-to-face teams. However, the chat teams also generated more recommendations (convergent task) that were not relevant to the topic (Kerr & Murthy, 2004). Electronic groups were found to take the same amount of time as the face-to-face groups in decision-making tasks when there was a clear majority preference in the group. Electronic groups took a longer time than face-to-face groups when there was no distinct majority preference (Shen, et al., 2015). Another study on strategic planning compared the use of GDSS with a facilitator with a more conventional approach (Adkins, Burgoon, & Nunamaker, 1995). The facilitated GDSS approach led to higher quality outcomes and reduced the time to generate the plan. However, this study confounded computer support with facilitator support. So it is not clear which factor was responsible for the positive outcomes.

Implications for E-Planning

It appears that EBS may be a useful technique for generation of ideas. We have provided a summary of the major findings and the factors that influence its effectiveness. These findings may provide useful information for those who practice e-planning that involves generation of ideas.

Although we could not find any published studies that have examined the role of EBS in e-planning, a variety of information exchange tools such as instant messaging, group chat and virtual groups provided by EBS systems can be useful in the context of e-planning. In e-planning, there is a complex interaction between information technology, services, stakeholders, policy–makers and citizens. The use of EBS technology could facilitate such interactions. More specifically this paradigm should enhance collaboration among stakeholders to come up with new perspectives and ideas regarding city planning. Research shows that the millennial generation considers computer mediated communication to be more effective in terms of removing the boundaries and enhancing group collaboration (Myers & Sadaghiani, 2010).

A dissertation by Suchocki (1995) examined the reactions of six groups of planners in a land use exercise using Group Systems. There was no systematic assessment of performance or decision-making, but it was reported that the groups perceived that the electronic tools helped organize data, provided some structure and support for equal contributions, and facilitated the decision making process. However, they were not as enthusiastic about the effectiveness of electronic system as their supervisors. One study of EBS in an industrial setting compared individuals with a group of 30 brainstormers generating ideas about organizational problems over a period of four days (Dornburg, Stevens, Hendrickson & Davidson, 2009). It was found that individuals and the group generated a similar number of ideas, but individuals came up with higher quality ideas. Thus contrary to prior research on group size in EBS, the large group did not outperform the individual brainstormers. However, past EBS studies on group size have used paradigms in which group members generate ideas simultaneously in a limited time period (Derosa, et al., 2007). Allowing group members to generate ideas over four days may have made the interaction processes less stimulating. Certainly the pace of idea sharing is slower in an asynchronous paradigm, and some participants may not contribute much until late in the session. So the stimulation provided by other group members' ideas may be less than in a synchronous paradigm. Michinov and Primois (2005) found that asynchronous brainstorming works better when participants are provided periodic feedback that allowed them to compare their performance with one another. Paulus, Larey, Putman, Leggett, and Roland (1996) and Roy et al. (1996) similarly found that performance feedback for groups or individuals enhances the generation of ideas in an EBS. Thus if the e-planning process will continue over a period of time in an asynchronous fashion, periodic feedback on the number of ideas generated by others may motivate participants to be more conscientious in making contributions.

Antunes, Sapateiro, Zurita, and Baloian (2010) conducted an experiment in which they instructed a team of three members to perform an e-planning task. The members were provided with a plan of the municipality of Santiago to merge the area occupied by the Faculty of Engineering of the Universidad de Chile with a park located close by which was separated by an avenue. The task was to generate ideas on how to create a continuous area from the Faculty of Engineering location to the park. The task was divided in to two sessions. In the first session, two of the team members went to the location and generated ideas in their PCs on the given problem. In the second session the two team members joined the third member at the third member's office where the third member used Smartboard to generate ideas while the other two members continued using their laptops. Finally the team collaboratively came up with a solution. The authors noted that the divergent collaboration mode was beneficial especially in the idea generation session. Thus this study demonstrated the importance of divergent tendencies in teams and its impact on e-planning in the context of urban planning. Since urban planning is a complex process requiring a variety of technical and human efforts, it requires the collaborative effort of various stakeholders such as architects, urban designers, city planners and public administrators (Antunes, Pedro, Claudio Sapateiro, Gustavo Zurita, & Nelson Baloian, 2010). Hence one can predict that functional diversity of various stakeholders would contribute towards the generation of creative ideas through the varied experiences and domains of knowledge of the team of stakeholders.

A new body of research suggests the importance of crowdsourcing platform to generate ideas for the development of products and services. Crowdsourcing is a form of online activity in which an undefined group of individuals brainstorm to solve a problem via a flexible open call (Howe 2006). Surowiecki (2004), in his book *The Wisdom of Crowds* claims that under the right circumstances, groups can be smarter than the smartest people in them. The use electronic brainstorming through the platform of crowdsourcing in the context of e-planning could be helpful since the wisdom and ideas of common citizens can be used to solve problems. Crowdsourcing can provide some solutions to urban problems based on available resources and citizen preferences (Anttiroiko, 2012). Some studies report that citizen oriented information systems are considered to be a crucial part of e-planning (e.g., Lenk, 1999). A recent study reported a strong positive relationship between the motivation of the community crowdsourcing users and their active contribution towards solution of problems (de Vreede, de Vreede & Reiter-Palmon, 2017).

Our review suggests that involving a large group of people in the e-planning process may not be a problem. There is some evidence that increasing group size may stimulate more ideas, although the increase in average number of ideas per person is rather small. Moreover, when one is faced with a large pool of ideas, one may not carefully process them. Thus in large groups there may not be as much building on the ideas of others as in smaller groups. For example, in a study of two organizations using EBS, de Vreede, et al. (2010) found that when multiple subgroups (ranging in size from 4-11) built on the work of previous groups, they produced more elaborations. However, groups that did the entire process without input from other groups generated more unique ideas. So the situation is rather complex and further study is needed on the best way to generate unique ideas that also build on the contributions of others (cf. Kohn, et al., 2011).

In applying electronic brainstorming or group decision systems to e-planning, it should be clear that, as with any brainstorming technique, its overall effectiveness will be affected by a number of factors. Some training and the use of facilitators may be beneficial (Cornelius & Boos, 2003). To help structure the ideation process, it is also beneficial if participants are instructed to focus on one aspect of the problem at a time. Providing brief breaks during the process may also be helpful, but it is not clear how long they should be or what kinds of break activities would be most beneficial. Osborn (1953) proposed four rules that he claimed to be beneficial in group ideation process. The rules are do not criticize ideas, freewheeling is welcome, quantity is wanted, combination and improvement are sought. The emphasis on following the standard of Osborn instructions seems to be beneficial (Parnes & Meadow, 1959), although the individual importance of each of the four rules have not been clarified. For example, the rule on freewheeling facilitates divergent thinking whereas the rule on combination promotes convergent thinking (Goldenberg, et al., 2013). Some evidence suggests that the *do not criticize rule* may not be as important as assumed (Ormiston, 2007). However such a rule may help reduce brainstormers' feeling of evaluation apprehension to some extent. Groups often tend to get distracted and generate irrelevant ideas (Ziegler, et al., 2000) or prematurely stop generation of ideas (Paulus, et al., 2006). Therefore adding some additional rules such as "do not tell stories", "do not explain ideas", "keep the brainstorming going", do appear to significantly enhance the number of ideas generated in part because they reduce the sharing of irrelevant material (Paulus, et al., 2006; Putman & Paulus, 2009). Setting goals for groups or providing feedback during the ideation process can also increase idea generation (Paulus & Dzindolet, 1993; Paulus, Larey, Putman, Leggett, & Roland, 1996). Some task relevant diversity in the group may also enhance the number and quality of ideas generated (Paulus & van Der Zee, 2015).

RECOMMENDATIONS FOR E-PLANNING

Based on our review of the literature on collaborative creativity we make the following recommendations for e-planning that is concerned with generating novel ideas:

1. Urban planning is a collaborative outcome of stakeholders from various fields of expertise. In order to facilitate e-planning, sharing and integration of ideas that include visual and spatial information (Wang, Song, Hamilton & Curwell, 2007) from various fields is important. Hence, our recommendation is to include in the group, individuals that represent different areas of expertise relevant to the problem being discussed.
2. If possible use a synchronous process. This may increase the degree of engagement since ideas are being generated in real time and receiving feedback on those ideas may increase the motivation to generate ideas and build on ideas.
3. If one is using a synchronous process, inserting some sessions in which participants generate ideas alone or off-line may allow the group to more effectively tap the stimulation potential of the group sharing process.
4. When using an asynchronous process, it is important for participants to regularly examine the contributions, respond to them and build on them.
5. It may be useful to have participants focus on one aspect of the problem at one time to increase the degree to which each of those are tapped deeply.
6. Brief breaks in the collaborative e-planning process may help participants overcome fixation on a particular train of thought and allow for additional associations to develop.
7. Brainwriting could be another approach to generate ideas for co-located groups. In order to reduce evaluation apprehension, enhance attention and promote cognitive stimulation, one can use a combination of brainwriting and EBS sessions to generate ideas.
8. Use of crowdsourcing in generating innovative ideas in urban-planning context could be a promising approach. Involvement of the community members through electronic media in generating ideas in terms of solving issues related to urban planning where the members generate and build on each others' ideas, provide critical comments and feedback, and vote for the best ideas could enable creation of more practical ideas and decisions.
9. Although there is some evidence that large groups may be more stimulating, such groups may lead to cognitive overload and the failure to attend carefully to the full range of ideas. Thus, it may be better to have planning occur in subsets

of small groups. These groups can then select their own best ideas and share them with the other groups for further elaboration.

10. The process of coming to a collective decision by electronic means can be difficult. One alternative is to have a co-located group discuss the shared ideas, select the best ones, and develop a comprehensive plan based on those ideas.
11. If decision-making needs to do be done electronically, participants should vote for the best ideas. The top ideas should be presented to the group for further discussion and elaboration. At that point each participant can individually suggest a comprehensive plan based on the elaboration and discussion process and share these with the group. The group can then vote on their preferred plan(s)
12. Some degree of facilitation and training for effective collaborative creativity will increase the extent to which group members will function effectively in e-planning. Specialized training to leverage performance of EBS groups with functional diversity may be especially beneficial. It is noteworthy that specialized trainings within specific areas in urban planning do exist, but such specialized training may lack the breadth of skills necessary to address the broader community needs (Duhl & Sanchez, 1999). Duhl and Sanchez reported that since the entities in the urban planning system are interconnected and interdependent, there need to be training programs with an increased emphasis on coalition building or collaboration. In addition, we suggest that training that focuses on collaboration of groups engaged through electronic media will prove to be effective in enhancing e-planning.
13. The use of mobile applications is a common trend because of its user-friendly features. Use of a group decision-making mobile application should foster convenient participation in collaborative brainstorming and decision-making for virtual groups.
14. Since the millennial generation perceives computer-mediated communication to be effective for group interactions (Myers & Sadaghiani, 2010), the use of social-media such as Facebook, Twitter and Instagram to engage the young generation in the process of e-planning could be an effective approach. However, age and familiarity with technology may have an impact on the usability of such paradigm. Gilson, Maynard, Jones Young, Vartiainen, and Hakonen (2015) reported that the variables of age and familiarity with technology have not received enough attention in research. Special training programs targeted towards enhancing the familiarity of such technologies across various age groups could be beneficial.

REFERENCES

Adkins, M., Burgoon, M., & Nunamaker, J. F. Jr. (2003). Using group support systems for strategic planning with the United States Air Force. *Decision Support Systems*, *34*(3), 3, 315–337. doi:10.1016/S0167-9236(02)00124-0

Adla, A., Zarate, P., & Soubie, J. L. (2011). A proposal of toolkit for GDSS facilitators. *Group Decision and Negotiation*, *20*(1), 57–77. doi:10.100710726-010-9204-8

Aiken, M., Vanjani, M., & Paolillo, J. (1996). A comparison of two electronic idea generation techniques. *Information & Management*, *30*(2), 91–99. doi:10.1016/0378-7206(95)00048-8

Amabile, T. M. (1996). *Creativity in context.* Boulder, CO: Westview Press.

Ancona, D. G., Okhuysen, G. A., & Perlow, L. A. (2001). Taking time to integrate temporal research. *Academy of Management Review*, *26*(4), 512–529.

Anttiroiko, A. V. (2012). Urban planning 2.0. *International Journal of E-Planning Research*, *1*(1), 16–30. doi:10.4018/ijepr.2012010103

Antunes, P., Sapateiro, C., Zurita, G., & Baloian, N. (2010). Integrating spatial data and decision models in an e-planning tool. In G. Kolfschoten, T. Herrmann, & S. Lukosch (Eds.), Lecture notes in computer science: Vol. 6257. *Collaboration and technology* (pp. 97–112). Berlin: Springer. doi:10.1007/978-3-642-15714-1_8

Baruah, J., & Paulus, P. B. (2008). Effects of training on idea generation in groups. *Small Group Research*, *39*(5), 523–541. doi:10.1177/1046496408320049

Baruah, J., & Paulus, P. B. (2009). Enhancing creativity in groups: The search for synergy. In M. Neale, B. Mannix, & J. Goncalo (Eds.), Research on managing groups and teams. Oxford, UK: Elsevier Science Press. doi:10.1108/S1534-0856(2009)0000012005

Baruah, J., & Paulus, P. B. (2011). Category assignment and relatedness in the group ideation process. *Journal of Experimental Social Psychology*, *47*(6), 1070–1077. doi:10.1016/j.jesp.2011.04.007

Baruah, J., & Paulus, P. B. (2016). The role of time and category relatedness in electronic brainstorming. *Small Group Research*, *47*(3), 333–342. doi:10.1177/1046496416642296

Bell, S. T., Villado, A. J., Lukasik, M. A., Belau, L., & Briggs, A. (2011). Getting specific about demographic diversity variable and team performance relationships: A meta-analysis. *Journal of Management*, *37*(3), 709–743. doi:10.1177/0149206310365001

Branon, R. F., & Essex, C. (2001). Synchronous and asynchronous communication tools in distance education. *TechTrends*, *45*(1), 36–36. doi:10.1007/BF02763377

Brown, V., Tumeo, M., Larey, T. S., & Paulus, P. B. (1998). Modeling cognitive interactions during group brainstorming. *Small Group Research*, *29*(4), 495–526. doi:10.1177/1046496498294005

Brown, V. R., & Paulus, P. B. (2002). Making group brainstorming more effective: Recommendations from an associative memory perspective. *Current Directions in Psychological Science*, *11*(6), 208–212. doi:10.1111/1467-8721.00202

Camacho, L. M., & Paulus, P. B. (1995). The role of social anxiousness in group brainstorming. *Journal of Personality and Social Psychology*, *68*(6), 1071–1080. doi:10.1037/0022-3514.68.6.1071

Chen, Y., Liu, C., & Wong, R. (2007). The adoption of synchronous and asynchronous media in the teaching of a second language. *Issues in Information Systems*, *3*(1), 217–223.

Cooper, W. H., Gallupe, R. B., Pollard, S., & Cadsby, J. (1998). Some liberating effects of anonymous electronic brainstorming. *Small Group Research*, *29*(2), 147–178. doi:10.1177/1046496498292001

Cornelius, C., & Boos, M. (2003). Enhancing mutual understanding in synchronous computer-mediated communication by training: Trade-offs in judgmental tasks. *Communication Research*, *30*(2), 147–177. doi:10.1177/0093650202250874

Coskun, H. (2000). *The effects of outgroup comparison, social context, intrinsic motivation, and collective identity in brainstorming groups* (Unpublished doctoral dissertation). The University of Texas at Arlington.

Coskun, H., Paulus, P. B., Brown, V., & Sherwood, J. J. (2000). Cognitive stimulation and problem presentation in idea generation groups. *Group Dynamics*, *4*(4), 307–329. doi:10.1037/1089-2699.4.4.307

Coursey, L. E., Kenworthy, J. B., Paulus, P. B., & Doboli, S. (2017). *Diversity and group creativity in an online, asynchronous environment* (Unpublished manuscript). University of Texas at Arlington.

Cramton, C. D. (2001). The mutual knowledge problem and its consequences for dispersed collaboration. *Organization Science*, *12*(3), 346–371. doi:10.1287/orsc.12.3.346.10098

Daily, B., Whatley, A., Ash, S. R., & Steiner, R. L. (1996). The effects of a group decision support system on culturally diverse and culturally homogeneous group decision making. *Information & Management*, *30*(6), 281–289. doi:10.1016/S0378-7206(96)01062-2

de Vreede, G.-J., Briggs, R. O., & Reiter-Palmon, R. (2010). Exploring asynchronous brainstorming in large groups: A field comparison of serial and parallel subgroups. *Human Factors*, *52*(2), 189–202. doi:10.1177/0018720809354748 PMID:20942250

de Vreede, T., de Vreede, G.-J., & Reiter-Palmon, R. (2017, January). Antecedents of engagement in community-based crowdsourcing. *Proceedings of the 50th Hawaii International Conference on System Sciences*. 10.24251/HICSS.2017.027

Dennis, A. R. (1996). Information exchange and use in group decision making: You can lead a group to information, but you can't make it think. *Management Information Systems Quarterly*, *20*(4), 433–457. doi:10.2307/249563

Dennis, A. R., Minus, R. K., & Williams, M. L. (in press). Creativity in computer-mediated virtual roups. In P. B. Paulus & B. A. Nijstad (Eds.), *Handbook of group creativity*. New York: Oxford University Press.

Dennis, A. R., Valacich, J. S., Connolly, T., & Wynne, B. E. (1996). Process structuring in electronic brainstorming. *Information Systems Research*, *7*(2), 268–277. doi:10.1287/isre.7.2.268

Dennis, A. R., & Williams, M. L. (2003). Electronic brainstorming: Theory, research, and future Directions. In P. B. Paulus & B. A. Nijstad (Eds.), *Group creativity: Innovation through collaboration* (pp. 160–180). New York, NY: Oxford University Press. doi:10.1093/acprof:oso/9780195147308.003.0008

Derosa, D. M., Smith, C. L., & Hantula, D. A. (2007). The medium matters: Mining the long-promised merit of group interaction in creative idea generation tasks in a meta-analysis of the electronic group brainstorming literature. *Computers in Human Behavior*, *23*(3), 1549–1581. doi:10.1016/j.chb.2005.07.003

Deuja, A., Kohn, N. W., Paulus, P. B., & Korde, R. M. (2014). Taking a broad perspective before brainstorming. *Group Dynamics*, *18*(3), 222–236. doi:10.1037/gdn0000008

Dickson, G. W., Lee-Partridge, J. E., Limayem, M., & Desanctis, G. L. (1996). Facilitating computer-supported meetings: A cumulative analysis in a multiple-criteria task environment. *Group Decision and Negotiation*, *5*(1), 51–72. doi:10.1007/BF02404176

Diehl, M., & Stroebe, W. (1987). Productivity loss in brainstorming groups: Toward the solution of a riddle. *Journal of Personality and Social Psychology*, *53*(3), 497–509. doi:10.1037/0022-3514.53.3.497

Dornburg, C. C., Stevens, S. M., Hendrickson, S. M., & Davidson, G. S. (2009). Improving extreme-scale problem solving: Assessing electronic brainstorming effectiveness in an industrial setting. *Human Factors*, *51*(4), 519–527. doi:10.1177/0018720809343587 PMID:19899361

Dugosh, K., Paulus, P. B., Roland, E. J., & Yang, H.-C. (2000). Cognitive stimulation in brainstorming. *Journal of Personality and Social Psychology*, *79*(5), 722–735. doi:10.1037/0022-3514.79.5.722 PMID:11079237

Dugosh, K. L., & Paulus, P. B. (2005). Cognitive and social comparison processes in brainstorming. *Journal of Experimental Social Psychology*, *41*(3), 313–320. doi:10.1016/j.jesp.2004.05.009

Duhl, L. J., & Sanchez, A. K. (1999). *Healthy cities and the city planning process: A background document on links between health and urban planning*. Copenhagen, Denmark: World Health Organization.

Dzindolet, M. T., Paulus, P. B., & Glazer, C. (2012). Brainstorming in virtual teams. In C. N. Silva (Ed.), *Online research methods in urban and planning studies: Design and outcome* (pp. 138–156). Hershey, PA: IGI Global. doi:10.4018/978-1-4666-0074-4.ch009

Faure, C. (2004). Beyond brainstorming: Effects of different group procedures on selection of ideas and satisfaction with the process. *The Journal of Creative Behavior*, *38*(1), 13–34. doi:10.1002/j.2162-6057.2004.tb01229.x

Gallupe, R. B., Bastianutti, L. M., & Cooper, W. H. (1991). Unblocking brainstorms. *The Journal of Applied Psychology*, *76*(1), 137–142. doi:10.1037/0021-9010.76.1.137 PMID:2016214

Geschka, H. (1993). The development and assessment of creative thinking techniques: A German perspective. In S. G. Isaksen, M. C. Murdock, R. L. Firestien, & D. J. Treffinger (Eds.), *Nurturing and developing creativity: The emergence of a discipline* (pp. 215–236). Norwood, NJ: Ablex.

Gilson, L. L., Maynard, M. T., Jones Young, N. C., Vartiainen, M., & Hakonen, M. (2015). Virtual teams research. *Journal of Management*, *41*(5), 1313–1337. doi:10.1177/0149206314559946

Goldenberg, O., Larson, J. R. Jr, & Wiley, J. (2013). Goal instructions, response format, and idea generation in groups. *Small Group Research*, *44*(3), 227–256. doi:10.1177/1046496413486701

Goodman, G. (1995). Brainwriting: What to do when there's not a cloud in the brainstorming sky. In J. Heilbrunn (Ed.), *Marketing encyclopedia: Issues and trends shaping the future* (pp. 40–46). Lincolnwood, IL: NTC Business Books.

Grise, M., & Gallupe, R. B. (2000). Information overload: Addressing the productivity paradox in face-to-face electronic meetings. *Journal of Management Information Systems*, *16*(3), 157–185. doi:10.1080/07421222.1999.11518260

Heslin, P. A. (2009). Better than brainstorming? Potential boundary conditions to brainwriting for idea generation in organizations. *Journal of Occupational and Organizational Psychology*, *82*(1), 129–145. doi:10.1348/096317908X285642

Hiltz, S. R., Johnson, K., & Turoff, M. (1991). Group decision support: The effects of designated human leaders and statistical feedback in computerized conferences. *Journal of Management Information Systems*, *8*(2), 81–108. doi:10.1080/07421222.1991.11517922

Hinds, J. M., & Payne, S. J. (2016). Collaborative inhibition and semantic recall: Improving collaboration through computer mediated communication. *Applied Cognitive Psychology*, *30*(4), 554–565. doi:10.1002/acp.3228

Howe, J. (2006). Crowdsourcing: A Definition. In *Crowdsourcing: Tracking the Rise of the Amateur*. Retrieved from http://crowdsourcing.typepad.com/cs/2006/06/crowdsourcing_a.htm

Hülsheger, U. R., Anderson, N., & Salgado, J. F. (2009). Team-level predictors of innovation at work: A comprehensive meta-analysis spanning three decades of research. *The Journal of Applied Psychology*, *94*(5), 1128–1145. doi:10.1037/a0015978 PMID:19702361

Jessup, L. M., Connolly, T., & Galegher, J. (1990). The effects of anonymity on group process in automated group problem solving. *Management Information Systems Quarterly*, *14*(3), 313–321. doi:10.2307/248893

Johnson, G. M. (2006). Synchronous and asynchronous text-based CMC in educational contexts: A review of recent research. *TechTrends*, *50*(4), 46–53. doi:10.100711528-006-0046-9

Johnson, G. M., Howell, A. J., & Code, J. R. (2005). Online discussion and college student learning: Toward a model of influence. *Technology, Pedagogy and Education*, *14*(1), 61–76. doi:10.1080/14759390500200193

Kanawattanachai, P., & Yoo, Y. (2007). The impact of knowledge coordination on virtual team performance over time. *Management Information Systems Quarterly*, *31*(4), 783–808. doi:10.2307/25148820

Kay, G. (1995). Effective meetings through electronic brainstorming. *Journal of Management Development*, *14*(6), 4–25. doi:10.1108/02621719510086147

Kerr, D. S., & Murthy, U. S. (2004). Divergent and convergent idea generation in teams: A comparison of computer-mediated and face-to-face communication. *Group Decision and Negotiation*, *13*(4), 381–399. doi:10.1023/B:GRUP.0000042960.38411.52

Kiesler, S., Siegel, J., & McGuire, T. W. (1984). Social psychological aspects of computer-mediated communication. *The American Psychologist*, *39*(10), 1123–1134. doi:10.1037/0003-066X.39.10.1123

Kohn, N., & Smith, S. M. (2009). Partly versus completely out of your mind: Effects of incubation and distraction on resolving fixation. *The Journal of Creative Behavior*, *43*(2), 102–118. doi:10.1002/j.2162-6057.2009.tb01309.x

Kohn, N. W., Paulus, P. B., & Choi, Y. (2011). Building on the ideas of others: An examination of the idea combination process. *Journal of Experimental Social Psychology*, *47*(3), 554–561. doi:10.1016/j.jesp.2011.01.004

Korde, R., & Paulus, P. B. (2017). Alternating individual and group idea generation: Finding the elusive synergy. *Journal of Experimental Social Psychology*, *70*, 177–190. doi:10.1016/j.jesp.2016.11.002

Kramer, M. W., Fleming, G. P., & Mannis, S. M. (2001). Improving face-to-face brainstorming through modeling and facilitation. *Small Group Research*, *32*(5), 533–557. doi:10.1177/104649640103200502

Larey, T. S., & Paulus, P. B. (1995). Social comparison and goal setting in brainstorming groups. *Journal of Applied Social Psychology*, *26*(18), 1579–1596. doi:10.1111/j.1559-1816.1995.tb02634.x

Larey, T. S., & Paulus, P. B. (1999). Group preference and convergent tendencies in groups: A content analysis of group brainstorming performance. *Creativity Research Journal*, *12*(3), 175–184. doi:10.120715326934crj1203_2

Laurini, R. (2001). *Information systems for urban planning: a hypermedia cooperative approach*. London: Taylor & Francis.

Lenk, K. (1999). Electronic support of citizen participation in the planning process. In B. N. Hague & B. D. Loader (Eds.), *Digital democracy: discourse and decision making in the information age* (pp. 87–95). London: Routledge.

Limayem, M., Banerjee, P., & Ma, L. (2006). Impact of GDSS: Opening the black box. *Decision Support Systems*, *42*(2), 945–957. doi:10.1016/j.dss.2005.08.004

MacDonald, P., Amparo, G.-C., & José Miguel, C.-S. (2013). Computer learner corpora: Analysing interlanguage errors in synchronous and asynchronous communication. *Language Learning & Technology*, *17*(2), 36–56.

Martins, L. L., & Shalley, C. E. (2011). Creativity in virtual work: Effects of demographic differences. *Small Group Research*, *42*(5), 536–561. doi:10.1177/1046496410397382

McLeod, P. L., & Liker, J. K. (1992). Electronic meeting systems: Evidence from a low structure environment. *Information Systems Research*, *3*(3), 195–223. doi:10.1287/isre.3.3.195

Michinov, N. (2012). Is electronic brainstorming or brainwriting the best way to improve creative performance in groups? An overlooked comparison of two idea generation techniques. *Journal of Applied Social Psychology*, *42*, 222–243. doi:10.1111/j.1559-1816.2012.01024.x

Michinov, N., & Primois, C. (2005). Improving productivity and creativity in online groups through social comparison process: New evidence for asynchronous electronic brainstorming. *Computers in Human Behavior*, *21*(1), 11–28. doi:10.1016/j.chb.2004.02.004

Munkes, J., & Diehl, M. (2003). Matching or competition? Performance comparison processes in an idea generation task. *Group Processes & Intergroup Relations*, *6*(3), 305–320. doi:10.1177/13684302030063006

Myers, K. K., & Sadaghiani, K. (2010). Millennials in the workplace: A communication perspective on millennials' organizational relationships and performance. *Journal of Business and Psychology*, *25*(2), 225–238. doi:10.100710869-010-9172-7 PMID:20502509

Nagasundaram, M., & Dennis, A. R. (1993). When a group is not a group: The cognitive foundation of group idea generation. *Small Group Research*, *24*(4), 463–489. doi:10.1177/1046496493244003

Nemeth, C. J., & Nemeth-Brown, B. (2003). Better than individuals? The potential benefits of dissent and diversity for group creativity. In P. B. Paulus & B. A. Nijstad (Eds.), *Group creativity: Innovation through collaboration* (pp. 63–84). New York: Oxford University Press. doi:10.1093/acprof:oso/9780195147308.003.0004

Nemeth, C. J., & Ormiston, M. (2007). Creative idea generation: Harmony versus stimulation. *European Journal of Social Psychology, 37*(3), 524–535. doi:10.1002/ejsp.373

Nemeth, C. J., Personnaz, B., Personnaz, M., & Goncalo, J. A. (2004). The liberating role of conflict in group creativity: A study in two countries. *European Journal of Social Psychology, 34*(4), 365–374. doi:10.1002/ejsp.210

Nijstad, B. A. (2009). *Group performance*. New York: Psychology Press.

Nijstad, B. A., & Stroebe, W. (2006). How the group affects the mind: A cognitive model of idea generation in groups. *Personality and Social Psychology Review, 10*(3), 186–213. doi:10.120715327957pspr1003_1 PMID:16859437

Nijstad, B. A., Stroebe, W., & Lodewijkx, H. F. M. (2002). Cognitive stimulation and interference in groups: Exposure effects in an idea generation task. *Journal of Experimental Social Psychology, 38*(6), 535–544. doi:10.1016/S0022-1031(02)00500-0

Nunamaker, J. F. Jr, Dennis, A. R., Valacich, J. S., Vogel, D. R., & George, J. F. (1991). Electronic meetings to support group work. *Communications of the ACM, 34*(7), 40–61. doi:10.1145/105783.105793

Offner, A. K., Kramer, T. J., & Winter, J. P. (1996). The effects of facilitation, recording, and pauses on group brainstorming. *Small Group Research, 27*(2), 283–298. doi:10.1177/1046496496272005

Osborn, A. F. (1953). *Applied imagination: Principles and procedures of creative thinking*. New York, NY: Scribner.

Oxley, N. L., Dzindolet, M. T., & Paulus, P. B. (1996). The effects of facilitators on the performance of brainstorming groups. *Journal of Social Behavior and Personality, 11*, 633–646.

Parnes, S. J., & Meadow, A. (1959). Effects of" brainstorming" instructions on creative problem solving by trained and untrained subjects. *Journal of Educational Psychology, 50*(4), 171–176. doi:10.1037/h0047223

Paulus, P. B., & Brown, V. (2003). Ideational creativity in groups: Lessons from research on brainstorming. In P. B. Paulus & B. A. Nijstad (Eds.), *Group creativity: Innovation through collaboration* (pp. 110–136). New York, NY: Oxford University Press. doi:10.1093/acprof:oso/9780195147308.003.0006

Paulus, P. B., & Brown, V. R. (2007). Toward more creative and innovative group idea generation: A cognitive-social-motivational perspective of brainstorming. *Social and Personality Psychology Compass, 1*(1), 248–265. doi:10.1111/j.1751-9004.2007.00006.x

Paulus, P. B., & Coskun, H. (2012). Group creativity. In J. M. Levine (Ed.), *Group Processes* (pp. 215–239). Amsterdam: Elsevier.

Paulus, P. B., Dugosh, K. L., Dzindolet, M. T., Coskun, H., & Putman, V. L. (2002). Social and cognitive influences in group brainstorming. Predicting production gains and losses. *European Review of Social Psychology, 12*(1), 299–325. doi:10.1080/14792772143000094

Paulus, P. B., & Dzindolet, M. T. (1993). Social influence processes in group brainstorming. *Journal of Personality and Social Psychology, 64*(4), 575–586. doi:10.1037/0022-3514.64.4.575

Paulus, P. B., & Dzindolet, M. T. (2008). Social influence, creativity and innovation. *Social Influence, 3*(4), 228–247. doi:10.1080/15534510802341082

Paulus, P. B., Dzindolet, M. T., & Kohn, N. W. (2011). Collaborative creativity-Group creativity and team innovation. In M. D. Mumford (Ed.), *Handbook of organizational creativity* (pp. 327–357). New York, NY: Elsevier.

Paulus, P. B., Kohn, N. W., Arditti, L., & Korde, R. (2013). Understanding the group size effect in electronic brainstorming. *Small Group Research, 44*(3), 332–352. doi:10.1177/1046496413479674

Paulus, P. B., Kohn, N. W., & Arditti, L. E. (2011). Effects of quantity and quality instructions on brainstorming. *The Journal of Creative Behavior, 45*(1), 38–46. doi:10.1002/j.2162-6057.2011.tb01083.x

Paulus, P. B., Korde, R. M., Dickson, J. J., Carmeli, A., & Cohen-Meitar, R. (2015). Asynchronous brainstorming in an industrial setting; Exploratory studies. *Human Factors, 57*(6), 1076–1094.

Paulus, P. B., Larey, T. S., Putman, V. L., Leggett, K. L., & Roland, E. J. (1996). Social influence processes in computer brainstorming. *Basic and Applied Social Psychology, 18*(1), 3–4. doi:10.120715324834basp1801_2

Paulus, P. B., Nakui, T., Putman, V. L., & Brown, V. R. (2006). Effects of task instructions and brief breaks on brainstorming. *Group Dynamics*, *10*(3), 206–219. doi:10.1037/1089-2699.10.3.206

Paulus, P. B., & Nijstad, B. A. (Eds.). (2003). *Group creativity: Innovation through collaboration*. New York: Oxford University Press. doi:10.1093/acprof:oso/9780195147308.001.0001

Paulus, P. B., & van der Zee, K. I. (2015). Creative processes in culturally diverse teams. In S. Otten, K. I. van der Zee, & M. Brewer (Eds.), *Towards inclusive organizations: Determinants of successful diversity management at work* (pp. 108–131). New York: Psychology Press.

Paulus, P. B., & Yang, H. C. (2000). Idea generation in groups: A basis for creativity in organizations. *Organizational Behavior and Human Decision Processes*, *82*(1), 76–87. doi:10.1006/obhd.2000.2888

Putman, V. L., & Paulus, P. B. (2009). Brainstorming, brainstorming rules and decision making. *The Journal of Creative Behavior*, *43*(1), 23–39. doi:10.1002/j.2162-6057.2009.tb01304.x

Reiter-Palmon, R., Wigert, B., & de Vreede, T. (2012). Team creativity and innovation: The effect of group composition, social processes, and cognition. In M. D. Mumford (Ed.), *Handbook of organizational creativity* (pp. 295–326). Amsterdam: Elsevier. doi:10.1016/B978-0-12-374714-3.00013-6

Rietzschel, E. F., Nijstad, B. A., & Stroebe, W. (2006). Productivity is not enough: A comparison of interactive and nominal brainstorming groups on idea generation and selection. *Journal of Experimental Social Psychology*, *42*(2), 244–251. doi:10.1016/j.jesp.2005.04.005

Rietzschel, E. F., Nijstad, B. A., & Stroebe, W. (2007). Relative accessibility of domain knowledge and creativity: The effects of knowledge activation on the quantity and originality of generated ideas. *Journal of Experimental Social Psychology*, *43*(6), 933–946. doi:10.1016/j.jesp.2006.10.014

Rietzschel, E. F., Nijstad, B. A., & Stroebe, W. (2010). The selection of creative ideas after individual idea generation: Choosing between creativity and impact. *British Journal of Psychology*, *101*(1), 47–68. doi:10.1348/000712609X414204 PMID:19267959

Roy, M. C., Gauvin, S., & Limayem, M. (1996). Electronic group brainstorming: The role of feedback on productivity. *Small Group Research*, *27*(2), 215–247. doi:10.1177/1046496496272002

Salas, E., DiazGranados, D., Klein, C., Burke, C. S., Stagl, K. C., Goodwin, G. F., & Halpin, S. M. (2008). Does team training improve team performance? A meta-analysis. *Human Factors*, *50*(6), 903–933. doi:10.1518/001872008X375009 PMID:19292013

Santanen, E. L., Briggs, R. O., & de Vreede, G. (2004). Causal relationship in creative problem solving: Comparing facilitation interventions for ideation. *Journal of Management Information Systems*, *20*(4), 167–197. doi:10.1080/07421222.2004.11045783

Scott, G., Leritz, L. E., & Mumford, M. D. (2004). The effectiveness of creativity training: A quantitative review. *Creativity Research Journal*, *16*(4), 361–388. doi:10.1080/10400410409534549

Shen, Z., Lyytinen, K., & Yoo, Y. (2015). Time and information technology in teams: A review of empirical research and future research directions. *European Journal of Information Systems*, *24*(5), 492–518. doi:10.1057/ejis.2014.8

Shepherd, M. M., Briggs, R. O., Reinig, B. A., Yen, J., & Nunamaker, J. F. Jr. (1995). Invoking social comparison to improve electronic brainstorming: Beyond anonymity. *Journal of Management Information Systems*, *12*(3), 155–170. doi:10.1080/07421222.1995.11518095

Shirani, A. I., Tafti, M. H., & Affisco, J. F. (1999). Task and technology fit: A comparison of two technologies for synchronous and asynchronous group communication. *Information & Management*, *36*(3), 139–150. doi:10.1016/S0378-7206(99)00015-4

Siegel, J., Dubrovsky, V., Kiesler, S., & McGuire, T. W. (1986). Group processes in computer-mediated communication. *Organizational Behavior and Human Decision Processes*, *37*(2), 157–187. doi:10.1016/0749-5978(86)90050-6

Smith, S. (2003). The constraining effects of initial ideas. In P. B. Paulus & B. A. Nijstad (Eds.), *Group creativity: Innovation through collaboration* (pp. 15–31). New York: Oxford University Press. doi:10.1093/acprof:oso/9780195147308.003.0002

Stasser, G., Abele, S. P., & Parsons, V. S. (2012). Information flow and influence in collective choice. *Group Processes & Intergroup Relations*, *15*(5), 619–635. doi:10.1177/1368430212453631

Suchocki, S. S. (1995). The impact of electronic tools for support of groups engaged in land use planning. *Dissertation Abstracts International Section A: Humanities and Social Sciences*, *56*(3-A), 0742

Surowiecky, J. M. (2004). *The wisdom of crowds*. New York: Random House.

van Dijk, H., van Engen, M. L., & van Knippenberg, D. (2012). Defying conventional wisdom: A meta-analytical examination of the differences between demographic and job-related diversity relationships with performance. *Organizational Behavior and Human Decision Processes*, *119*(1), 38–53. doi:10.1016/j.obhdp.2012.06.003

van Knippenberg, D., De Dreu, C. K. W., & Homan, A. C. (2004). Work group diversity and group performance: An integrative model and research agenda. *The Journal of Applied Psychology*, *89*(6), 1008–1022. doi:10.1037/0021-9010.89.6.1008 PMID:15584838

VanGundy, A. B. (1981). *Techniques of structured problem solving*. New York: Van. Nostrand–Reinhold.

Wang, H., Song, Y., Hamilton, A., & Curwell, S. (2007). Urban information integration for advanced e-planning in Europe. *Government Information Quarterly*, *24*(4), 736–754. doi:10.1016/j.giq.2007.04.002

Ziegler, R., Diehl, M., & Zijlstra, G. (2000). Idea production in nominal and virtual groups: Does computer mediated communication improve group brainstorming? *Group Processes & Intergroup Relations*, *3*(2), 141–158. doi:10.1177/1368430200032003

Chapter 8
SketchBoard:
Design and Evaluation of Interactive Tools for the Design Review Process

Rojin Vishkaie
Ball State University, USA

Richard Levy
University of Calgary, Canada

ABSTRACT

This chapter seeks to investigate the subject of the design review process and its embedment in digital media, with a primary focus on mobile interactive surfaces. Despite the growing body of academic research on the topic, there is a gap in evaluating how aspects of the design review process can be performed with digital media. The main point of this study is to combine empirical and conceptual design components to evaluate a new communication medium called SketchBoard that uses interactive surfaces to perform selected tasks of the design review process. This study specifically contributes to the state-of-the-art of visualization, communication, and participation aspects of the design review process and mobile interactive surfaces.

1. INTRODUCTION

The zoning ordinance is a prime text-based source of information on land use and district regulations. It is regularly referred to and used by urban planners, property owners, developers, attorneys, local governing bodies, and members of zoning appeals boards. However, there is a lack of comprehensibility of the technical information and concepts on the zoning ordinance document that causes frustration and bewilderment

DOI: 10.4018/978-1-5225-5999-3.ch008

for urban planning stakeholders. It is also uncommon to have accompanying maps, graphics, and diagrams that support the readability and intelligibility of the zoning ordinance document. Thus, computing and visualizing Floor Area Ratio (FAR) are difficult activities when referring to the text-based document. FAR refers to the ratio of a building's total floor area compared to the size of a building's parcel, as shown in Figure 1. In addition, setbacks that are specified under a zoning ordinance can impose constraints which can make it impossible to achieve the permitted FAR. Even though these definitions are somewhat clear, without illustrations they are cumbersome to comprehend (Vishkaie, 2014; Kaufman, 1963).

Furthermore, the work conducted within the design review process is complex because it requires enhancements to the participatory communication aspects; participatory communication refers to creating shared understanding, perception, and knowledge between urban planning stakeholders within a collective decision-making environment. Moreover, participatory communication within the design review process should empower the voice and visibility of urban planning stakeholders through simple visualization, thus discovering solutions to their complex design problems (Vishkaie, 2014; Tufte & Mefalopulos, 2009).

However, these various aspects of the design review process significantly rely on conventional tools such as paper, pen, pencil, tape measure, large binders, multiple stacks of bylaws, and so on, as well as desktop computer applications such as Computer Aided Design (CAD), Planning Support System (PSS), and Geographic Information System (GIS) software. Yet, these existing tools and technologies used in the design review process lack simplicity, usefulness, and mobility (Vishkaie, 2014).

Although, mobile, small interactive surfaces are proliferating, there remains a gap in determining whether computer aided planning can fully capitalize on the benefits of interface and interaction design of such surfaces. Specifically, there is an absence of empirical research studies to identify how urban planning stakeholders can benefit from using mobile, small interactive surfaces within the design review process. Yet, by characterizing the contemporary trend of interactive surfaces, perhaps as one solution amongst many, this exploratory study speculates that integration of a mobile, small interactive surface such as a tablet with SketchBoard could potentially enhance the participatory communication and visualization needs within the design review process (Vishkaie, 2014).

The goal of this exploratory study was to design and evaluate SketchBoard, a communication medium that uses interactive surfaces to perform selected tasks for the design review process. This study explored how technology can be used to support communication, participation, visualization, and analysis throughout the design review process. As part of the user-centered design process, the primary goal of this exploratory study was to create a minimal interface and interaction design for SketchBoard that will allow urban planners with limited computer skills to take part

Figure 1. Floor Area Ratio (FAR)

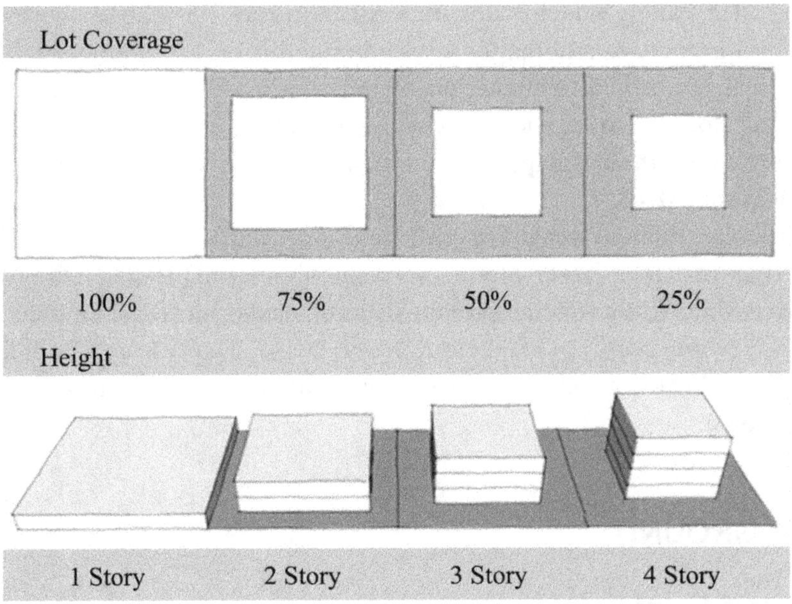

in a co-design process with other urban planners and the researcher. In the absence of such an interactive communication medium, a digital prototype was created to help test its potential use within the design review process. The interface and interaction design of SketchBoard were designed using an interactive storyboard technique on a tablet (iPad). In this exploratory study, the primary data collection techniques relied on semi-structured interviews together with demonstrations of the digital prototype (Vishkaie, 2014; Rogers, Y., Sharp, H., & Preece, J., 2011; Warfel, 2009).

This chapter was emanated from a PhD thesis entitled *SketchBoard: Conceptualizing Interactive Communication Media for the Design Review Process,* published by this chapter's first author. The PhD thesis consists of a four-part empirical and design research study (Studies I-IV) that investigated the communication between urban planners and mobile interactive surfaces for the participatory production of mutual intelligibility within the design review process. These studies were all conducted by this chapter's first author in Calgary, Canada and included participants from the United States, Canada, and most European countries. Ultimately, the PhD thesis focuses on informing the design specifications for creating SketchBoard, an interactive communication medium that is participatory, procedural, spatial, and encyclopedic. Participatory affordance that represents shared dialogue with others, and procedural affordance that represents rule-based behavior assist in making SketchBoard interactive. Spatial affordance that represents navigable space, and encyclopedic affordance that represents logical, symbolic representations of a process assist in making SketchBoard immersive (Murray, 1997; 2012). Furthermore, SketchBoard introduces a ubiquitous, pseudo intelligent, and geo-located method for 3D visualization and computation of the zoning ordinance, specifically the FAR. SketchBoard also provides a vision for simple, intuitive, and intelligible interactions within the participatory communication around the future shaping of cities (Vishkaie, 2014).

In particular, this chapter draws on Pilot Study and Focus Group which were conducted as part of a larger Study IV of the above PhD thesis. Study IV also consisted of Main Study which is previously published in an article entitled *Making Design Review Interactive,* (Vishkaie & Levy, 2014). The research findings from Pilot Study, Main Study, and Focus Group contributed to the creation and evaluation of SketchBoard.

2. BACKGROUND

2.1 Urban Planning Practice

Urban planning practice involves considerable technical and communicative activities. Technical actions include architectural and aesthetic techniques and design, as well as drawing, computing, calculating, making plans, interpreting, and developing policy, while, communicative actions involve communicating with the public, granting permissions, and administrating guidelines and regulations (Brail & Klosterman, 2001). As negotiators, urban planners must work closely with their respective city councils, mayors, and the developers. Their work often requires them to submit proposals on behalf of the communities they represent. The review of urban development plans and proposals is a political process that requires participation and communication amongst all stakeholders. During this process or reviews, building proposals are analyzed, negotiated, investigated, and delayed. Oftentimes proposals are sent back to the developer with requested changes. In many cases, the decisions of urban planners must deem issues of aesthetics, politics, law, and design. Even the most straightforward conversations can raise complex issues for assessment (Forester, 1989; Vishkaie, 2014).

However, the technical character of urban planning actions can be difficult for the layperson to comprehend. This can entrench an intellectual divide between urban planners and non-urban planners on a range of issues. Limited public access to information can also increase misinformation. A reliance on sophisticated technical tools such as CAD, PSS, and GIS also contributes to this divide between urban planners and the public. However, developing shared understanding of the issues and creative problem-solving through communication and participation amongst stakeholders is key to the practice of urban planning. Therefore, attention must be made to the strategic shaping and planning of communicative action and the building of a cooperative working relationship in order for the developers, the community, and bureaucrats to trust the technical information (Forester, 1989; 1987; Vishkaie, 2014).

2.2 Urban Planning Review

The urban planning review process involves the parallel activities of design and decision. In the process of reviewing a plan, alternative solutions are often created. A plan can change in response to government regulations or the need to generate a specific level of profit. At each stage in the design process, experts need to consult with the public. Ideally, the end product of this participatory process should be a plan that respects the aesthetics and socio-economic-political context of a proposed development project (Brail & Klosterman, 2001). Furthermore, the urban planning review process must resolve conflicts which arise out of a diversity of opinions among stakeholders. Conflicting demands in urban planning review mainly arise when private profit and the public good clash. Depending on the political situation surrounding the design review process, one of the parties involved in the political endeavor wins or loses (Forester, 1989; Vishkaie, 2014).

Design Review Scenario and Persona

To provide concrete context, a typical design review scenario from the perspective of an urban planner is described. This scenario incorporates three concrete elements of communication and participation around technical activities that is based on the researcher's day-to-day engagement with the urban planners in a city planning department's offices. A precise description of pretend users who are fictional characters and what they want to accomplish is provided, and the design concept is developed for them. The cast of characters including Don, a developer; Pat, an urban planner; and Sydney, a member of the public who is also the primary persona, are created to narrow down the spectrum of users for a design (Murray, 1997; Cooper, 2004; Vishkaie, 2014).

Idea Generation

Don is a developer who plans to apply for a permit for a multi-building, multi-family development. He requests a three-hour pre-application workshop with Pat, an urban planner who works for the city. At their first meeting, a number of design concepts are discussed. Pat provides feedback, anticipating issues that Don may encounter within the design review process. Together they work on potential design solutions. This design charrette session involves the use of paper and pencil to draft and discuss possibilities. Because Don is an experienced developer, he does not need information about the design review process, however, Pat would provide this for a less experienced developer (Vishkaie, 2014).

Application Refinement

Following this session, Don refines the design and submits the initial development application to the urban planning department as a large set of physical and electronic documents. Pat shepherds the application through the city's design review process. With his team, Pat reviews the development application by cross-referencing the design against bylaw and policy requirements. Through several iterations, Pat identifies details of Don's design that need to be clarified or may need modification. Some of these changes could conflict with Don's original plans, and he will need to refine several features of his design. Often, this involves intense phone conversations with Pat; other times, they meet to discuss the city's concerns, and how best to resolve this issue. Sometimes this process involves marking up printed plans with markers and pens. Together, Pat and Don meet many times to negotiate, identifying opportunities for exceptions on the formal building code to address the city's concerns. Don's project is not too far out of the ordinary, but communications continue for well over three years. Once communications wrap up, and changes are satisfactory to the city, as well as Don, and he is granted a provisional development permit (Vishkaie, 2014).

Public Consultation

Sydney is a civic-minded citizen who attends every event at his local community center. One of these is a Public Consultation session run by Pat. Here, Pat gives an electronic slide-based presentation to Sydney about the new proposed development, perhaps with a scale model provided by Don. This discussion is aimed at understanding the impact of the development on the community. Sydney, one of the few members of the public at the session, feels very strongly against the plan, as it will impede his view of the park. He voices his concerns to Polly who is an elected official. If these concerns are taken into account, the process may bounce back to the Application Refinement phase. When they are finally resolved or ignored, if the political consequences are not major, Pat releases the final building permit to Don (Vishkaie, 2014).

2.3 Urban Planning Technology

There is a dichotomy between visual and verbal cues of urban planning Computer Mediated Communication (CMC) versus urban planning face-to-face communication. Pattern of user interaction is largely dictated by the urban planning technology rather than diversity and creativity of the users. Traditionally, urban planning technologies include substantial use of paper-based plans, maps, and text documents. Currently, urban planners rely on digital technologies, which lack mobility. Because they require specialized software, urban planners tend to stay in their offices to use software, such as CAD and GIS. Further, urban planners who cannot leave their offices require visualization aids for informational and communicational needs, which allow them to conceptualize the potential impact of new building projects. In addition to visualization, urban planning decisions are made in a participatory fashion, another function that technology should support for participatory communication (Brail & Klosterman, 2001; Flew, 2005; Short, J., Williams, E., & Christie, B., 1976; Vishkaie, 2014).

Everyday life of the design review work is shaped by corporate and bureaucratic systems where contemporary capitalist practice entails inequality, differences, and power relationships. Optimistically, participation technology for urban planning communication could potentially re-shape social and cultural differences evident in the structure of the design review process. Technologies that support participation comprise an increasingly important part of the Information Communication Technology (ICT) infrastructure in organizations, including process improvement and teamwork. Electronic participation (e-participation) that encompasses computer mediated collaborative work refers to use of digital media among different individuals to accomplish a common task. E-participation is increasingly being used in industry to promote collaborative activities, tools, and functionalities embedded in large enterprise systems. Implementation of e-participation exercise in the context of urban planning technology could move towards enhancing the participatory decision-making process in democratic participation. Accordingly, in automotive industry, the application of e-participation is used for effective coordination of product development in real-time. 2D drawings are not easy to understand, but automotive industry still relies heavily on them to access technical product information. By trading 2D drawings for 3D visualization, the automotive industry uses technological substitutions for traditional exchange of information to replace face-to-face participation and paper-based product development with e-participation (Flew, 2005; Short, J., Williams, E., & Christie, B., 1976; O'Doherty & Einsiedel, 2012; Kock, 2005; Kock & Nosek, 2005; Lefebvre, E., Lefebvre, L. A., Hen, G. L., & Mendgen, R., 2006; Vishkaie, 2014).

2.4 Adaptation of Digital Media in the Design Review Process

There is now a wide array of immobilized, single-user digital media that could support information and visualization needs of the design review process (Brail & Klosterman, 2001). However, mobility is fundamental for visual perception (Gibson, 1986). Specifically, the design review process requires mobility that supports spontaneous interaction and participation, exchange of useful information, local communication, and mutual awareness. Technology for this process should support self-explanatory operations without significant training through mutual intelligibility of individuals and technological interactions (Suchman, 2007; Norman, 2010). Yet, these aspects of the design review process are not supported by exiting technology that is tied to desktop computers and afford explicit communication (Bellotti & Bly, 1996; Luff & Heath, 1998). There is, an absence of intelligent and intelligible 3D visualization media that allows participatory communication and ubiquitous, interactive behavior. Having such media could reduce the time needed to examine development proposals within the design review process. This media may also result in the preparation of development proposals that does not require knowledge of CAD, PSS, and GIS. Furthermore, the proliferation of interactive surfaces and their chameleon-like quality have pervaded both the public and work environments, and in education and entertainment, however, the role of representational affordances and properties provided by such surfaces within the design review process is still largely unexplored (Rogers & Lindley, 2004; Murray, 2012; Vishkaie, 2014).

Furthermore, the rapid evolution of ICT has influenced the socio-economic-political structure of the design review process through decentralized coordination of activities as well as reshaping the cultural technologies of this process. In particular, the availability of information through constant mobile communication has transformed the way individuals interact, communicate, and participate. Moreover, the participation paradigm in the design review process assumes that a technology for group work should allow the urban planning stakeholders to actively participate and engage in the design process. The integration of digital media and ICT with the emergence of groupware into the structure of the design review process could increase the opportunity for participation. E-participation in conjunction with ICT can provide tools for engagement via delivering better quality information to geographically dispersed locations (O'Doherty & Einsiedel, 2012; Hanzl, 2007; Klabbers, 1999; Shiode, 2000; Evans-Cowley, 2010).

More advanced technological support should be given to support an achievement in essential tasks of the design review process such as visualization and communication of plans for future cities. This could also promote an efficient and a shared environment for the urban planning stakeholders. Yet, the use of conventional urban planning tools and technology is not likely to be suppressed. In addition, cultural technologies for the design review process where media, context, and power relations are assembled has not been matched to potential forms of education and training necessary to acquire socio-economic-political knowledge and skills to make use of the digital media. Thus, it is not yet evident whether digital media in urban planning can evolve enough to become a mainstream technology that could potentially take advantage of advances in computing technology (Norman, 2010; Brail & Klosterman, 2001; Klabbers, 1999; Gaver, 1996; Norman, 1993).

3. SCOPE OF THE STUDY

The objective of this exploratory study was to design and evaluate SketchBoard for visualizing the zoning ordinance. This study further provided an understanding of the participatory, communicative, and technical activities around the computation and visualization of the permitted FAR (Vishkaie, 2014).

3.1 Method and Study Design

Integrating ethnography and contextual design as empirical approaches and minimalism as a design technique framed the overall methodology of research in this exploratory study. Cognitive ethnography uses a multiple data collection and analysis method to provide an understanding of the interaction among urban planners, tools and technologies, as both urban planner and tools and technologies are culturally immersed and socially distributed in the environment. Contextual design was used to help urban planners crystallize and articulate their work experiences. Contextual design was combined with the cognitive ethnography technique to develop an understanding about the work practices of urban planners. This understanding formed a basis for developing SketchBoard to support urban planners' work within the design review process. Minimalism, as a design technique employed in this study, aims to accommodate the design review's workflow process (Vishkaie, 2014; Hollan, J., Hutchins, E., & Kirsh, D., 2000; Hutchins, 2001).

This exploratory study consisted of two phases specifically a Pilot Study and a Focus Group, each of which was conducted in two steps. Step one involved the demonstration of SketchBoard. Step two consisted of the semi-structured interviews. In the first step, participants were given a walk-through of SketchBoard, then asked to interact with SketchBoard. In the second step, participants had to explain how they thought SketchBoard could potentially support the participatory communicative and technical activities within the design review process, and how SketchBoard could be improved. Both steps - the demonstration of digital prototypes and interviews

- lasted approximately for half an hour for each participant. The demonstration of SketchBoard and interviews were entirely anonymous (Vishkaie, 2014).

Furthermore, to define the sample size, this study followed the general guideline in qualitative research. The intent was to focus on a few individuals and elucidate particular and specific aspects, events, experiences, and incidents. This allowed for the collection of extensive detail which deepened the researcher's understanding of each participant, their process of work, and the artifacts used in the course of their work. Moreover, to decrease inevitable biases generated by the researcher, participants' original quotes are also included in the data analysis sections (Vishkaie, 2014; Beyer & Holtzblatt, 1998; Creswell, 2007; Crabtree, 2003; Obendorf, 2009; Murray, 2012; Rogers, Y., Sharp, H., & Preece, J., 2011).

3.2 Pilot Study and Focus Group

Initially, a Pilot Study was conducted as a small-scale exploration designed to gather information prior to a larger study and to evaluate the feasibility of the Focus Group. The results of the Pilot Study were useful in improving the quality and efficiency of the Focus Group by revealing the deficiencies in the design of SketchBoard, and useful information was gathered for the later iteration of SketchBoard demonstrated for the Focus Group. Furthermore, in the Pilot Study, a total of five participants were recruited by using a purposeful sampling technique, and via an email distribution list sent to urban planning and urban design students at the University of Calgary. Participants were given a demo of SketchBoard and interviews sessions in which they were questioned. Participants were interviewed individually; in addition, they had the freedom to explain their background, existing knowledge, and prior experience (Vishkaie, 2014; Creswell, 2007; Rogers, Y., Sharp, H., & Preece, J., 2011).

In the Focus Group, a total of five participants were recruited by using a purposeful sampling technique, and an email distribution list that reached urban planners, urban designers, and architects in land use planning and policy; development and building infrastructure; and infrastructure and information services at the City of Calgary. Participants for the Focus Group had between 15 to 34 years of experience in their professions as senior architects, urban planners, and urban designers, as shown in Table 1. The sampling strategy employed in this exploratory study was meant to discover extensive details about specific contexts, and as a result, even a small sample size will generate a large quantity of data. However, while participants who are experts in the field were purposefully selected to fulfill informational needs, the small sample size for this study may not meet the criteria for statistical representation.

Ideally, conducting a survey involving a larger sample size would certainly improve the comprehensiveness of the Focus Group. However, as an initial study, given the strict means, the Focus Group needed to be limited to one city (Calgary). Although,

Table 1. Participants' Biographical Information

Participant	Position	Years of Experience
P1	Policy Maker: Land Use Planning & Policy	34
P2	Coordinator Architect: Building Infrastructure	16
P3	Director: Infrastructure & Information Services	22
P4	Senior Planner: Land Use Planning & Policy	15
P5	Senior Coordinator Architect: Urban Design & Heritage	24

the process, procedures, and context will change from city to another city, the base issues faced by planners remains largely the same in both mid-size cities and larger cities. The Focus Group can be considered as a precursor to a larger study if there is interest in further development of the application.

Focusing on uncontrolled conditions, personal characteristics, and events uncovered the uniqueness of the scenarios in this study, which is novel and differs in context from other situations. Generalization generally modulates context-free, deterministic values with the aim of prediction and control, which was not intended in this study (Vishkaie, 2014; Creswell, 2007; Rogers, Y., Sharp, H., & Preece, J., 2011). Furthermore, the discussion in the Focus Group was led by the researcher - author of the above PhD thesis. The Focus Group was held to identify potential conflicts in terminology arising out of the communicative and technical activities or expectations from different individuals within the land use planning and policy department at the City of Calgary. To enable participants to put forward their own opinions in a supportive environment, they were allowed to give their opinions and discuss their ideas about SketchBoard with the other participants in the Focus Group (Vishkaie, 2014; Creswell, 2007; Rogers, Y., Sharp, H., & Preece, J., 2011).

The first step in Pilot Study and Focus Group was the digital prototyping that was used to demonstrate the potentials of SketchBoard which could assist in the design review process. The design was used to investigate how well SketchBoard could potentially support the goals of this study. Participants were asked to navigate through SketchBoard by interacting with the digital prototype. During the course of the demonstrations, they provided feedback on the flow of interaction and interaction functionality. Audio recording was employed as a data gathering technique (Vishkaie, 2014).

The second step in Pilot Study and Focus Group was a semi-structured interview with the purpose of engaging participants in a conversation. The purpose of this interview was to gain impressions about how urban planners might react to SketchBoard. This interview included four open-ended and evolving questions (Vishkaie, 2014).

Specifically, SketchBoard targeted the design charrette that was part of the early phase of the design review process. SketchBoard did not aspire to replace modelers targeting photorealistic rendering, but to be a visualization and sketching tool for the design review process. Furthermore, SketchBoard supports the visualization of the zoning ordinance and the calculation of FAR by manipulating different dimensions on a 2D plan. SketchBoard can be further enhanced by automatically defining the geo-location of the parcel, illustrating the setbacks associated with the parcel, providing a number-based dialogue for manipulating dimensions such as setbacks, heights, and number of floors, and then creating a three-dimensional building that represents the proposed building on the prospective parcel (Vishkaie, 2014).

In particular, design specifications were defined to create a minimal interface and interaction design for SketchBoard that were unobtrusive and differential to the current tools and technologies for the design review process. The interface and interaction design complexity of SketchBoard were functionally reduced, mainly because it only offers a minimal number of tools that focus on simple sketching and visualization functions; also it does not include complex features of CAD or GIS interface. The interface and interaction design of SketchBoard are more radically tailored to the notion of functional minimalism that concentrates on the requirements of individual functionalities such as a visualization of the zoning ordinance and computation of FAR (Vishkaie, 2014; Warfel, 2009; Obendorf, 2009; Murray, 2012).

The SketchBoard interface and interaction design were created to be easy-to-use, and especially could be promoted to urban planners, and urban planning stakeholders who are novices and interested in visualization and calculation speed and variability. Architectural minimalism in the constituents of SketchBoard denotes that the interface and interaction design should be modularized. This can be attained through the development of a gird system that allows for the achievement of a much more harmonious relationship between individual elements on the interface and interaction design through coherence, asymmetrical balance, alignment, and distinct dimensions. The interface and interaction design of SketchBoard also attempt to accentuate its visual characteristics through designing all the icons, refining typography, and incorporating a whole new palette of colors (Vishkaie, 2014; Warfel, 2009; Obendorf, 2009; Murray, 2012).

The interface and interaction design of SketchBoard tend to attain a minimal access structure for the user. This can be accomplished by using distinct and functional buttons and menus that help establish hierarchy and order and increase a sense of simplicity. Structural minimalism can also be achieved through representing the important features as a focal point, as well as reducing unnecessary elements and contents that are less noticeable and cause distraction and cognitive load for the user. Moreover, the SketchBoard interface and interaction design apply a compositional minimalism approach to the situationally intelligent features associated with the interactivity and mobility of the communication and participation needs within the design review process (Vishkaie, 2014; Warfel, 2009; Obendorf, 2009; Murray, 2012). However, in a real world practical sense, it is still to be concluded if the features of SketchBoard can realistically operate as projected in the following design specifications: pseudo visualization and computation of the zoning ordinance and FAR; pseudo three-dimensional, intelligent, and geo-located interaction; mobile, participatory, intelligible, simple, and intuitive interaction; as well as minimalist aesthetics (Vishkaie, 2014). Figure 2 shows a storyboard that consists of six illustrations of SketchBoard.

In the Pilot Study and Focus Group, extensive, manual note-taking and audio recording were used to document interviews. The interview notes and audio recordings were written up and transcribed verbatim. Moreover, the affinity diagraming technique was then used to analyze the data (Vishkaie, 2014; Beyer & Holtzblatt, 1998; Creswell, 2007; Crabtree, 2003; Rogers, Y., Sharp, H., and Preece, J., 2011). It organized ideas and statements gathered during the interviews into four categories: SketchBoard and Technical Activities; SketchBoard and Communicative Activities; SketchBoard and Education; and SketchBoard Improvement. For each category appropriate subcategories were ascribed during the detailed content analysis.

4. FINDINGS

The next sections, the analysis of the interviews within two phases (Pilot Study and Focus Group) are presented.

4.1 Interviews for Pilot Study

SketchBoard and Technical Activities

All participants mentioned that SketchBoard supports the visualization of the zoning ordinance that explains why it would be difficult to achieve the maximum FAR on any particular site. These opinions are discussed in the section below.

Comprehension

Participants believed that whenever zoning regulations pose restrictions to achieve the maximum FAR on a site, SketchBoard provides a great degree of comprehension of the technical concepts. Participants saw value in SketchBoard because it supports the

Figure 2. SketchBoard

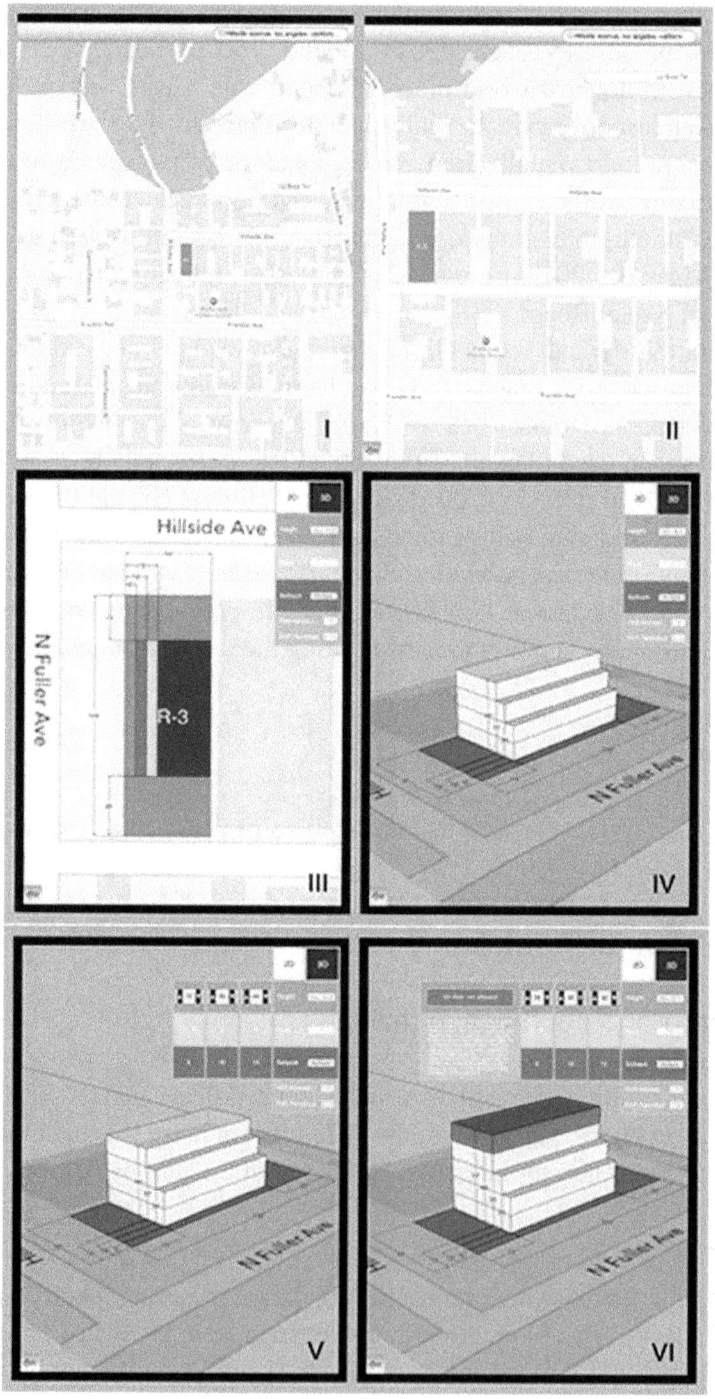

visualization of the zoning ordinance and computation of the FAR in a comprehensible format. Participants said that, on a positive note, SketchBoard offers the opportunity to change parameters, such as floor height, also to locate the blocks and parcels geographically. Furthermore, being able to access the exact height, setbacks, number of floor, and permitted and allowed FAR, also to visualize the impact on developable area were stated as being SketchBoard's forte. Participants also mentioned it would be useful to explore potential variances that might be granted to the developers.

Visual Cues

Participants stated that the use of red to color restrictions and limitations was unclear. Participants said that the use of a 3D model, graphics, and visual forms makes SketchBoard very straightforward and easy to understand. Participants further elaborated that representing the height limits through the use of 3D modeling might propose a good option. As three participants noted,

- ***(P3):*** *[SketchBoard] gives a really good graphic and puts it into an exact spatial reference, so in that sense gives the viewer a more accurate description rather than simply telling them or showing blueprints even.*
- ***(P4):*** *[SketchBoard] is very easy to use. It is visually appealing. It is straightforward and quite simple. The blocks are simple. It is kind of like Google SketchUp almost. I like that it has all the setbacks, which is important as well. I really like it.*
- ***(P5):*** *[SketchBoard] obviously shows that you can or cannot go over a certain limit, and the 3D model is probably the best way to represent that. You can visibly see it. It does not make sense if you sometimes put it into words. So I thought that was pretty good.*

SketchBoard and Communicative Activities

All participants believed that SketchBoard would be helpful in communication between urban planners and developers, and some of these opinions are detailed in the section below.

Intelligibility and Readability

Participants thought that the information in urban planning, particularly the traditional zoning concept is communicated through a complex paper and text-based document that is not clear, intelligible, or readable. However, participants said that SketchBoard explicitly visualizes the required information using graphics rather than text documents. They also stated that SketchBoard is more readable, comprehensible, and easy to use while negotiating technical contents. Some of the participants also mentioned that SketchBoard could be useful while negotiating plans with the public. As three participants said,

- ***(P2):*** *It is always nice to have visual information instead of written, but most of the time this information is conveyed through text. So, showing it digitally and visually is good.*
- ***(P4):*** *[SketchBoard] helps both sides to better understand the impact of the particular tower, for instance. If there was a requirement that all applications must have a model that fits, that would allow the city to input this into their system, that would really help both sides when coming up with a more comprehensive plan rather than just looking side by side; this is more cohesive with the whole community.*
- ***(P5):*** *[SketchBoard] could help to get the information across more easily to the developers of the building. You can visibly see the information.*

SketchBoard and Education

All participants agreed that SketchBoard could be used to help educate the public on how FAR is considered by the zoning ordinance, and certain specific comments are elaborated in the section below.

Public Education and Spatial Thinking

Participants said that although urban planning involves spatial thinking, urban planning concepts are mostly conveyed through paper and text-based documents. However, SketchBoard offers a very simple and visual way to represent spatial and mathematical concepts such as FAR. Furthermore, the use of tablet versus desktop computers makes the information more accessible to anyone. As two participants said,

- ***(P2):*** *Urban planning is spatial. We are talking about spaces. And spaces can sometimes be hard to describe, so the best way to represent the space is spatially with graphics.*
- ***(P4):*** *[SketchBoard] helps to breakdown the FAR which is a bit of a hard concept to understand. There is a bit of a math to it and most laymen and the public will not really get into that. So, this makes it easy if they can [visualize this] in open houses and it would be particularly useful.*

SketchBoard and Improvement

Participants particularly expressed their opinion on how to improve SketchBoard, and some of these opinions are emphasized in the section below.

Various Zoning Ordinances

All participants mentioned that SketchBoard could be improved by including all zoning ordinances, uncommon zoning ordinances, irregularly shaped buildings and different building types into the working version of SketchBoard.

Graphics

Most participants said that the red color is the only indication that shows the restriction on the extra floor. They suggested that it would be useful to provide more information indicating what is/is not allowed. As two participants noted,

- ***(P1):*** *It would be helpful if there is a sign saying that this floor is not permitted.*
- ***(P2):*** *If there is a little information box that I could click and it would give me like a short paragraph explaining what is allowed, that would help me.*

Modifiable Variables

One participant indicated that it would be useful to allow the user to change the FAR. This would allow the urban planner to see the impact of increasing or decreasing the FAR on the built form. This participant also mentioned that it would be useful to have access to statistics on census data and building permit data. As a participant observed,

- ***(P4):*** *If you could actually drop down the FAR as well; currently, you can only increase it, but if you could click down and show if it was only one story, what the FAR would be or if it was two stories or three stories, what the FAR would be, and then change and break down some statistics on the side, that might be useful.*

4.2 Interviews for Focus Group

SketchBoard and Technical Activities

All participants believed that SketchBoard helps to visualize why zoning would make it difficult to achieve the maximum FAR on a site. These opinions are noted in the section below.

Land Use Designations

One participant mentioned that SketchBoard is very helpful because it presents four key variables from the land use district and designation such as FAR, height, setbacks, and floor height. However, this participant suggested that floor height, setbacks, and FAR variables should be modifiable. This would allow the developers to input their proposed variables and to create their own development proposals. The urban planners can also input the allowed variables and create what is allowed on the site. Then, the proposed and allowed development application can be compared, hence this would help to envision the differences between what a developer wants to achieve and what the bylaws allow. This participant said,

- ***(P2):*** *Urban planners know applicants do not like what the rules say, but with [SketchBoard] urban planners can ask applicants show us what they would like.*

One participant mentioned that SketchBoard shows only one individual site, however rules are more complex when there are multiple compounds of developments. All participants also believed that SketchBoard could be used for calculating the when a proposal exceeds the allowed FAR. As a participant said,

- ***(P5):*** *Depending on what variables are present, it can give an idea of the allowable, but may be tricky when broken down into various forms. The City of Calgary has many more complex variables.*

Contradictions

Even though participants were told that the mock-up shows a red color when the building exceeds the permitted height, however, one participant thought that the red color should also show when the building has a FAR greater than what is allowed. On the other hand, one participant mentioned that FAR and density cannot be relaxed.

Two participants were skeptical that urban planners and developers would know the FAR, height, and setback restrictions. One participant mentioned specifically that not necessarily everybody knows about these rules and restrictions. This participant said,

- ***(P1):*** *There is an occasional architect who could not figure out how to calculate FAR; most of them have already figured it out, and they know that they are already exceeding it, but they are looking for ways to get relaxations.*

Interestingly, the researcher noticed that even the participants who were senior urban planners expressed the rules of relaxation about FAR and density in a contradictory manner; this then caused some degree of confusion and uncertainty during the meeting, as well as spending unnecessary time to solve a fairly evident issue. It appeared that even urban planners' opinions vary when interpreting whether FAR or density are/are not allowed.

One participant said that SketchBoard could be a good tool for creating massing models. However, another participant felt that people always respond very negatively to massing models, because they are not aesthetically attractive-looking models. This participant noted,

- ***(P3):*** *People think you are going to get them chunky buildings like those on the massing models.*

Efficient Technical Activities

One participant provided an example where they had to spend a significant amount of time in SketchUp, a 3D modeling computer program, trying to figure out the various scenarios in order to achieve what was expected on the development application (@ Last Software, 2000). Participants believed that SketchBoard can present different FAR in various forms and scenarios that are easier and quicker than what can be achieved in SketchUp. Participants said that they would like to have SketchBoard because it is simpler to use than SketchUp. As a participant argued,

- ***(P1):*** *We spent quite a bit of time in SketchUp trying to figure out what the various scenarios for the twenty FAR were with the building and without the existing buildings, with the park and without the park.*

SketchBoard and Communicative Activities

All participants believed that SketchBoard would improve communication between urban planners and developers, and specific comments are discussed in the section below.

Efficient Communication

Participants believed that SketchBoard can help them in explaining and having discussions with the developers, and even the public, thus making it easier to draw quicker conclusion. Most participants mentioned that SketchBoard could also be used for discussions surrounding the relaxation of setbacks, FAR, and height variables.

They agreed that SketchBoard could be used specifically for graphical comparison between the allowed and proposed envelopes. As a participant said,

- ***(P1):*** *[SketchBoard] can quickly generate what is allowed internally, and then your applicant gives you the sketches of what they are proposing; you can [then] visually compare those pretty quickly without having to input the new variables and spending time yourself creating a new model because that could take some time.*

Initial Communication and Concept Development

One participant stated that the initial analysis of a development or land use application should compare what the developers propose against what is allowed; this is an important conversation between the urban planners and developer. This participant thought that being able to visualize those differences would facilitate discussion between urban planners and developers and could advance the discussion to reach an agreement. This participant noted,

- ***(P4):*** *You could just pop-up the image of what is allowed, put in what developers are asking for, and then start to have those conversations and talk about, and if you need to go and talk to other people, immediately they get a sense of what you are looking for.*

SketchBoard and Improvement

All participants provided some suggestions on how SketchBoard could be improved. These opinions are stated in the section below.

Superimposed Envelope

All participants suggested that SketchBoard should be able to show the differences between the allowable and proposed envelopes as a transparent-object. As a participant observed,

- ***(P2):*** *If we were to have the technology to take the crude FAR massing and show it on a 3D [model] and [transparent].*

Comparing Envelopes

One participant suggested that it would be very useful if SketchBoard could be accessible at a public meeting or online. This would then allow for viewing and comparing what the applicant proposed, and what is permitted.

Sunlight Restrictions

A participant suggested that being able to show the shadowing path and sunlight restrictions on SketchBoard would be very useful.

Modifiable Variables

Participants felt that the variables should be modifiable to allow either the urban planner or developer to input their own variables for each zoning district. As a participant noted,

- ***(P3):*** *Many zoning rules can be relaxed. A static model cannot reflect various options. The model would be more useful if this was possible.*

Various Building Forms and the Surrounding Context

Participants said that they want to be able to make various forms and types of buildings, compare existing land use versus proposed, as well as show the surrounding context on a site. Participants thought that these options should be accessible on SketchBoard. As a participant said,

- ***(P4):*** *Being able to look at that tower and the surrounding context is really important, and would be very helpful to have that on [SketchBoard].*

5. DISCUSSION AND FUTURE RESEARCH

5.1 Beyond Zoning Ordinance's Visualization and Computation

Findings of this exploratory study revealed that the problem of shared understanding in the design review process can be solved through the unrealized use of new digital media. Taking this further could possibility lead to the creation of an intelligent and interactive "expert-help-medium" that is capable of resolving complications in face-to-face human communication, while also providing solutions to technical problems within the design review process. Furthermore, findings of this exploratory study also suggested that visualization and computation of the zoning ordinance and the FAR should be enhanced. However, exploring beyond this revelation could provide alternative directions in interface and interaction design as a site for further research. For example, the results from this exploratory study revealed that urban planners and their stakeholders have difficulties both in visualizing future developments on a site, as well as in interpreting bylaws and these activities can be somewhat confusing for them. Also, currently missing from the development proposals is the comparison between the proposed and allowed building envelopes that is often significantly important, and mostly an imaginary process which happens in the mind of urban planners and their stakeholders. Thus, as part of future research, incorporating SketchBoard with the capabilities to view the contextual building envelope and visualize the surrounding areas of proposed buildings would be worthwhile. In addition, the findings from this exploratory study recognized that there is also a need for future research to explore possibilities to store and retrieve the history of a design, as well as multiple design solutions within the design review process. Thus, the question of how technology could potentially support the work of comparing the proposed and allowed building envelops, as well as showing the surrounding contexts of a particular building envelope, is also another encouraging direction for future research (Vishkaie, 2014).

5.2 Promoting Participatory Communication Around Technical Activities Within the Design Review Process

While this exploratory study provides insights about the range of participatory communication activities and how they are influenced by the interface and interaction design of SketchBoard within the design review process, however, there are still open-ended questions in this domain that need to be addressed. A particular challenge is to find a balance between enabling participatory design around technical issues on interactive surfaces. As this exploratory study has shown, this interaction style comes with many benefits and prevents situations where the knowledge level and expertise of experts and novices could interfere with the optimal outcome for the design review process. However, the introduction of such system-based constraints comes with the restricting action of prescribing interactions and thus hampering the notion of participation and shared understanding in the design review process. To some degree, creating new digital media would help define the boundaries of urban planning stakeholder's imaginations and idea generation, which could make for participatory communication in a timely fashion. The question, therefore, is how would it be possible to introduce design elements on the interface and interaction design of SketchBoard in order to potentially prevent endless and unproductive discussion, while enhancing participatory communication with experts and novices around technical problems. This exploratory study has begun to explore these ideas on an empirical level, but they have not been explored in much depth from a design perspective (Vishkaie, 2014).

Furthermore, often for the developers and mostly for the public, the design review process traditionally involves a collection of paper-based and digital rules, guidelines, and 2D/3D material that have been arranged and compiled in a certain way. Historically, urban planners have active roles as gatekeepers of information and access, while, developers and the public are passive actors. However, effective communication and acquiring knowledge and expertise are not passive process, but involve the dynamic and constant participation of all actors engaged in the process. From this point of view, developers, and particularly the public are considered to be actors who are encouraged to actively participate in the design review process through participatory communication and involvement. Moreover, keeping in mind the notion of participation, the public, in particular, should be able to add their own ideas, experiences, and contents to the process of design to which other members have already added their own contributions. However, unfortunately, this is not always the case at the present time. Thus, what types of technology can be invented to facilitate this participation is a kind of open-ended question that can be explored as part of future research. In addition, what a perceivable interactive communication medium can do to enable participatory active roles and disable misinterpretation of information and access, are yet largely unexplored questions from a political perspective (Vishkaie, 2014).

5.3 The Influence of SketchBoard on the Design Review Process

This exploratory study mainly focused on the relation between urban planning stakeholders' communication and technical activities in the design review process using SketchBoard on a small display. While this exploratory study has covered a large and important area of the entire design review process, it does not elaborate much about the interaction between different urban planning departments and other organizations involved in urban planning processes, and how this could influence the larger context of these processes. Taking this to an even higher level, this exploratory study examines how the use of both small and large interactive surfaces can transform the urban planning processes. This would introduce a broad yet important research question that will require studies to be done that focus on the urban planning processes as a whole, rather than concentrating on a particular design review process (Vishkaie, 2014).

5.4 The Future of Mobile, Small Interactive Surfaces Within the Design Review Process

While this exploratory study has exclusively focused on mobile, small interactive surfaces, there are also a wider range of technologies that could also be adapted to the process of design review. Research directions that could be explored include the adaptation of large interactive surfaces in the design review process or even within a larger context of urban planning processes. Throughout this exploratory study, the results of the exploration of urban planning stakeholders' participatory interactions with mobile, small interactive surfaces reveal that there are both challenges and benefits to these devices. Moreover, large interactive surfaces are not accessible and compatible with the current systems, but they could possibly enhance a shared view and understanding of the presented information together in the same place. Yet, mobile, small devices have the potential in the design review process to enable mobility and ease of access for the information required, while communicating ideas. The question about what role large interactive surfaces, such as tabletop and large wall displays, could potentially play in the design review process, and how they could be integrated with the existing desktop computer aided planning systems therefore guides another promising research direction (Vishkaie, 2014).

5.5 Rapid 3D Visualization of Buildings and Objects on Interactive Surfaces

Inspirations from research findings suggest that rapid and crude 3D visualization and the construction of buildings and objects on interactive surfaces is the future generation of 3D modeling. By tapping on the surface, this technique envisions the integration of 2D sketches to 3D blocks by sketching the border lines of the cross-section of a building; sketches of buildings can then be selected and extruded. In addition, dimensions and form of blocks can be manipulated freely by pinch and zoom gestures on interactive surfaces. Although, the following idea is not directly suggested by the research findings, however, rapid and crude 3D visualization and construction of products such as a phone, chair, or car on interactive surfaces is identical to the rapid 3D visualization and construction of buildings on interactive surfaces. Yet, considering buildings as complex products/objects, the fundamentals of rapid 3D visualization and construction of buildings and products on interactive surfaces are much the same. This technique envisions the integration of 2D sketches to 3D models of products. This can be done by tapping on the surface to sketch the border lines of the cross-section of a product; these product sketches can then be selected and extruded. Additionally, dimensions and form of products can be manipulated freely by pinch and zoom gestures on interactive surfaces (Vishkaie, 2014).

6. CONCLUSION

The rapidly advanced computer aided planning imposes limitations to the work of urban planners. These limitations derive from impoverished concepts of participatory, intelligibility, mobility, automation, flexibility, and efficiency of interface and interaction design of urban planning technology. For example, CAD and GIS require an extensive technical training before the user becomes proficient in their use. This requirement for training has prevented the development of non-professional genre of urban planning technology. This example offers a starting point for exploring the potential of digital media for the design review process, one which suggests that the user's competence should be broadened and improved. It further suggests that the digital media used in the design review process should require user's rudimentary or experimental knowledge. In addition, the use of interactive digital media within the design review process should provide communication platforms, which inhibit the problem of software non-professionalism (Vishkaie, 2014; Evans-Cowley, 2010; Hanzl, 2007; Shiode, 2000).

Yet, there is a wide range of PSS, and each system has a set of unique specifications. But, the goal of this exploratory study was not to measure the effectiveness or efficiency of each technology. In fact, it would be cumbersome to evaluate the usefulness of all software urban planners utilize. However, that was not the purpose of this exploratory study. The goal was the creation of a vision of a new generation of communication media that could be supported by interactive surfaces. Thus, combining both design and empirical research should provide a useful case study for urban planners, interaction designers, and researchers (Vishkaie, 2014).

REFERENCES

@Last Software. (2000). SketchUp. *Trimble Navigation*. Retrieved August 30, 2015, from http://www.sketchup.com

Bellotti, V., & Bly, S. (1996). *Walking Away from the Desktop Computer: Distributed Collaboration and Mobility in a Product Design Team.* Paper presented at the CSCW, Cambridge, MA. 10.1145/240080.240256

Beyer, H., & Holtzblatt, K. (1998). *Contextual Design - Defining Customer-Centered Systems*. San Francisco, CA: Morgan Kaufmann Publisher.

Brail, R. K., & Klosterman, R. E. (Eds.). (2001). *Planning Support Systems: Integrating Geographic Information Systems, Models, and Visualization Tools*. Redlands, CA: ESRI Press.

Cooper, A. (2004). *Inmates Are Running the Asylum, The: Why High-Tech Products Drive Us Crazy and How to Restore the Sanity*. Indianapolis, IN: Sams Publishing.

Crabtree, A. (2003). *Designing Collaborative Systems: A Practical Guide to Ethnography*. London: Springer.

Creswell, J. W. (2007). *Qualitative Inquiry and Research Design: Choosing Among Five Approaches*. SAGE Publishing.

Evans-Cowley, J. (2010). Planning in the Real-Time City: The Future of Mobile Technology. *Journal of Planning Literature*, *25*(2), 136–149. doi:10.1177/0885412210394100

Flew, T. (2005). *New media: an introduction*. South Melbourne, Australia: Oxford University Press.

Forester, J. (1989). *Planning in the Face of Power*. Los Angeles, CA: University of California Press.

Forester, J. (1987). *Critical Theory and Public Life*. Cambridge, MA: The MIT Press.

Gaver, W. W. (1996). Situating Action II: Affordances for Interaction: The Social Is Material for Design. *Ecological Psychology*, *8*(2), 111–129. doi:10.120715326969eco0802_2

Gibson, J. J. (1986). *The Ecological Approach to Visual Perception*. Hillsdale, NJ: Lawrence Erlbaum Associates.

Hanzl, M. (2007). Information technology as a tool for public participation in urban planning: A review of experiments and potentials. *Design Studies*, *28*(3), 289–307. doi:10.1016/j.destud.2007.02.003

Hollan, J., Hutchins, E., & Kirsh, D. (2008). Distributed Cognition: Toward a New Foundation for Human-Computer Interaction Research. *ACM Transactions on Computer-Human Interaction*, *7*(2), 174–196. doi:10.1145/353485.353487

Hutchins, E. (2008). *Distributed Cognition*. International Encyclopedia of the Social and Behavioral Sciences.

Kaufman, J. L. (1963). *Illustrating the Zoning Ordinance*. Chicago: American Planning Association.

Klabbers, J. H. (1999). A paradigm shift in planning: information technology and self-steering in complex systems. *International conference; 6th, Computers in urban planning and urban management.*

Kock, N. (2005). Media Richness and Media Naturalness? The Evolution of Our Biological Communication Apparatus and Its Influence on Our Behvaiour Toward E-Communication Tools. *IEEE Transactions on Professional Communication*, *48*(2), 117–130. doi:10.1109/TPC.2005.849649

Kock, N., & Nosek, J. (2005). Expanding the Boundaries of E-Collaboration. *IEEE Transactions on Professional Communication*, *48*(1), 1–9. doi:10.1109/TPC.2004.843272

Lefebvre, E., Lefebvre, L. A., Hen, G. L., & Mendgen, R. (2006). Cross-Border E-Collaboration for New Product Development in the Automotive Industry. *International Conference on System Science.*

Luff, P., & Heath, C. (1998). *Mobility in Collaboration.* Paper presented at the CSCW, Seattle, WA. 10.1145/289444.289505

Murray, J. H. (2012). *Inventing the medium: principles of interaction design as a cultural practice*. Cambridge, MA: MIT Press.

Murray, J. H. (1997). *Hamlet on the Holodeck – The Future of Narrative in Cyberspace*. Cambridge, MA: MIT Press.

Norman, D. A. (2010). *Living with Complexity*. Cambridge, MA: MIT Press.

Norman, D. A. (1993). *Things that Make Us Smart*. Addison-Wesley Publishing Company.

O'Doherty, K., & Einsiedel, E. (2012). Public Engagement and Emerging Technologies. Vancouver: University of British Columbia Press.

Obendorf, H. (2009). *Minimalism: Designing Simplicity*. London, UK: Springer. doi:10.1007/978-1-84882-371-6

Rogers, Y., Sharp, H., & Preece, J. (2011). *Interaction Design: Beyond Human - Computer Interaction*. Sussex, UK: Willey.

Rogers, Y., & Lindley, S. (2004). Collaborating around vertical and horizontal large interactive displays: Which way is best? *Interacting with Computers*, *16*(6), 1133–1152. doi:10.1016/j.intcom.2004.07.008

Shiode, N. (2000). Urban Planning, Information Technology, and Cyberspace. *Journal of Urban Technology*, *7*(2), 105–126. doi:10.1080/713684111

Short, J., Williams, E., & Christie, B. (1976). *The Social Psychology of Telecommunications*. London: Wiley.

Suchman, L. A. (2007). *Human-Machine Reconfigurations: Plans and Situated Actions*. Cambridge, UK: Cambridge University Press.

Tufte, T., & Mefalopulos, P. (2009). *Participatory Communication: A Practical Guide*. Herndon, VA: World Bank Publications.

Vishkaie, R. S. (2014). *SketchBoard: Conceptualizing Interactive Communication Media for the Design Review Process* (Doctoral dissertation). Available from ProQuest Dissertation and Theses database. (URI No. 11023/1426)

Vishkaie, R. S., & Levy, R. M. (2015). Making Design Review Interactive. In G. Mura (Ed.), *Analyzing Art, Culture, and Design in the Digital Age* (pp. 234–257). Hershey, PA: IGI Global. doi:10.4018/978-1-4666-8679-3.ch018

Warfel, T. Z. (2009). *Prototyping: A Practitioner's Guide*. Brooklyn, NY: Rosenfeld Media.

Chapter 9
Residential Resource Consumption:
A New Approach to Collection of Dwelling Data

Maryam Saydi
University of Melbourne, Australia

Ian Bishop
University of Melbourne, Australia

ABSTRACT

Residential energy and water consumption depend on dwelling structure and the behaviour of residents. Aspects of residential behaviour can be derived from census data. Dwelling information is harder to obtain. Using both aerial and street-level views from Google mapping products, exterior dwelling characteristics were captured in each of 40 postal areas in and around Melbourne, Australia. This approach saved the time and cost of travelling to the widely spread suburbs and provided data not otherwise available. The census and dwelling data were compared with resource usage statistics in linear regression models. It was found that energy and water use are highly correlated, with socio-economic variables better explaining water consumption and dwelling structure factors better explaining energy consumption. Nevertheless, the proportions of households that include a couple with children and have a swimming pool provided useful models of variations in both energy and water use. Applications to planning through spatially explicit scenario testing were developed in ArcGIS ModelBuilder.

DOI: 10.4018/978-1-5225-5999-3.ch009

INTRODUCTION

Rapid urban population growth and the consequent increase in urban resource demand are putting pressure on limited natural resources. Water use has been restricted in many parts of the world (Kenney, Goemans, Klein, Lowrey & Reidy, 2008; Worthington & Hoffman, 2008) and the broader implications of energy use are now widely recognised (Gram-Hanssen, 2014; Newton & Meyer, 2010; O'Callaghan, Green, Hyde, Wadley & Upadhyay, 2012). Over the past decade, a nexus between energy and water resources has been increasingly described (Marsh, 2008; Stillwell, King, Webber, Duncan & Hardberger, 2011; U.S. Department of Energy, 2006) and governments are devising policies and beginning to take action (DHI, 2008; Fisher & Ackerman, 2011; Proust, Dovers, Foran, Newell, Steffen & Troy, 2007; Stillwell, A. S., King, C. W., Webber, M. E., Duncan, I. J., & Hardberger, A., 2011) targeting the factors influencing the consumption of both energy and water in various sectors of the economy. This chapter focuses on the residential sector that, in Australia, consumes 11.7% of energy (Syed, Melanie, Thorpe & Penney, 2007) and 10.7% of non-agricultural water (ABS, 2013).

Saydi, Bishop and Rajabifard (2015) reported their use of online resource instruments to capture dwelling structure data for urban resource management, in particular residential energy and water demand modelling. The purpose of that research was to develop a novel *Virtual Identification of Dwelling Characteristics Online (VIDCO)* technique to support analysis of the influence of dwelling structure characteristics on urban resource consumption. In this chapter, we summarize that development and then describe how it was combined with other data sources and modelling to support a scenario testing platform for e-planning. In particular, we provide an overview of our data collection methods – especially VIDCO, and then discuss VIDCO application issues and evaluate the data produced. We then bring the VIDCO derived dwelling data together with socio-economic data and water and energy consumption data to model the respective influences of dwelling and behavioural characteristics on urban residential resource consumption. Finally, we illustrate the use of the derived relationships in spatially explicit scenario testing for resource planning purposes and report some conclusion.

BACKGROUND

Urban Resource Consumption

Residential energy and water consumption depend on both the structure of the dwelling (and any associated garden) and the number the residents, and their behaviour:

we refer to these as dwelling characteristics and socio-economic characteristics respectively. The influence of dwelling characteristics on residential consumption of energy and water has emerged as an essential component of urban resource reduction strategies. For example, separate dwellings consume more residential energy and water than other dwelling types such as semi-detached, or flats and apartments (Crawford & Fuller, 2011; Druckman & Jackson, 2008; Fox, McIntosh & Jeffrey, 2009; Troy & Holloway, 2004). While social-economic factors are also influential in household resource consumption (Arbués, Villanúa, & Barberán, 2010; Domene, Saurí & Parés, 2005; Druckman & Jackson, 2008; O'Callaghan, B., Green, H. J., Hyde, R. A., Wadley, D., & Upadhyay, A., 2012), dwelling characteristics are a key element for planners and policy makers to devise urban resource management strategies (Guerra Santin, Itard & Visscher, 2009). Also, while the socio-economic characteristics of an area can commonly be extracted from census data, identifying the most significant dwelling structure factors is more problematic and these are not routinely surveyed. However, analysis in support of policy development requires detailed and high-quality data about the structure of dwellings.

Sources of Dwelling Data

Traditionally, there have been two broad approaches to collecting dwelling structure data: interview-based surveys of users, and spatial data analysis. Acquiring interview data with a large sample size is usually conducted by large organisations. For example, Residential Energy Consumption Surveys (RECS) have been conducted by the US Energy Information Administration (EIA) to collect data about the type and construction year of dwellings (EIA, 2011). Alternatively, spatial analysis has been used by some researchers to extract dwelling structure characteristics. For example, Chang, Parandvash & Shandas (2010) used a geographic information system (GIS) for geo-coding billing data using linked street addresses. This data was then linked to a tax-lot spatial layer to provide a dataset of dwelling structure characteristics including lot size, building size, building density, and building age.

Satellite images have also been used to extract dwelling structure data. Siebrits (2012) used high quality satellite images (red-green-blue (RGB) orthophotos) and GIS analysis to identify dwellings with a swimming pool for a sample of 14,233 properties. Furthermore, in a study on identifying determinants of water demand in Phoenix, Arizona, Wentz and Gober (2007) captured the percentage of the garden cover for each census tract from a land cover database classified from a 1998 Landsat TM image using automated satellite image data extraction.

Data about dwelling characteristics can be categorised as either observable or non-observable. Several physical variables, such as insulation status, are not observable. However, there are many factors that can be collected by direct observation (Table 1).

Table 1. Studies on traditional dwelling structure data capture techniques

Dwelling Characteristics		Dwelling Structure Data Capturing Techniques		
Examples		Interview-Based	GIS Method	Satellite Image
Observable	Type of dwelling Size of dwelling Construction year Garden type Garden style Swimming pool presence	(Adua, 2010), (Domene & Saurí, 2006), (Randolph & Troy, 2008), (Newton & Meyer, 2010), (Syme, Shao, Po, & Campbell, 2004), (Liao & Chang, 2002), (Guerra Santin, O., Itard, L., & Visscher, H., 2009), (Gram-Hanssen, 2014)	(Siebrits, 2012), (Chang, H., Parandvash, G. H., & Shandas, V., 2010), (HousePeters, Pratt, & Chang, 2010)	(Siebrits, 2012), (Wentz & Gober, 2007)
Non-Observable	Ownership Insulation Dwelling value	(Adua, 2010),(Liao & Chang, 2002), (Randolph & Troy, 2008), (Newton & Meyer, 2010),(O'Callaghan, B., Green, H. J., Hyde, R. A., Wadley, D., & Upadhyay, A., 2012), (Guerra Santin, O., Itard, L., & Visscher, H., 2009),(Gram-Hanssen, 2014)		

Characteristics of Online Resources

In recent years, due to the high cost of collecting observable data at street level, there has been increasing interest in utilizing online sources, such as Google Earth (GE), Google Map (GM) and Google Street View (GSV), to remotely capture such data (Badland, Opit, Witten, Kearns & Mavoa, 2010; Clarke, Ailshire, Melendez, Bader & Morenoff, 2010; Rundle, Bader, Richards, Neckerman & Teitler, 2011; Taylor, Fernando, Bauman, Williamson, Craig & Redman, 2011). These sources allow users to remotely obtain quantitative and qualitative information about the urban environment, costly field work is not required.

GM provides two-dimensional maps of buildings and streets with labels which make it easy to identify addresses on the map. In addition, its 'Measure distance' tool helps users to measure the features such as buildings, streets and sidewalks. The ability to zoom in and out as well as to pan the maps makes it convenient for users to observe and compare different levels of detail on the map.

GE combines satellite imagery and aerial photography of the earth's surface, as well as map data. It provides a three-dimensional interactive image of the world. GE views the Earth as a virtual globe which allows the user to spin it and 'fly' from space down to the Earth's surface. Other functionalities that are useful in this context include: a tool to create a 'Placemark' to represent a specific location on the Earth; the 'Ruler' tool to measure the distances; and the availability of historical imagery that makes it possible for us to observe changes over time.

GSV, which is available from within both GM and GE, provides 360° horizontal and 180° vertical panoramic views at street level. This allows users to view streets at roughly eye level and virtually walk through them. Its zoom function makes it possible for users to interpret more qualitative and detailed information about urban features such as a building's wall material or number of stories – factors not directly observable in satellite imagery.

To date, only a few studies, mainly in public health assessment, have explicitly adopted a 'virtual' field audit to inventory and measure relevant environmental characteristics. These studies have included audit of land-use, sidewalk amenities, pedestrian safety, motorized traffic and parking, infrastructure for active travel and aesthetics, and compared these with 'in-person' or 'direct' field observations (Badland, H. M., Opit, S., Witten, K., Kearns, R. A., & Mavoa, S., 2010; Rundle, A. G., Bader, M. D., Richards, C. A., Neckerman, K. M., & Teitler, J. O., 2011; Taylor, B. T., Fernando, P., Bauman, A. E., Williamson, A., Craig, J. C., & Redman, S., 2011).

In-person audits are costly and extremely resource intensive, making them prohibitive for many studies (Clarke, P., Ailshire, J., Melendez, R., Bader, M., & Morenoff, J., 2010). For example, Taylor et al. (2011) used both GE and direct observation to capture physical-activity-related environment data for 50 parks in Sydney, Australia. They reported saving 38 hours: 4 hours using GE in comparison to 42 hours for direct observation.

Moreover, in a study capturing the attributes of the built environment for four neighbourhoods in New Zealand, Badland et al. (2010) found that using the GSV tool was faster than direct observation using a specially developed New Zealand Systematic Pedestrian and Cycling Environment Scan (NZ-SPACES) tool (χ=115.3 min (virtual), χ=148.5 min (direct)). They showed that the time taken to complete the audits decreased as familiarity with the measures and tools increased; this was most evident for the virtual audit.

In addition to requiring less time, virtual audit instruments such as GSV and GE, have been shown to be reliable (Badland, H. M., Opit, S., Witten, K., Kearns, R. A., & Mavoa, S, 2010; Clarke, P., Ailshire, J., Melendez, R., Bader, M., & Morenoff, J., 2010; Taylor, B. T., Fernando, P., Bauman, A. E., Williamson, A., Craig, J. C., & Redman, S., 2011; Wilson, Kelly, Schootman, Baker, Banerjee, Clennin & Miller, 2012). For example, in a study carried out on public health in the city of Chicago, Clarke et al. (2010) demonstrated that in using GSV for housing assessment on a virtual walk, accuracy rates of 97.3% for high-rise buildings, 83.3% for detached housing and 73.0% for low-rise private apartments were achieved.

Consistent with this study, Badland et al. (2010) showed that using GSV had perfect agreement (100%) between virtual and in-person measures with a few elements, such as cycling surface and cycling infrastructure, and acceptable levels of reliability (≥70%) for a majority of streetscape features such as path safety and traffic safety. Moreover, in assessing the agreement between GE instruments and in-person observation in obtaining public open space data, Taylor et al. (2011) showed that the results of these two methods were highly correlated ($r=0.90$, $p < 0.0001$).

While virtual audit seems to be a reliable method to obtain data, caution should be exercised when trying to gather more finely detailed observations or those that typically exhibit temporal changes, such as using GSV to collect data about items related to social and physical disorder (garbage, litter, or broken glass in the street or on sidewalks) (Clarke, P., Ailshire, J., Melendez, R., Bader, M., & Morenoff, J., 2010; Rundle, A. G., Bader, M. D., Richards, C. A., Neckerman, K. M., & Teitler, J. O., 2011).

METHODS

Data Sources

Postal Areas (POA) were, at the time of data acquisition, the smallest spatial resolution available for resources consumption data. This was therefore used as the aggregation level for other data acquisition and modelling. POA boundaries for 2006 (the most recent census year at the commencement of this research) were sourced from the Australian Bureau of Statistics (ABS) in ESRI Shape file format (*.shp). This layer included 653 POAs of Victoria projected in the GDA_1994_MGA coordinate system (Zone 55) on the D_GDA_1994 datum. POA areas ranged from about 1 km^2 to more than 11000 km^2.

The data analysis based on energy and water consumption was undertaken per household (HH), rather than per capita, because a household is the usual unit for social and economic measurements and people in a household often follow the same culture and lifestyle. Moreover, the benefits of energy and water consumption are typically shared between the members of a household regardless of age or income level.

The then Department of Environment and Primary Industries (DEPI) in Victoria provided 2006 data for residential gas and electricity consumption. These data were originally provided to government by the distribution companies: Envestra and Multinet for gas, and SPAusnet, Citipower, Powercore, Jemena and United Energy for electricity. Data for residential energy consumption was the result of converting electricity and gas data to the same unit (Gigajoule (GJ)) and then combining them together. This data was divided by the number of HHs in each POA to achieve per household data for energy consumption (GJ/HH). Pacific Micromarketing Company (http://www.pacmicro.com.au/) provided data for the number of HHs in each POA in 2006. There was no POA level water consumption (l/HH) data available for 2006, however data for 2008 was provided by the Victorian Government.

To ensure that the input data was appropriate for analysis, we applied several filters on both residential energy and water consumption, such as excluding POAs which had low residential population and POAs which did not have both electricity and gas data available. After filtering, data for both residential energy and water consumption were compiled for 203 POAs across Victoria.

The socio-economic data were all sourced from ABS, specifically from the 2006 census, and aggregated by POA. In general, the raw numbers provided by ABS were converted to percentages based on the number of households in each POA. We were then in a position to compare household resource use with household demographics and dwelling characteristics.

The ABS provided limited data on dwelling characteristics at POA level in 2006. These data categorised the type of dwellings only. We reclassified these data to make them more appropriate to our research. For example, similar types relating to multi-unit dwellings (e.g. 'flat, unit or apartment'), were aggregated. Moreover, data for temporary dwellings and dwelling types that were seen as not primary dwellings, such as 'caravan, cabin, houseboat' and 'improvised home, tent, sleepers out', as well as 'not stated' and 'not applicable', were ignored. The percentage of the main dwelling types in each POA was then calculated.

However, this was not seen as sufficient to assess the influence of dwelling structure on residential energy and water demand; it was therefore necessary to acquire additional dwelling structure data by using VIDCO. Forty randomly selected POAs, where data for both residential energy and water consumption were available, Figure 1 and Table 2, were used to develop and test the VIDCO process.

Defining Dwelling Structure Variables for VIDCO

As VIDCO was based on inspecting online images, only exterior visual characteristics could be considered. After preliminary inspection of dwellings in the sample POAs using GE and GSV, several major dwelling characteristic groups and sub-groups were defined (Table 3). For example, in the garden group, the size and the style of the gardens were defined as sub-groups. Variables were then defined in each sub-group; the dwelling structure variables were based on the most frequently observed characteristics of dwellings. For example, for roof material, four variables were defined as the percentages of 'tile', 'corrugated iron', 'others' and 'unknown'.

Most of the variables were categorical and were simply given a numerical code for ease of data entry: e.g., 1, 2, 3 and 4 were assigned respectively to 'separate', 'semi-attached', 'attached' and 'other'. Only variables of the roof edge sub-group

Figure 1. VIDCO study area: 40 randomly-selected POAs-distributed across inner and outer parts of Melbourne. The city is at the north end of Port Phillip Bay.

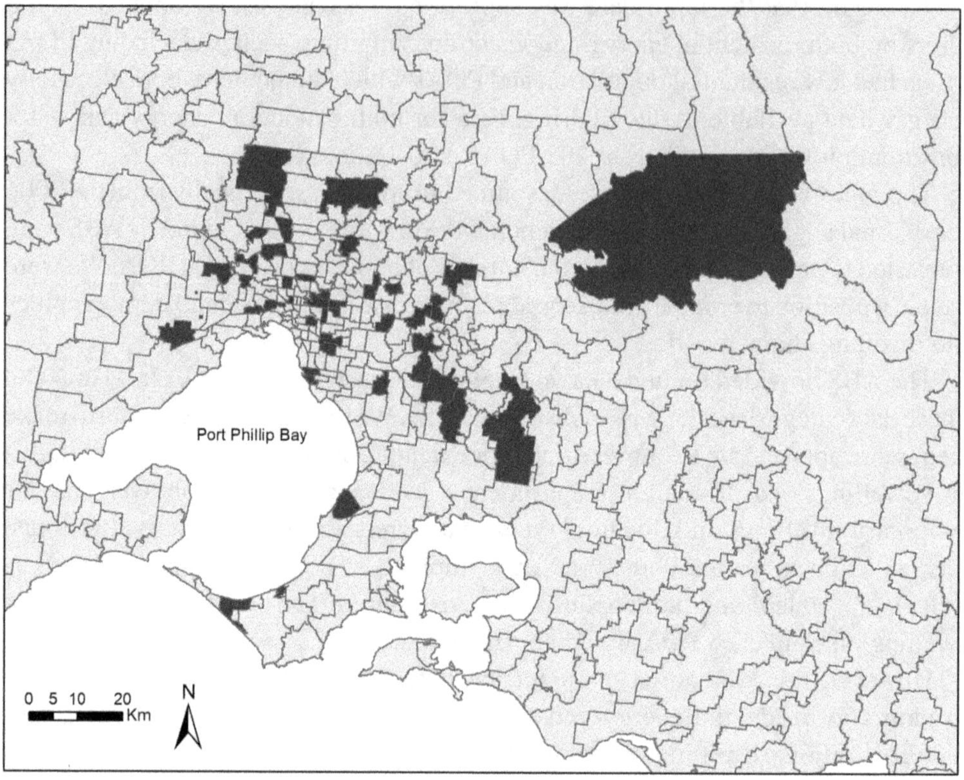

Table 2. Descriptive statistics for the 40 sample POAs in Victoria

	Min.	Max.	Mean	Standard Deviation
Energy use (GJ/HH)	29.99	131.71	81.83	22.08
Water use (l/HH)	93075	247105	162039	25991
Number of households (HH)	1053	17570	6968.6	4171.8
POA boundary (km^2)	1.75	1210.5	49.99	188.14

were ordinal: the percentages of 'no eave', 'with eaves' and 'veranda' variables corresponded to increasing levels of protection of the dwellings from sun and rain. A simple Roof Edge Index (REI) was therefore created by weighting the categories, based on their degree of shading:

$$REI = 1 * \left(Noeaves\%\right) + 2 * \left(witheave\%\right) + 3 * \left(verandah\%\right)$$

Using VIDCO for Accruing Dwelling Structure Data

VIDCO is based on using vertical and horizontal images to determine dwelling characteristics, such as the presence of swimming pool, the garden type, building materials and dwelling dimensions. Two online image-based instruments (GSV and GE) were used to capture the exterior characteristics of dwellings from front (horizontal) and top (vertical) views. GM was also used to search the addresses on the map. VIDCO also uses GIS (in this case ArcGIS v10) for several spatial analyses such as identifying the sample areas.

We acquired dwelling structure data from 12th April to 14th June 2011 using the classic versions of GM and GSV. Version 6.1.0.5001 was available for GE. At this period, the available satellite images (top view) provided by GE's historical imagery tool dated from 2000 to 1st January 2009. For GSV, most of the street view images were captured in either 2009 or 2010 (for the newly built areas). Later, Google upgraded the classic versions of GM and GSV and added the option to view historical street view images; this facility was not available at the time of our data collection. Images from 2009 were used, through the steps below, to collect dwelling structure data even though other data used in the research was from 2006. These images (in GE) had better image quality than the older images. However, the existence in 2006 of the selected dwellings was checked before data collection.

Random locations for dwelling structure data collection were based on random points created using ArcGIS 'Spatial Analyst Tools': first, a raster layer (1 m resolution) with the same extent as the 40 sample POAs was created using 'Create Random Raster' function. The generated raster layer had continuous cell values across a gradual range from 0 to 1. We used the 'Set Null' function to eliminate those with a value below 0.96. The remaining 4% were converted to a vector point layer in (*.shp) file format using the 'Raster to Point' function in 'Conversion Tools'.

The two layers containing the 40 sample POA boundaries as well as the randomly created points in each POA, were converted to (*.KML) format using ArcGIS's 'Conversion Tool' and loaded into GE. The next step was to identify sample locations from among the random points in each sample POA. This, as well as the sampling of dwellings for application of the VIDCO instrument, involved several criteria:

- We selected sample locations only in residential areas and not in other land-use areas such as business, industrial, commercial or parkland.
- Where possible, the locations selected for data acquisition were well distributed across the residential areas of each sample POA.

Table 3. Variables defined and acquired using VIDCO

Group	Sub-Group	Dwelling Structure Variables						
	Code	1	2	3	4	5	6	7
Building	Type	Separate	Semi-detached, row and terraced	Attached	Others			
	Size	Large	Medium	Small	Unknown			
	Storey	One	Two	Three	More than 3	1.5		
	Age	After 2000	1960-2000	Pre 1960	Unknown			
Windows	Type	Outside awnings or shutters	Tinted glass	Transparent	Louvre	Unknown		
Walls and Roof	Wall Material	Brick	Concrete	Timber	Fibro-cement	Stone	Steel	Unknown
	Roof Material	Tile	Corrugated Iron	Others	Unknown			
	Roof Shape	Hip & Valley	Gable	Shed	Flat	Butterfly	Unknown	
	Roof Edge	No overhang	With eaves	Veranda	Other			
Garden	Garden Size	Large	Medium	Small	No Garden			
	Garden Style	High (mostly trees)	Short (a few trees)	All grass and flowers	No planting	Unknown		
Pool	Swimming Pool	No	Yes					
Car	Car Park	Undercover	Open air	None	Unknown			
Apartment	Number of Levels	One	Two	Three	Four or more			
	Size	Large	Medium	Small				
	Flats Size	Large	Medium	Small	Unknown			

- To make it possible to visually inspect dwelling structure characteristics, only locations which were close to the roads and had the street view images available in GSV were selected.
- Dwelling structure data was acquired around each location only when both street view and top view images were 'clear'; this means that both images had reasonable quality and there were no barriers, such as clouds or trees and cars, to cover either the street-view or top-view images.

Both GE and GM resources provide top view and street (front) view images. However, for street view, we chose to access GSV through GM as it had the capability of 'zooming in' on the street view images. When capturing dwelling structure data from the top view, GE was preferred to GM as it was quicker to navigate.

GE provided us with several other advantages for acquiring dwelling structure data: we used GE's ability to import the (*.KML) files of POA boundaries and randomly created points to help determine the appropriate locations for data acquisition. With the help of GE's 'Placemark' tool, we marked the locations and coordinates (latitude and longitude) of the appropriate locations for extracting dwelling structure data. In addition, using the 'Ruler' tool, we measured the size of dwellings and gardens.

Building size, roof material, garden size and swimming pool existence are examples of data which were collected from top view. Garden style, building storey number, wall material and window type are examples of data which were extracted from the front view.

Inspecting the top and front view characteristics of the dwellings with VIDCO allowed us to collect data for ten dwellings around each of five different locations (totally 50 dwellings) for each sample POA. In most cases, five dwellings were selected on each side of the street. The distance of the dwellings to the data collection locations depended on the urban form (density of the dwellings) and the availability of the dwelling structure data as mentioned in the criteria section. We entered the relevant numerical codes, which were assigned to each dwelling structure variable, in an Excel sheet designed to record the dwelling characteristics (Figure 2).

The apparent size of the dwellings could be misleading. For example, size could be overestimated when dwellings had a large eave size or an undercover car park attached to the house. We were not always able to distinguish these situations. A dwelling foot print was coded as small (Less than 50 m^2), including flats and apartments and attached dwellings, medium (between 50 to 100 m^2), including either separate, semi-detached, row or terraced houses, or large (>100 m^2), including many separate houses and mansions (Figure 3). In cases where the dwellings had both front and back yards, we measured both gardens and added them together.

Figure 2. Dwelling structure data acquisition sheet and list of collection locations

Postcode: 3046 Glenroy, Hadfield and Oak Park **Date :** 27 May 2011

Approximatle Location: West

Dwelling structre data of residential areas														
No.	Type	Size	Storey	Age	Window	Material (Wall)	Material (Roof)	Roof Shape	Roof Egde	Garden Size	Garden Style	Pool	Car Park	Point ID
1	1	1	1	3	3	1	1	1	2	2	1	1	1	a
2	1	1	2	3	3	1	1	1	2	2	1	1	1	a
3	1	2	2	3	3	1	1	1	3	2	1	1	1	a
4	1	2	2	3	3	1	1	1	2	2	2	1	1	a
5	1	2	2	2	3	1	2	4	2	1	3	1	1	a
6	1	2	2	3	3	1	1	1	2	1	1	1	1	a
7	1	1	5	3	3	1	2	2	2	1	1	1	1	a

ID	Latitude	Longitude	Street
a	37°42'48.58"S	144°54'30.99"E	New Road, Start from the Avenue
b	37°41'59.94"S	144°54'36.97"E	Nene Avenue
c	37°42'25.30"S	144°56'3.75"E	Glenroy Road, Start from Farview St
d	37°41'54.65"S	144°56'29.28"E	Evell Street
e	37°42'47.79"S	144°57'2.66"E	Tristan Court

Figure 3. Dwelling sizes: (a) Large, (b) Medium, (c) Small

Paved areas were excluded. For garden style, we defined four categories based on the quantity and the size of vegetation in the garden; (1) High garden: mostly trees; (2) Short garden: with a few trees; (3) Almost all grass and flowers (rarely trees); (4) Not planted.

After collecting dwelling structure data, we filtered out several 'unnecessary' variables. For example, variables with null values, such as all cases referred to as 'unknown' and 'others'; fibro-cement, stone and steel wall materials and louver window, were removed because of their low frequency of occurrence. In addition, several other variables related to the roof shape, car park and apartment style sub-groups were ignored, as our preliminary data analysis showed that these variables were not significant. Ultimately, variables for only ten dwelling characteristic sub-groups remained. We calculated the percentage of the variables in each sub-group for all 40 sample POAs.

Statistical Analysis

Statistical analysis was undertaken using IBM SPSS software (v22). Correlations between variables were run first but are not reported here except to the extent that they help with interpretation of the results of linear regression. Multiple linear regression, using a stepwise approach was the main analytical tool. Residential energy and water consumption were separate dependent variables and the socio-economic and dwelling variables were the independent variables, and thus potential resource use predictors.

Scenario Testing

ArcMap (the mapping component of ArcGIS software) version 10.1 was used as an appropriate GIS environment for developing and implementing scenario testing. New toolboxes for each of the statistical regression models were created in the ArcToolbox window using ModelBuilder application. In each Toolbox, four interfaces were created to define spatial scenarios. The first interface added several new fields to the table of attribute data to store the values of socio-economic or dwelling structure indicators as well as dependent variables in each scenario. The other three interfaces provided different ways to define the scenarios: specific values, ratios of current values or fixed amounts of increase or decrease relative to current values. Users (e.g. urban planners or policy makers) could choose an interface depending on which of three different methods were preferred for manipulating the socio-economic and dwelling structure indicators.

ISSUES IN USING VIDCO

While capturing dwelling structure data using VIDCO, we confronted issues of two different types:

- **Technical Issues:** Arising from the functionality of GSV and GE instruments, as well as the resolution, quality or availability of images from both top and street views.
- **Data Sampling Issues:** Related to the individual dwelling instances or their locations.

Technical Issues

- **Street View Data is not Available:** This happened in newly built areas where not all the roads had been captured by Google. Such points were not included in the sample, which was appropriate given that we were using 2006 data, but would have been a problem with more recent resource consumption data.
- **GSV Image Separation:** The distance between consecutive GSV images sometimes made it hard to collect the dwelling structure data, especially in high density areas with small dwelling sizes.
- **Street View and Top View were not Synchronous:** A time difference between images provided by GSV and GE could result in mismatching top and front views of dwellings.
- **Difficulties in Inspecting Brightly Coloured Materials:** Identifying the material of the wall or roof was difficult, especially walls with bright colours (e.g. it was hard to distinguish between white timber and white painted bricks).
- **Poor Quality GSV Images:** There were some streets where the street view was provided with low-resolution and poor quality. These images usually were taken in bad conditions like foggy weather.
- **Address did not Match:** In some cases, finding the same address in GSV and GE was difficult; the problem was that street types on GM are stored as abbreviations ('Kerr St.') whereas on GE they are fully named ('Kerr Street').

Data Sampling Issues

- **Tree Barrier:** It is hard to see the front view of the dwellings when high trees (or high plantations) obstruct the building at street level.
- **Topology Barrier:** The topography of some areas may conceal the whole front view of a building.
- **Object Barrier:** When an object such as a bus or truck blocks the front view of dwellings, it is not possible to collect data.
- **Long Distance between the Road and the Buildings:** Distinguishing the characteristics of the dwellings, such as window type, wall or roof material, is difficult when there is a long distance between the street view and a building

location. This occurs when dwellings are set well back in the block or in two-way roads where GSV is available from just one side of the road.

- **Not Enough Dwellings around Data Capture Locations:** The distance of the dwellings from the data capture locations depends on the urban form (density of the dwellings) and the availability of the dwelling structure data. In low density areas, there may be insufficient buildings close to the data collection points (less than 10 buildings).
- **Difficulties in Identifying a Reference Point:** In some locations, it can be difficult to identify the reference point necessary to ensure that the same dwelling is being coded from top and front views (Figure 4a). This is particularly problematic where dwellings are in the same architecture style and the same size. In other locations, street trees made recognition difficult (Figure 4b).

In this study, if it was not possible to be sure about the characteristics of a dwelling because of one or more of these issues, the dwelling was passed over and the next dwelling along the street used instead. In cases where these issues affected the majority of dwellings (e.g. image quality) an alternative dwelling collection location was selected (Table 4). We assumed that the issues occurred randomly (i.e. are not

Figure 4. Reference point problems, (a) similar dwelling style (b) dense street trees.

Table 4. Strategies applied to resolve VIDCO data collection issues

Strategies	Issues in Using VIDCO
Move to another data collection location	• Street view data is not available • Poor quality GSV images • Long distance between the road and the buildings • Difficulties in identifying a reference point
Use the next appropriate dwelling	• When GSV's arrow moved the camera point too far • Street view and top view were not synchronous • Difficulties in inspecting brightly coloured materials • Tree barrier • Topology barrier • Object barrier • Not enough dwellings around data capture locations
Use alternative key word for address	• Address did not match

correlated with specific dwelling variables) and therefore moving to the next house (or next location) does not reduce the representativeness of the sample.

Validity of VIDCO Data

As discussed in Section 1, several studies, with a focus on public health, have confirmed that data captured from virtual audits have acceptable agreement with in-person field observation (Badland, H. M., Opit, S., Witten, K., Kearns, R. A., & Mavoa, S., 2010; Kelly, Wilson, Baker, Miller & Schootman, 2013; Taylor, B. T., Fernando, P., Bauman, A. E., Williamson, A., Craig, J. C., & Redman, S., 2011). Exceptions only appeared when virtually capturing data for small or ephemeral features. In resource consumption modelling we were not concerned with such small or fast changing features but with the macro characteristics of the dwellings and their outdoor areas. This reduces the potential of serious validity issues, and so in this study we did not use on-site audit for confirmation. In any case, it would not be possible to precisely check variables such as dwelling and garden size by direct street-level observation, or to do so at the time (2008/9) when the images were acquired. Instead, we used the limited dwelling data sourced from ABS, in which dwelling type was the only characteristic available, as a partial check of the validity of VIDCO data.

For this, we applied bi-variate correlation (Pearson) analysis on dwelling type in both ABS and VIDCO for the 40 sample POAs. For separate dwellings, the correlation was 0.748, for semi-detached, terrace and row dwellings it was 0.628, and for attached units or apartments it was 0.695. All correlations were significant with $p < 0.01$.

This means that the highest agreement between VIDCO and ABS data was approximately 75% (for separate dwellings) and less for other types of dwellings. This relatively weak agreement may be because we restricted VIDCO data capture to residential areas, and not everyone lives in such areas. Many dwellings are situated in commercial zones or mixed zones, for example. This is more likely to be the case in older POAs closer to the city centre, where many people live above shops or in flat blocks within shopping precincts. We would therefore expect POAs closer to city centre to have higher disagreement in the percentage of separate dwelling type than those in outer areas. Figure 5 shows that this is precisely what was found with disagreement levels up to 60% in some inner POAs but below 10% in outer commuter areas.

Similarly, semi-detached, terrace and row dwellings as well as flats, apartments and attached dwellings had higher agreement (less than 10% difference) in outer POAs. The differences were all in the expected direction with separate dwelling being generally over estimated in VIDCO and semi-detached and attached units under

Figure 5. Difference between VIDCO and ABS in percentage of separate dwellings in POAs, Melbourne. The one POA with negative difference is shown with N.

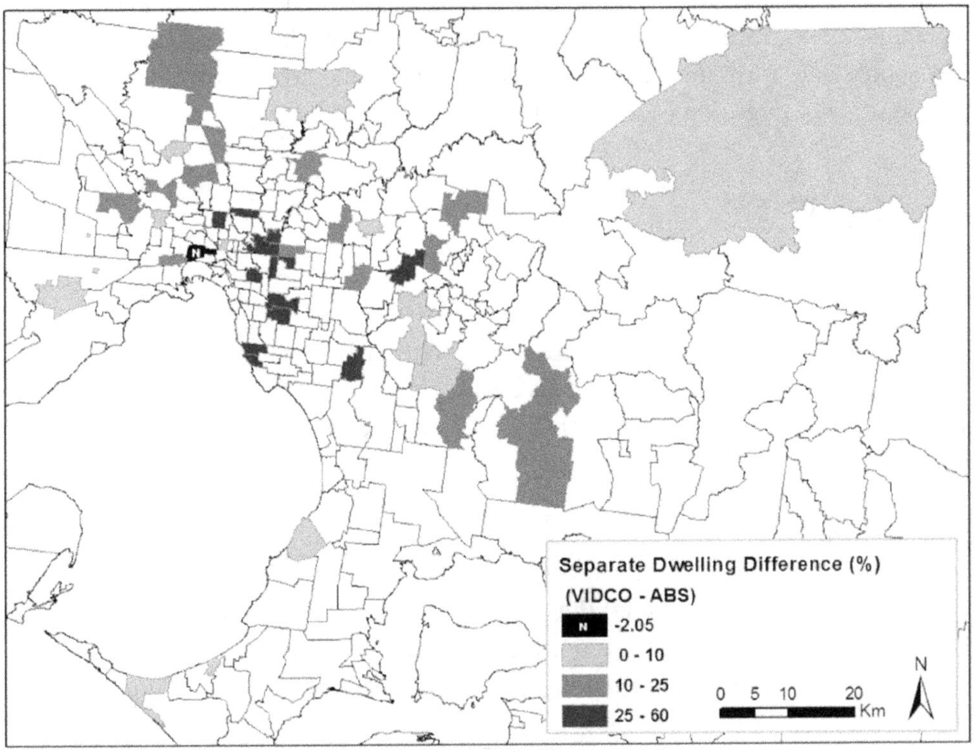

estimated. Since ABS data was available only for dwelling type, it was not possible to assess the level of agreement of VIDCO data with other dwelling characteristics.

When modelling urban resource consumption by aggregated spatial unit, rather than by individual household, the chosen spatial units should be as homogenous as possible. If the spatial units are too small then data collection is unnecessarily costly, and if the units are too large the modelling will not be valid. The question here is whether the spatial unit we used (POA level) was appropriate for capturing dwelling structure data? To investigate this, we assumed that if a dwelling structure characteristic is homogenous inside each POA, then this spatial unit is ideal for data capture. A second question is, was a sample of 50 dwellings sufficient for the proposed analysis? Again, this depends on the homogeneity, since a single sample is sufficient in a completely homogeneous spatial unit.

A variety of diversity indices have been developed to measure the evenness of samples. Diversity indices are used in many studies especially biology and ecology (Peet, 1974; Bush, Lafferty, Lotz & Shostak, 1997). We applied the 'Simpson index of diversity' to measure the degree of homogeneity of the characteristics of dwellings in our sample areas. The Simpson index is often used to quantify the biodiversity of a habitat by taking the number of species present and the abundance of each species into account. Simpson's index (D) measures the probability that two individuals randomly selected from a sample will belong to the same species (The Offwell Woodland & Wildlife Trust, 2008), or in our case one of the sub-group categories. The appropriate formula for D depends on the sample size. As the VIDCO data sample was not large, we used formula (1):

$$D = \frac{\sum_{i=1}^{r} n_i \left(n_i - 1\right)}{N\left(N - 1\right)} \tag{1}$$

With this index, 0 represents infinite diversity and 1, no diversity. That is, the bigger the value of D, the lower the diversity. This is not intuitive, and so D is often subtracted from 1 as in formula (2). In this case, the greater the value, the greater the sample diversity.

$$\text{Simpson index} = 1 - D \tag{2}$$

In this study, N represented the sample size (50) which is the total number of dwelling in each dwelling sub-group, r is the number of variables in each sub-group and n_i is the number of values collected for each variable.

Examining the level of evenness for ten dwelling characteristic sub-groups, we calculated the Simpson index inside each of the 40 sample POAs (Table 4). In addition, we combined dwelling structure data for the entire sample of POAs, as if they were a single spatial unit, and calculated the Simpson index for this 'super-unit'. If the POA's diversity estimates are smaller than the super-unit's, this means that the POA is more homogeneous than the city as a whole. The last column of Table 5 shows the percentage of POAs which had a diversity index higher than the super-unit. Ideally these numbers should be very small.

The results are quite widely spread, indicating that the POAs were very homogenous on some characteristics and quite diverse on others. Whether this makes POA level analysis valid for resource consumption modelling depends on whether the homogenous characteristics are also those that are the better predictors of resource use (see below). The mean value of the Simpson index for dwelling structure sub-groups illustrated that most POAs had a higher level of evenness than the super-unit in dwelling type, residential architecture style, wall material as well as the presence of swimming pool. POAs had the lowest homogeneity in garden size, garden style, and dwelling size.

The low diversity of residential architectural style suggests that in Melbourne the main architectural style has not changed substantially since the 1960s and many newer dwellings retain older characteristics. The low swimming pool diversity shows that swimming pools are widely dispersed in both close and far-distant POAs.

Table 5. Simpson index for dwelling structure characteristics, 40 sample POAs, Melbourne

Dwelling Structure Sub-Groups	Simpson Index		POAs with Higher Diversity Value to Super-Unit (%)
	Inside POAs (Mean)	Super-Unit	
Dwelling type	0.07	0.14	17.5
Dwelling size	0.38	0.48	40
Number of storey	0.29	0.35	32.5
Residential architecture style	0.10	0.11	35
Window type	0.23	0.25	40
Wall material	0.15	0.18	32.5
Roof material	0.26	0.35	27.5
Garden size	0.46	0.58	40
Garden style	0.53	0.70	2.5
Swimming pool presence	0.17	0.19	40

Dwellings use predominantly the same material (brick) for outer walls for most of the sample areas. The style and size of garden, as well as dwelling size, had low levels of evenness in both individual POAs and the whole sample area. Although POAs closer to city centre had more flats, unit and apartments, which do not have gardens, or have only a small garden, there was still considerable diversity.

As the Simpson index shows, not all the dwelling characteristics are homogenous inside POAs; however, more than 60% of POAs had less diversity than the whole sample area. This suggests that while POA level is not perfect for data capture, it is clearly better than using the city as a whole. Moreover, capturing 50 data samples appear to be adequate for POAs with low levels of diversity.

URBAN RESOURCE CONSUMPTION MODELLING

This section firstly reviews the relationship between energy and water consumption. We then describe the application of multiple regression statistical analysis, using IBM SPSS software (v22), with residential energy and water consumption as separate dependent variables and dwelling variables as the potential predictors. Next, we report similar analysis with socio-economic variables as the potential predictors. Finally, the two resources are considered together and their predictive variables compared for their similarities and differences.

Relating Energy and Water Consumption

Figure 6 Shows the scatter plot of per household energy and water use. Each point in the scatter plot represents annual per household energy and water consumption in one POA. As is expected from the positive correlation, the 'regression' line has an upward trend.

$$E = 0.000552W - 10.64$$

E : Energy use (GJ/HH) and W : Water use (Litre/HH)

Points near the line have a typical ratio of energy to water usage. Points above the regression line (positive residuals) are POAs with a high rate of per household energy usage relative to their water use and points below the reference line (negative residuals) are POAs with larger relative rates of water usage than energy.

Figure 7 shows the spatial distribution of positive and negative residuals. Most of the POAs from the eastern suburbs had positive residuals, which means that in

Figure 6. Scatter plot for per household energy and water use, 203 POAs. POAs with higher energy use rate are above the line. POAs with higher water use rate are below. (from Saydi, 2015)

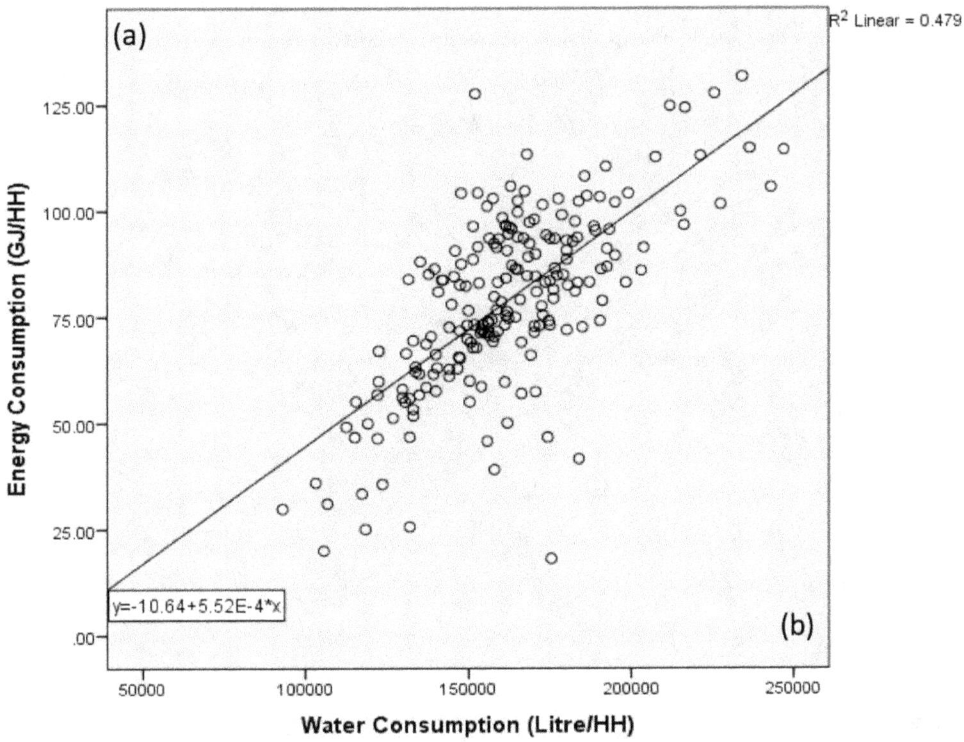

those areas per household energy use rate was comparatively higher than household water use.

Regression Modelling

The results showed that both energy (adjusted-R^2 0.68) and water (adjusted-R^2 0.55) consumption in a POA can be estimated with just two dwelling variables: areas with high proportion of dwellings without swimming pool and dwellings with the tile roofs (Table 4). Based on the derived Beta coefficients, when the roof material (Tile) variable is constant, a 1% decrease in the variable 'swimming pool (No)' – in other words a 1% increase in the proportion of houses with a swimming pool – would have increased annual energy consumption by between 0.757 and 1.125 (GJ/HH) in the year 2006 and would have increased water usage by between 1182 and 1792.5 (l/HH) in 2008 (Table 6).

Figure 7. Spatial distribution of residuals of energy use model relative to water use, 203 POAs, Melbourne. Darker areas are comparatively high energy use and lighter areas comparatively high water use. (from Saydi, 2015).

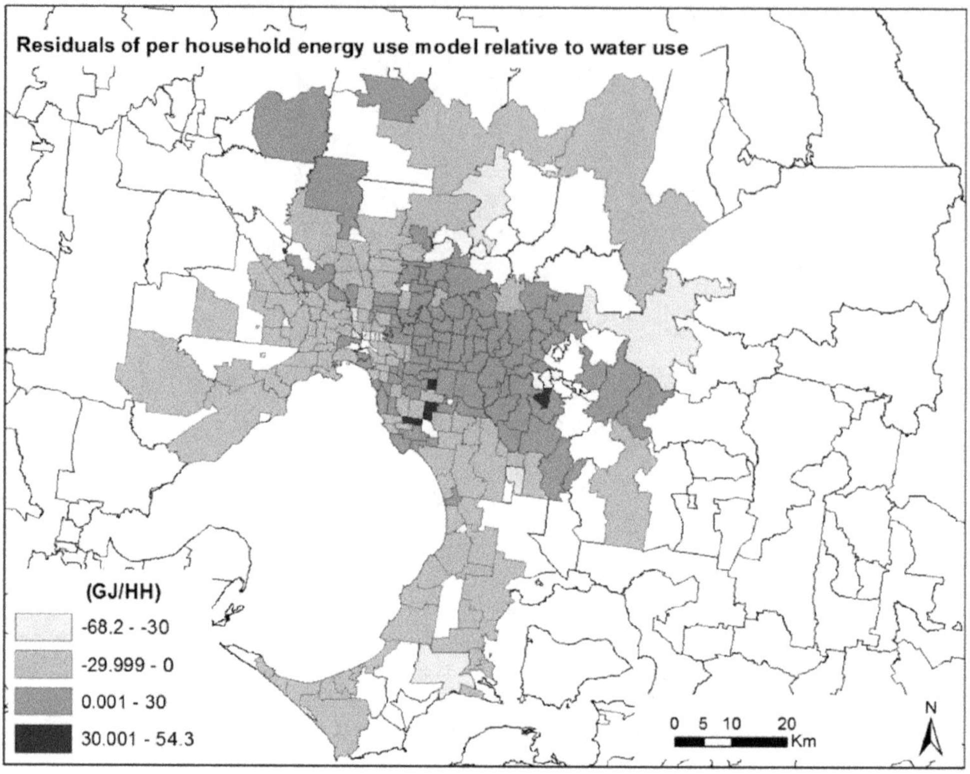

This result is in agreement with the findings of several studies (Shearer, 2009; Syme, G. J., Shao, Q., Po, M., & Campbell, E., 2004; Wentz & Gober, 2007). For example, Wentz and Gober (2007) demonstrated that an increase of one percentage point in the numbers of swimming pools raised average household demand annually by 4631 l (about 12.7 l/day). Although this is larger than the estimated range suggested here, a swimming pool is also an indicator for a large dwelling and a degree of household affluence – factors that also tend to increase resource consumption (March & Saurí, 2010; Wentz & Gober, 2007).

To explain why the percentage of dwellings with tiled roof in a POA was a strong predictor of a household's energy and water consumption, we applied bi-variate Pearson correlation analysis. Table 7 shows that these dwellings were likely to be large, separated and one storey houses with gardens. It is likely that households in these dwellings use more energy for heating and cooling as there is no other wall adjacent to insulate the house. In addition, because roof tile dwellings typically had

Table 6. Multiple regression with dwelling variables

VIDCO Variables		Unstandardized Coefficients		Standardized Coefficients	t	Sig.
		B	Std. Error	Beta		
Energy use *N=39* *Adj.R^2= 0.680*	(Constant)	140.908	18.935		7.442	.000
	Swimming pool (No)	-.941	.184	-.482	-5.117	.000
	Roof material (Tile)	.595	.102	.612	5.851	.000
Water use *N=39* *Adj.R^2= 0.546*	(Constant)	109757.280	11902.049		9.222	.000
	Swimming pool (No)	-1487.227	305.305	-.549	4.871	.000
	Roof material (Tile)	516.197	152.288	.382	3.390	.002

*Table 7. Correlation analysis (N = 40, ** p <0.01)*

Variable	Energy Usage (GJ/HH)	Water Usage (Litre/HH)	Separate Dwelling	Large Size Dwelling	Garden	One Storey Dwelling
Roof material (Tile)	0.651**	0.542**	0.611**	0.442**	0.672**	0.448**

gardens, these dwellings were likely to use more water than dwellings without garden. Indeed, an alternative model for energy use, with similar explanatory power is:

Energy use per household = 49.985 +0.818(Having swimming pool %) +0.366 (Large size dwellings %)

Moreover, the finding that large and separated dwellings (related to dwellings with tile roof) use more residential energy is consistent with other studies such as Druckman & Jackson (2008) and Crawford & Fuller (2011). More specifically, dwelling size especially influences the amount of energy consumed for heating (Bartusch, Odlare, Wallin, & Wester, 2012; Gram-Hanssen, 2014).

Models based on socio-economic variables explain nearly 68% of energy use and around 75% of water use variability. Because, in contrast to dwelling characteristics modelling, we had a higher number of cases with data available (196 for water and 270 for energy), models with a larger number of significant variables emerged from the analysis. Unlike the case with dwelling data, different variables contribute to explanation of water and energy usage.

For energy use the percentage of 'couple family with children' in a POA had the highest positive influence on household energy consumption (Beta=0.628). This was almost 1.5 times the influence of high income households (earning more than $3000 per week) (Beta=0.406). Areas with households with a family member working in trade industry (Beta=0.538) also positively influenced household energy usage. ABS define a 'trade' job as working in wholesale or retail trades. Correlation analysis showed that having two or more motor vehicles, as well as owning the dwelling with a mortgage, was likely in areas with large number of families with a member working in trade. It may be that these are the contributing behaviours, rather than the specific area of employment. The percentage of one parent families in a POA had also positive influence on per household energy consumption (Beta=0.267). ABS define 'community services' industry of employment as working in transport, postal and warehousing, public administration and safety, health care and social assistance and education and training industries. Households with a member in this type of work had a positive influence on the regression model (Beta=0.221). The percentage of households having children at primary school was the only variable in this model which had negative influence on energy use per household (Beta=-0.312). POAs with a high percentage of such families were likely to have a high proportion of households with two or more motor vehicles.

The reasons for the presence of some variables in the models is fairly clear, the interpretation of other is less clear. It appears however that high energy use tends to occur in households with:

- Children-- Especially if older than primary school age.
- Higher Incomes – Especially if household includes persons likely to be at home for longer periods as might apply to one parent families and people with 9-5 style jobs, and perhaps also if the dwelling is subject to a mortgage.

Water use is also highly influenced by a high proportion of 'couple families with children' (Beta=0.744). Although not significantly correlated with per household water consumption; more water was used in areas with large percentage of high income households (earning more than $3000 per week (Beta=0.660). Highly educated families (with bachelor or postgraduate degrees) as well as households with family member working in jobs such as 'Information media and telecommunications' were likely to live in these areas. However, the percent of households in a POA with a professional family member had negative influence on water usage per household (Beta=-0.606). Water use per household had a negative simple correlation with areas with many households having a professional member ($r=-0.243$, $p<0.01$). Small size (singles and couples without children), middle age families (aged 25-60 years old), high income households (earning $2000-$3000 p.w. or higher) as well as well-educated families (with bachelor or higher degree) were likely to live in these areas. While the percentage of households having young children studying at pre-school had significant influence on reducing water consumption per household (Beta=-0.327), it did not have a significant simple correlation with water use. These areas were likely to include many couple families with children, to have medium income families and to have households with two or more motor vehicles.

Again, it is apparent that higher water consumption is likely in households with:

- Children (post pre-school age)
- High Income (non-professional)

The results of applying multiple regressions on the combination of all socio-economic and dwelling structure variables on per household water consumption and energy use are detailed in Table 9.

By combining variables, we were able to account for 78.7% (Adj.R^2=0.787) of the variance in water consumption per household. POAs with the high percentage of couple family with children, areas with large proportion of two storey dwellings and areas with households having family member who did not go to school (Beta= 0.725, 0.474 and 0.388 respectively) had positive influence on water use per household. On

Table 8. Multiple regression with socio-economic variables

Multiple Regression Models		Unstandardized Coefficients		Standardized Coefficients	t	Sig.
		B	Std. Error	Beta		
Energy Use	(Constant)	-97.324	11.101		-8.767	.000
	Couple family with children	1.498	.111	.628	13.438	.000
N= 270	Industry of employment (Trade)	4.379	.368	.538	11.909	.000
	One parent family	1.416	.231	.267	6.118	.000
Adj.R^2= 0.677	Family income (> $3000 p.w)	1.231	.158	.406	7.783	.000
	Education type (Primary school)	-.844	.141	-.312	-5.968	.000
	Industry of employment (Community services)	.880	.162	.221	5.422	.000
Water Use *N=196* *Adj.R^2= 0.752*	(Constant)	121144.04	7641.64		15.853	.000
	Couple family with children	1930.755	117.696	.744	16.405	.000
	Education type (Pre-school)	-5149.810	634.597	-.327	-8.115	.000
	Family income ($3000-more p.w.)	2116.284	238.283	.660	8.881	.000
	Occupation (Professional)	-1416.917	187.805	-.606	-7.545	.000

Table 9. Socio-economic and dwelling structure determinants of household energy and water use

Multiple Regression Models (Urban Resource Use Analysis)		Unstandardized Coefficients		Standardized Coefficients	t	Sig.
		B	Std. Error	Beta		
Water use per household (Litre/HH) *N=36* ***Adj.R²= 0.787***	(Constant)	177833.10	35711.10		4.980	.000
	Couple family with children	2256.994	303.604	.725	7.434	.000
	Number of dwelling storey (Two)	768.461	154.036	.474	4.989	.000
	Swimming pool (No)	-1584.080	298.431	-.500	-5.308	.000
	Highest year of school completed (Did not go to school)	7010.139	1595.009	.388	4.395	.000
Energy use per household (GJ/HH) *N=36* ***Adj.R²= 0.868***	(Constant)	-67.579	15.188		-4.450	.000
	Couple family with children	.988	.175	.475	5.658	.000
	Family income (more than $3000 p.w.)	1.204	.143	.571	8.408	.000
	Industry of employment (Trade)	2.885	.658	.359	4.386	.000
	Occupied private dwelling	.441	.138	.223	3.204	.003

the other hand, and as expected, dwellings without swimming pool reduced water use per household (Beta=-0.500).

A combined model accounted for 86.8% (Adj.R^2=0.868) of the variance in energy use per household. All the variables in this model were significant and had positive influence on energy use per household: POAs with high percentage of high income families (Beta=0.571), areas with a large proportion of couple family with children (Beta=0.475), areas with many households working in trade (Beta=0.359), and areas with a high percentage of occupied private dwellings (Beta=0.223).

Water use per household is better explained by socio-economic (up to 75.2%) than dwelling structure variables, while dwelling structure variables better estimated the variation in energy use per household (up to 83.8% of variation). For energy use per household, the simple explanation for the greater dependence on dwelling variables is that in Victoria the largest proportion of energy use per household is for space and water heating followed by appliances (Petchey, 2010) which are mainly independent of the number of people in the house (or at least for the 'common' rooms – kitchen, living, dining). Therefore, per household energy use is primarily dependent on the characteristics of the dwelling itself (particularly dwelling size).

For water use, although the literature suggests that household water use is heavily influenced by outdoor activities (such as watering the lawns and gardens as well as filling the swimming pool) (Inman & Jeffrey, 2006; Syme, G. J., Shao, Q., Po, M., & Campbell, E., 2004; Wentz & Gober, 2007), and that this is related to the dwelling characteristics more than to the number of people at home, the results here indicated that variation of water consumption per household is somewhat better explained by the socio-economic characteristics of household members. This could be simply the result of Victorian water restrictions in 2008 which banned watering of lawns and restricted swimming pools filling. In these conditions, the number of people in the house, rather than the structure of the dwelling, had greater influence on water use, especially indoor water use such as showers, toilets, clothes and dish washing.

Identifying common indicators and making statistical models to estimate and link the variation of both per household energy and water use, multiple regression analysis (stepwise method) was applied on all six indicators contributing to the statistical models of energy and water use. Table 10 shows the results this analysis.

The results indicated that areas with a high proportion of two parent family with children and with large numbers of dwellings with a swimming pool were the most influential common indicators of residential energy and water use: both with positive influence. Variation in per household energy use was better explained (Adj. R^2 =0.733) than water use per household (Adj.R^2 =0.584).

Under the Melbourne water use restrictions in place in 2008, pools and spas over 2000 litres could not be filled without approval from the local water corporation (Melbourne Water, 2012). Even under these conditions swimming pools have been

Table 10. Common indicator statistical models of per household energy and water use

Multiple Regression Analysis Results Common Indicator Models		**Unstandardized Coefficients**		**Standardized Coefficients**	**t**	**Sig.**
		B	**Std. Error**	**Beta**		
Energy use per household *N=38, Adj.R^2= 0.733*	(Constant)	17.613	8.852		1.990	.054
	Couple family with children	1.142	.188	.538	6.063	.000
	Swimming pool(Yes)	.952	.162	.523	5.892	.000
Water use per household *N=39, Adj.R^2= 0.584*	(Constant)	85224.92	16137.88		5.281	.000
	Swimming pool(Yes)	1385.909	297.574	.512	4.657	.000
	Couple family with children	1375.309	345.725	.437	3.978	.000

found to be a significant indicator. This suggests that in a period without water restrictions their influence would be even stronger.

Although not all large dwellings had a swimming pool, dwellings with a swimming pool were generally large and had higher per household energy use. Thus, having swimming pool is also a good indicator of higher energy use.

SPATIAL SCENARIOS AND PLANNING APPLICATIONS

It is apparent that household resource consumption may change as a result of changes in the make-up of households and their dwellings, or through changes in technology and behaviour. Within the types of statistical models developed in this research, this corresponds to either variations in the values of the socio-economic or dwelling structure indicators themselves or changes in the coefficients of these indicators.

Modelling the former requires understanding of indicators' trends over time; for example, changing average dwelling size or household income level, which influences the percentage of these variables in each POA, would change the model outputs. The new values would be estimates of consumption levels simply based on continuation of existing trends.

A change in coefficients could happen, for example, through the application of new water restrictions or by devising conservation plans to encourage households to make more efficient use of residential energy and water such as using storm or recycled water, or improving home appliances. In Australia, it was estimated that replacing a traditional single flush toilet with a water efficient dual flush toilet saves about 51 l/person/day of water. This reduces household water use by 30,000 to 40,000 litres per year (Australian Government, 2014). Similarly, several strategies could influence household energy consumption. In 2009, roof insulation saved Victorian households in newer dwellings 12% on their energy bills compared to 5.8% for households in older dwellings (ABS, 2010).

Here are several examples of scenarios which planners and policy makers may wish to test for their influence on urban resource consumption:

1. What happens to per household energy and water consumption if nothing major changes in the world and no new resource policies are implemented and thus socio-economic and dwelling structure characteristics continue their existing trends, while coefficients remain the same?
2. What happens if these trends were modified by expected demographic changes (e.g. aging baby boomers moving to smaller dwellings and no long wanting a swimming pool)?

3. How would world or national level forces (or policies), beyond the control of the local decision-maker, such as climate change, a one-child policy, high carbon price, or national level incentives against separate dwellings influence urban resource consumption?
4. What happens if some local government policies (either whole Victoria or single LGA) influence socio-economic and housing life style trends? For example: heavily subsidizing childcare by a municipal (LGA) government; giving baby bonuses to the families living in far distant areas; not allowing new dwellings over 400 square meters; or relaxation of planning rules to allow high rise apartments in areas currently dominated by separate dwellings.

From these examples, it is clear that decisions of policy makers can influence both the trend of socio-economic and dwelling structure factors and their level of influence (as modelled by the coefficients). The scenario comparison which follows (further examples can be found in Saydi, 2015) tests the influence on per household water consumption of changes in socio-economic trends applied through the scenario testing module developed in ArcGIS ModelBuilder.

- **Scenario (1)***:* Socio-economic trends continue and no new policies apply?
- **Scenario (2)***:* The government promotes and invests in infrastructure development which requires 50% more professionals leading to a 10% increase in high income families, while other characteristics remain constant?

The distribution maps in Figure 8 show that scenario 2 has the effect of reducing per household water use in the majorities of POAs, especially those closer to the inner city. The percentage of couple families per household and families having pre-school children were held constant in this scenario. Variations in per household water consumption arose from the higher percentages of professionals and high-income families in and around inner cities areas. The total water use in the study area reduced by 8.9% under scenario 2.

CONCLUSION

Traditionally, there have been several impediments to capturing detailed data on dwelling structure (e.g. access). This study contributes to the growing literature on adoption of online resource instruments (such as GE and GSV) by presenting a novel technique called Virtual Identification of Dwelling Characteristics Online (VIDCO). Using this technique made it possible to capture data on exterior dwelling

Figure 8. Four conditions of per household water use; (a) Actual use (b) 2008 model output (c) after scenario 1 (d) after scenario 2. (from Saydi, 2015)

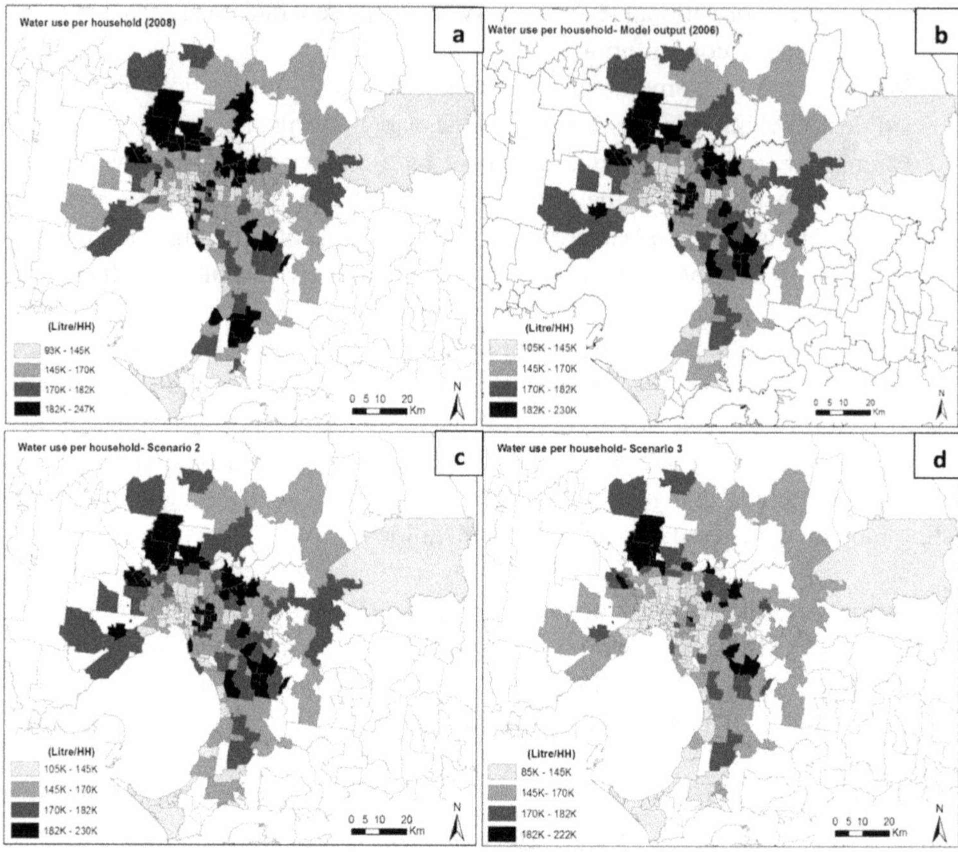

characteristics from both top and front views in order to assist urban resource consumption studies.

Using VIDCO successfully saved the time and cost of travelling to the entire 40 sample POAs, and also travelling inside POAs. Time spent acquiring data for 50 dwelling instances usually took from half a working day (4 hours) to two working days for each POA. Time for data capture decreased as familiarity with the online-resource instruments and measures increased, though there were still several issues adopting these instruments for VIDCO. For example, it took longer to capture dwelling structure data in POAs with high gardens, mainly because trees covered part of the dwellings. Also, in high density areas; it was difficult to find a reference point to make sure the same dwelling was observed from both top and street views. When this occurred, it was necessary to move to the next dwelling or choose another random starting location. Other difficulties in using VIDCO mainly related to the availability, resolution and quality of the images provided by GSV and GE tools at the time of carrying out this research, as low-quality street images made it difficult to capture some qualitative data such as roof or wall materials.

The VIDCO data sample was assessed for its agreement with other data sources, its homogeneity of spatial units and sampling size. Comparing VIDCO data with ABS data revealed that, generally, dwelling types (including separate, semi-attached and flat, unit and apartments) in outer suburbs had higher agreement (less than 10% difference) than in inner city suburbs. This difference probably occurred because our data collection zone was restricted to residential areas and ignored mixed land-use areas such as residential units on the top of business or commercial shops in inner city areas.

Applying statistical analysis (multiple linear regression) on VIDCO data, the presence of a swimming pool and tile roof material emerged as the best predictors of both energy (GJ/HH) and water (Litre/HH) consumption. Correlation analysis showed that dwellings with a tiled roof typically were large, separated and one storey houses. These results are broadly consistent with the literature on both residential energy and water demand, even though a variable such as 'tile roof' will not necessarily have the same implications in other cities or countries.

We have demonstrated the potential of using these online resources to conduct urban environment data collection for planning, providing evidence for researchers to consider dwelling structure characteristics during their strategic design phases. We did not compare outcomes to on-site survey results but suggest that any lack of validity in derived data is more related to the sampling process than to image interpretation. For example, restricting the sampling to residential zones caused a partial mismatch with other data in inner city areas. Other authors have tested the validity of image interpretation and found high levels of agreement except for small or ephemeral entities. As the frequency and quality of online imagery increases, even these issues should diminish, giving even greater reliability. We note however that this approach can only extract data about exterior aspects of residential property. Internal characteristics may also be important.

For comprehensive data compilation, a hybrid approach may be appropriate. VIDCO could be combined with either data available from government sources, or survey data. In Victoria, for example, property boundary data is available on-line and could be used with a GIS to determine the area of sample properties. If building footprint data were available, then this could be combined with VIDCO derived information on storeys to estimate building floor areas. Non-built area within the property could also be derived but this might not be helpful as the proportion that is paved, pool, lawn, etc. is only available visually. A telephone/questionnaire survey method would answer many different questions about dwelling structure such as indoor characteristics and tenure type and so would complement the exterior data gathered using VIDCO. The linkage between these two methods would be the addresses of the dwellings. It would first be necessary to check if data collection locations were visually appropriate for VIDCO application. If so, a survey could be conducted in those areas.

Bringing the VIDCO data together with census derived socio-economic data and resource consumption data, this study has found that POAs with higher per household energy use were also likely to consume more water per household indicating a clear link between residential energy and water demand. As a result of investigating the influence of key socio-economic and dwelling structure characteristics on residential energy and water use, it was found that, in general, socio-economic variables better explain water consumption (especially in a period of water use restrictions) while dwelling structure factors better explain energy consumption.

The preferred model for predicting water consumption per household included four socio-economic variables; percentages of couple families with children, percentages of children going to pre-school, percentages of high income families (earning $3000 or more per week) and percentages of household having family member working as a professional. The influences of family composition and income on residential water consumption are in agreement with other studies. The preferred model for predicting energy consumption per household included two dwelling structure variables: percentage of dwellings with swimming pool and percentage of large-size dwellings.

These two models, with no overlapping variables, suggest no clear link between policies for management of energy and water demand. However, a high proportion of couple family with children and high percentage of dwellings with swimming pool were common indicators of high residential consumption of energy and water. These common indicators better explained variation in energy use (Adj.R^2=0.733) than water use (Adj.R^2=0.584) per household.

By linking residential energy and water use, policy makers can define more complicated scenarios and investigate how changes in socio-economic and dwelling structure characteristics can influence both energy and water use per household. Further scenarios might include building town houses with a shared swimming pool in new development areas, or applying tax strategies which provide incentives to reduce dwelling sizes.

Applying the results of statistical analyses within spatial models can provide better understanding of possible changes in energy and water consumption over time through visual interpretation of the impacts of different planning scenarios, population growth policies and variations in people's life styles. This can lead to the design and scheduling of location specific programs to reduce urban resource demand.

However, the reliability of the spatial patterns of per household energy and water use is limited by the statistical models used for estimating the consumption of energy and water. Even should excellent statistical models exist for a particular date; it is difficult to judge their potential longevity without repetition of the analysis over an extended time period. The uncertainty associated with these statistical models therefore increases when applied to long-term scenarios.

REFERENCES

ABS. (2010). 1367.2 - State and Regional Indicators, Victoria, Jun 2010. *Feature article: Water and energy efficiency elements of households in older and newer dwellings.* Retrieved from http://www.abs.gov.au/AUSSTATS/abs@.nsf/Lookup/1367.2Feature+Article1Jun+2010

ABS. (2013). Water Supply and Use 2011–12 — Australia. *4610.0 - Water Account, Australia, 2011–12.* Retrieved 11 August 2014, 2013, from http://www.abs.gov.au/AUSSTATS/abs@.nsf/DetailsPage/4610.02011%E2%80%9312?OpenDocument

Adua, L. (2010). To cool a sweltering earth: Does energy efficiency improvement offset the climate impacts of lifestyle? *Energy Policy, 38*(10), 5719–5732. doi:10.1016/j.enpol.2010.05.021

Arbués, F., Villanúa, I., & Barberán, R. (2010). Household size and residential water demand: An empirical approach. *The Australian Journal of Agricultural and Resource Economics, 54*(1), 61–80. doi:10.1111/j.1467-8489.2009.00479.x

Australian Government. (2014). Water Efficiency Labelling and Standards (WELS) scheme. *Water efficiency.* Retrieved 22, February, 2015, from http://www.waterrating.gov.au/consumers/water-efficiency

Badland, H. M., Opit, S., Witten, K., Kearns, R. A., & Mavoa, S. (2010). Can virtual streetscape audits reliably replace physical streetscape audits? *Journal of Urban Health, 87*(6), 1007–1016. doi:10.100711524-010-9505-x PMID:21104331

Bartusch, C., Odlare, M., Wallin, F., & Wester, L. (2012). Exploring variance in residential electricity consumption: Household features and building properties. *Applied Energy, 92*, 637–643. doi:10.1016/j.apenergy.2011.04.034

Bush, A. O., Lafferty, K. D., Lotz, J. M., & Shostak, A. W. (1997). Parasitology meets ecology on its own terms: Margolis et al. revisited. *The Journal of Parasitology*, 575-583.

Chang, H., Parandvash, G. H., & Shandas, V. (2010). Spatial variations of single-family residential water consumption in Portland, Oregon. *Urban Geography, 31*(7), 953–972. doi:10.2747/0272-3638.31.7.953

Clarke, P., Ailshire, J., Melendez, R., Bader, M., & Morenoff, J. (2010). Using Google Earth to conduct a neighborhood audit: Reliability of a virtual audit instrument. *Health & Place, 16*(6), 1224–1229. doi:10.1016/j.healthplace.2010.08.007 PMID:20797897

Crawford, R., & Fuller, R. (2011). *Energy and greenhouse gas emissions implications of alternative housing types for Australia.* Paper presented at the State of Australian Cities National Conference, Melbourne, Australia.

DHI. (2008). Linking Water, Energy & Climate Change: A proposed water and energy policy initiative for the UN Climate Change Conference, COP15. In *Copenhagen 2009*. Draft Concept Note.

Domene, E., & Saurí, D. (2006). Urbanisation and water consumption: Influencing factors in the metropolitan region of Barcelona. *Urban Studies (Edinburgh, Scotland), 43*(9), 1605–1623. doi:10.1080/00420980600749969

Domene, E., Saurí, D., & Parés, M. (2005). Urbanization and sustainable resource use: The case of garden watering in the metropolitan region of Barcelona. *Urban Geography, 26*(6), 520–535. doi:10.2747/0272-3638.26.6.520

Druckman, A., & Jackson, T. (2008). Household energy consumption in the UK: A highly geographically and socio-economically disaggregated model. *Energy Policy, 36*(8), 3177–3192. doi:10.1016/j.enpol.2008.03.021

EIA. (2011). *Residential Energy Consumption Survey (RECS).* Retrieved 11,August, 2014, from http://www.eia.gov/consumption/residential/data/2009/index.cfm?view=characteristics

Fisher, J., & Ackerman, F. (2011). *The Water-Energy Nexus in the Western States: Projections to 2100 U.S.* Stockholm Environment Institute.

Fox, C., McIntosh, B., & Jeffrey, P. (2009). Classifying households for water demand forecasting using physical property characteristics. *Land Use Policy, 26*(3), 558–568. doi:10.1016/j.landusepol.2008.08.004

Gram-Hanssen, K. (2014). New needs for better understanding of household's energy consumption–behaviour, lifestyle or practices? *Architectural Engineering and Design Management, 10*(1-2), 91–107. doi:10.1080/17452007.2013.837251

Guerra Santin, O., Itard, L., & Visscher, H. (2009). The effect of occupancy and building characteristics on energy use for space and water heating in Dutch residential stock. *Energy and Building*, *41*(11), 1223–1232. doi:10.1016/j.enbuild.2009.07.002

HousePeters, L., Pratt, B., & Chang, H. (2010). Effects of urban spatial structure, sociodemographics, and climate on residential water consumption in Hillsboro, Oregon. The American Water Resource Association, 46(3).

Inman, D., & Jeffrey, P. (2006). A review of residential water conservation tool performance and influences on implementation effectiveness. *Urban Water Journal*, *3*(3), 127–143. doi:10.1080/15730620600961288

Kelly, C. M., Wilson, J. S., Baker, E. A., Miller, D. K., & Schootman, M. (2013). Using Google Street View to audit the built environment: Inter-rater reliability results. *Annals of Behavioral Medicine*, *45*(1), 108–112. doi:10.100712160-012-9419-9 PMID:23054943

Kenney, D. S., Goemans, C., Klein, R., Lowrey, J., & Reidy, K. (2008). Residential Water Demand Management: Lessons from Aurora, Colorado. JAWRA Journal of the American Water Resources Association, 44(1), 192-207.

Liao, H.-C., & Chang, T.-F. (2002). Space-heating and water-heating energy demands of the aged in the US. *Energy Economics*, *24*(3), 267–284. doi:10.1016/S0140-9883(02)00014-2

March, H., & Saurí, D. (2010). The suburbanization of water scarcity in the Barcelona metropolitan region: Sociodemographic and urban changes influencing domestic water consumption. *The Professional Geographer*, *62*(1), 32–45. doi:10.1080/00330120903375860

Marsh, D. M. (2008). *The Water-Energy Nexus: A comprehensive analysis in the context of New South Wales (PhD)*. Sydney, Australia: University of Technology.

Melbourne Water. (2012). *Permanent water use rules and targets*, from http://www.melbournewater.com.au/getinvolved/saveandreusewater/Pages/Water-restrictions.aspx

Newton, P., & Meyer, D. (2010). The determinants of urban resource consumption. *Environment and Behavior*, *44*(1), 107–135. doi:10.1177/0013916510390494

O'Callaghan, B., Green, H. J., Hyde, R. A., Wadley, D., & Upadhyay, A. (2012). Exploring the influence of housing design and occupant environmental attitudes on energy and water usage. *Architectural Science Review*, *55*(3), 176–185. doi:10.1080/00038628.2012.693813

Peet, R. K. (1974). The measurement of species diversity. *Annual Review of Ecology and Systematics*, *5*(1), 285–307. doi:10.1146/annurev.es.05.110174.001441

Petchey, R. (2010). *End use energy intensity in the Australian economy. Australian Bureau of Agricultural and Resource Economics – Bureau of Rural Sciences (ABARE-BRS)*. Australian Government.

Proust, K., Dovers, S., Foran, B., Newell, B., Steffen, W., & Troy, P. (2007). *Climate, energy and water: Accounting for the links.* The Australian National University, Occasional Paper. Retrieved from http://www.crdc.com.au/uploaded/File/E-Library/E-ENVIRO/Climate_Energy_and_Water_May_2007.pdf

Randolph, B., & Troy, P. (2008). Attitudes to conservation and water consumption. *Environmental Science & Policy*, *11*(5), 441–455. doi:10.1016/j.envsci.2008.03.003

Rundle, A. G., Bader, M. D., Richards, C. A., Neckerman, K. M., & Teitler, J. O. (2011). Using Google Street View to audit neighborhood environments. *American Journal of Preventive Medicine*, *40*(1), 94–100. doi:10.1016/j.amepre.2010.09.034 PMID:21146773

Saydi, M. (2015). *An integrated spatial model for understanding residential energy and water consumption* (PhD Thesis). The University of Melbourne.

Saydi, M., Bishop, I. D., & Rajabifard, A. (2015). Virtual identification of dwelling characteristics online for analysis of urban resource consumption. *International Journal of E-Planning Research*, *4*(3), 1–28. doi:10.4018/IJEPR.2015070101

H. Shearer (Ed.). (2009). Using Geographical Information Systems to explore the determinants of urban household water consumption and response to the Queensland government demand-side policy measure imposed during the drought of 2006-2008. *Proceedings of the 4th National Conference on the State of Australian Cities.*

Siebrits, R. (2012). Swimming pools and intra-city climates: Influences on residential water consumption in Cape Town. *Water S.A.*, *38*(1), 133–144. doi:10.4314/wsa.v38i1.16

Stillwell, A. S., King, C. W., Webber, M. E., Duncan, I. J., & Hardberger, A. (2011). The Energy-Water Nexus in Texas. *Ecology and Society*, *16*(1), art2. doi:10.5751/ES-03781-160102

Syed, A., Melanie, J., Thorpe, S., & Penney, K. (2007). *Australian energy national and state projections to 2029-30, ABARE research report 07. 24*. Australian Bureau of Agricultural and Resource Economics (ABARE). Retrieved from http://data.daff.gov.au/brs/data/warehouse/pe_abare99001416/rr07.24_energy_projections_full_report.pdf

Syme, G. J., Shao, Q., Po, M., & Campbell, E. (2004). Predicting and understanding home garden water use. *Landscape and Urban Planning*, *68*(1), 121–128. doi:10.1016/j.landurbplan.2003.08.002

Taylor, B. T., Fernando, P., Bauman, A. E., Williamson, A., Craig, J. C., & Redman, S. (2011). Measuring the quality of public open space using Google Earth. *American Journal of Preventive Medicine*, *40*(2), 105–112. doi:10.1016/j.amepre.2010.10.024 PMID:21238857

The Offwell Woodland & Wildlife Trust. (2008). *Simpson's Diversity Index*. Retrieved 1st,July, 2014, from http://www.countrysideinfo.co.uk/simpsons.htm

Troy, P., & Holloway, D. (2004). The use of residential water consumption as an urban planning tool: A pilot study in Adelaide. *Journal of Environmental Planning and Management*, *47*(1), 97–114. doi:10.1080/0964056042000189826

U.S. Department of Energy. (2006). *Energy Demands on Water Resources: Report to Congress on the interdependency of energy and water*. U.S. Department of Energy.

Wentz, E. A., & Gober, P. (2007). Determinants of small-area water consumption for the city of Phoenix, Arizona. *Water Resources Management*, *21*(11), 1849–1863. doi:10.100711269-006-9133-0

Wilson, J. S., Kelly, C. M., Schootman, M., Baker, E. A., Banerjee, A., Clennin, M., & Miller, D. K. (2012). Assessing the built environment using omnidirectional imagery. *American Journal of Preventive Medicine*, *42*(2), 193–199. doi:10.1016/j.amepre.2011.09.029 PMID:22261217

Worthington, A. C., & Hoffman, M. (2008). An empirical survey of residential water demand modelling. *Economic Surveys*, *22*(5), 842–871. doi:10.1111/j.1467-6419.2008.00551.x

Section 3
Wireless and Social Media

Chapter 10
Public Space and the Development of Wireless Media

Alex Lambert
Monash University, Australia

Scott McQuire
University of Melbourne, Australia

Nikos Papastergiadis
University of Melbourne, Australia

ABSTRACT

This chapter builds on research into Australian free Wi-Fi initiatives conducted in late 2012 and early 2013. It tours through a range of global developments in wireless internet delivery, focusing on how these influence the character of public spatiality, participation, and social inclusion. While there have been numerous technical and commercial advances, the authors argue that free public services narrowly focus on constructing public spaces of consumption and spectacle, and valorising public activities through increasingly granular sensor surveillance. The authors offer an expanded conception of what it means to value public space and to participate socially, culturally, and politically in public. The chapter concludes with the concerning gap between small scale projects that experiment with these concepts and the large-scale institutions that ignore them.

DOI: 10.4018/978-1-5225-5999-3.ch010

INTRODUCTION

This chapter builds on research we conducted in late 2012 and early 2013 on the free Wi-Fi initiatives of Australian municipalities and cultural institutions. In the original article, we explored the drivers behind Wi-Fi services, the way that visitors used them, and how they changed the character of public spaces (Lambert, McQuire, & Papastergiadis, 2014a). At the time Australia was experiencing a significant transformation in its broadband delivery systems with the auctioning of analogue television spectrum and the roll out of our National Broadband Network, which delivers fibre to the home and local node. Going beyond the common focus on private internet use, we sort to understand how these transformations were influencing public spaces.

Since this research there has been a much greater mobilisation of free Wi-Fi both in Australia and across the globe (Gabriel & Fellah, 2016; McShane, Wilson, & Meredyth, 2014). In this chapter we tour through some of these developments, finding that, while many new advances have been made, our original argument retains its persuasiveness. Wi-Fi services are generally planned and managed in a technocratic fashion that delimits the nature of public participation and the capacity for social inclusion. Wi-Fi, along with other wireless technologies, is becoming a central mechanism in the commodification, surveillance and social sorting of public space. This circumscribes the way that people value being in public, and even circumscribes the very concept of 'value' within an economic framework that lacks personal, social, cultural and political richness (Frow, 1995). For example, public spaces increasingly focus on commerce, rather than cosmopolitan cultural expression or social cohesion.

This chapter builds on an evolving interest in digital technologies, space and everyday life within media studies. People draw a great deal of meaning from a sense of place. Public spaces often mediate our experience of being part of a broader community, and hence our sense of cultural belonging. They are also significant for the mediation of difference in contemporary societies, such as the expression of multicultural diversity and political resistance (Carr, Francis, Rivlin, & Stone, 1992). For this reason, some argue that public spaces should be imbued with an ethic of cosmopolitanism that is open and respectful to difference (Papastergiadis, 2012). A long history of critical scholarship takes aim at the way that the technocratic and neo-liberal control of public spaces leads to gentrification, inequality, and break downs in cosmopolitan cohesion (for an overview see Harvey, 2009). Meanwhile, discourses of urban development and technological progress tend to mythologise certain places and either disguise or trivialise the differences that make up urban communities.

These political issues are becoming inextricably linked with digital media. For example, within the context of the emerging Internet of Things, Kitchen and Dodge (2011) explore the way that software and various 'coded objects' and infrastuctures have come to play an essential role in the governance of everyday spatial practices. When these media enter a space they 'transduce' it into something radically different. A café, when linked with Wi-Fi and smart devices, becomes a workplace. The authors advocate a sensitivity to the discursive regimes that justify and naturalise what they call 'code/space'. These include safety, security, efficiency, anti-fraud, empowerment, productivity, reliability, flexibility, and competitive advantage. Code/spaces are essentially systems of social control, yet the allure of these discourses allow people to willingly accept their disciplinary effects. A nice example of this is the relationship between urban surveillance and social sorting, or what Graham (2005) calls 'software sorted geographies'. Software comes to regulate access to spaces (such as toll roads, libraries, and Wi-Fi networks). Algorithmic filter bubbles and persuasive computing 'nudge' us toward particular experiences that reaffirm our beliefs and desires (Lambert, 2017). Consequently the experience of serendipity – if we understand this as the discovery of something valuable that we did not expect (Olma, 2016) – becomes highly regulated. This applies to the serendipitous encounter with cultural difference, and hence relates to the ethic of cosmopolitanism discussed above.

This framework encourages a conception of what various authors call 'hybrid' spaces (Antoniadis & Apostol, 2014; de Souza e Silva, 2006). Consider the way that everyday practices interlace and coproduce both online and offline senses of spatiality. When one imagines the diverse practices that produce public spaces – the practices of managers, technicians, service workers, and visitors – a complex, hybrid spatial assemblage emerges. Yet, as Easterling (2016) argues, far from producing a radically diverse menagerie of spaces, the current infrastructure paradigm is defined by its formulaic reproducibility, and hence its global standardisation of 'urban arrangements'. The integration of information communication with infrastructure via computer networks and sensors only amplifies the capacity for a global 'infrastructure space' as a kind of interlinked, updatable platform. While the design and governance of this global system does continue to be enacted by governments and inter-government agencies, it is also standardised within technical and economic systems to the point of operating almost autonomously, acting as what Easterling calls 'extrastatecraft', and what Thrift (2004) would call the 'technological unconscious'. This has serious consequences for effective planning, discussed in the conclusion.

In what follows we stay close to our expertise, which involves Australian case studies. However we also draw on important international comparisons to illustrate global trends. In many cases Australian wireless initiatives are highly typical of these trends.

THE ORIGINAL STUDY

In late 2012 we surveyed the extant and planned free public Wi-Fi initiatives of 30 Australian municipalities and 5 publically funded cultural institutions (Lambert, McQuire, & Papastergiadis, 2014a). Our methodology included 35 interviews of relevant public servants, an archival review of public documents such as council minutes and planning reports, short qualitative surveys with visitors to particular places, and ethnographic research in 5 public spaces. We identified various drivers for free Wi-Fi, including: the curation of digital art spaces; a means for marketing and communications via splash screens; a multi-channel system in which free services would be subsidised by an increase in efficiency for city services; the desire to support communities; a means for social sorting to attract tech savvy visitors; and most of all a desire for economic benefits through smart city branding, investment, tourism, competition over tenders, innovation in the local ICT industry, and creative digital production.

From the perspective of visitors, we conceptualised four key forms of Wi-Fi use. Wi-Fi was used to support the travel plans and communication practices of travellers. It was used for entertainment such as video watching in high-bandwidth spaces. It supported mobile workers who either did not have an office or who sort a more vibrant atmosphere to do work in, a finding also apparent in other ethnographic literature (Fortano, 2008). Finally, Wi-Fi was used to stay in contact on social media. Yet while users socialised via the Internet, they rarely engaged in serendipitous interactions with strangers. We discovered that Wi-Fi alone could not draw people to a particular place. Rather, it played a part in an ensemble of factors that make a place attractive.

A central finding of our study was that institutions were rarely willing to provide the operating costs for a 'user-centric' service that indulged the needs and desires of most visitors for internet services such as video and voice. Municipalities in particular engaged in what Anttiroiko (2012) calls 'technocratic planning' processes that often excluded the perspectives of their constituents and led to unpopular or defunct services. While we could identify a small number of participatory planning initiatives, we suggested that there was a dire need for more user-centric planning and service provision if Wi-Fi was to serve a social 'good'.

Shortly following the publication of this article a series of high profile public-private partnerships between telcos and large Australian cities began to roll out that we identified in a follow up article (Lambert, McQuire, & Papastergiadis, 2014b). This marked an important sea-change in the provision of Wi-Fi in Australia that reflected global trends. Following these changes a survey conducted by the Royal Melbourne Institute for Technology's Centre for Urban Research found that more than one in five Australians are frequent users of free public Wi-Fi, accessing a network at least once a week (McShane, Gregory, & Wilson, 2016). The most popular places to access free Wi-Fi are commercial in nature, such as cafes, restaurants, and shopping centres, although one in six people surveyed had accessed a network in a municipal space, such as a park or square. These findings reaffirm the important distinction between private and public ownership of Wi-Fi networks, and encourage us to probe how these different models are continuing to influence service design and public participation.

DEVELOPMENTS

Traditionally, Wi-Fi refers to the collection of 801.11 technological standards that allow devices to network across unlicensed spectra, predominantly the 2.4 GHz band. Since our original article, a variety of wireless technological innovations have emerged that have been recently detailed in the Wireless Broadband Alliance's 2016 industry report (Gabriel & Fellah, 2016). A 'next generation' of Wi-Fi technologies are emerging that include: second wave 802.11ac standards that boost speed and peak data rates; hyper-fast 802.11ad 'WiGig', capable of 7Gig download speeds and useful for short distance home, business and public Wi-Fi networks; and 802.11ah 'HaLow', a low-power, large range transmission technology useful for machine-to-machine networking and wireless backhaul for the Internet of Things and hence Smart City sensors. To support these new technologies, the International Telecommunications Union is lobbying governments and industry to open up new unlicensed spectra, such as the 60GHz band for WiGig and sub-GHz bands for HaLow.

Importantly, licensed mobile broadband, unlicensed Wi-Fi spectra and new technologies are converging together to form a heterogeneous network that will undergird the much anticipated 5G ecology. Industries are focusing on creating seamless protocols to network various forms of wireless radio transmission, as well as system virtualisation platforms that enable the control of heterogeneous infrastructure simultaneously, such as multiple Wi-Fi and cellular hotspots. Wi-Fi – particularly in urban areas – will continue to play a dramatic part in this evolution. It will be relied upon to take up the data hungry tasks of mobile broadband handoff, the Internet of Things, location services, smart city services (including transport and utility management), big data analytics, enterprise services and free public internet, all while maintaining an increasing expectation for a high quality of experience. Unsurprisingly, then, the global shipment of Wi-Fi radio units has exceeded 15 billion (Wi-Fi Alliance, 2016).

Within these developments we can identify a variety of cases that illustrate the diverse ways in which Wi-Fi is being utilised. Overall these uses suggest a highly complex assemblage of hybrid public spaces constituted by diverse practices. Yet for most of these cases a discourse of free, cheap or convenient Wi-Fi disguises hidden costs and modes of surveillance. Many of them also provoke critical reflection on how wirelessness transforms public space.

Telecommunication Operators and Mobile Internet

Telecommunication operators are significant supporters of next generation Wi-Fi technologies. Telcos face broadband congestion problems in dense urban areas in their mobile internet spectra. There is an increasing user demand for Voice-Over-Internet and Video-Over-Internet services, as well as business demand for data hungry enterprise services. A solution is the 'handoff' of internet traffic onto urban Wi-Fi networks with surplus bandwidth. This innovation has significantly driven telco investment in Wi-Fi, with Australia being a prime case study. For example, the largest Australian telco, Telstra (2017), has constructed a heterogeneous Wi-Fi network composed of homes, Telstra stores, and over 200,000 public telephones re-fitted with Wi-Fi routers to produce a network that has over a million hotspots. As well as enabling Telstra to handoff traffic, the network adds value to Telstra's mobile, home and pre-paid internet subscription products, and thus serves to attract new customers. As is the case with many incumbent telcos that were originally publically owned, Telstra has a significant advantage in the market with its control over infrastructure such as phone booths and copper and fibre wires. Telstra brands this network as 'free', when in reality it is subsidised by subscriptions.

Public-Private Partnerships

Significant public-private partnerships have also emerged in Australia, with large cities such as Adelaide, Perth, Canberra and Melbourne partnering with the internet service provider TPG to provide free public Wi-Fi in large urban areas. A typical case is the State of Victoria's (2017) partnership, which seeks to roll out 61000 access points over 600 square kilometres in Melbourne, Bendigo, and Ballarat at a cost of $6.7 million AUD. The service is already operating in multiple high population traffic areas in the Melbourne CBD, such as Flinders Street Station, the Queen Victoria Markets and the Melbourne Central mall. With these services TPG are able to capitalise on municipal spaces and infrastructure to build their brand, push innovation and open up new urban user markets. More importantly, they are able to access large volumes of valuable personal data. We conducted a week-long ethnography of Adelaide's CBD free Wi-Fi service in late 2014, when it was leading Australia in terms of network size, continuity and quality of experience. Interestingly, we found an important difference between small, localised, user-centric services such as those at Federation Square in Melbourne, and city-wide services. The former are capable of attracting visitors due to their boundedness and user-centric qualities. The latter do not create visitor clustering, as their boundaries are large, and effectively cover multiple forms of public space and practice.

These public-private partnerships are also occurring on impressive scales elsewhere in the world. A key example is the European WiFi4EU plan. The plan emerged from the Juncker Commission's *Digital Single Market Strategy*, which encourages telecommunications and digital services that cross-country borders, increased interoperability with common regulations and standards, and the development of higher internet capacity (European Union 2015). The discourse here is very much one of maximising the 'digital economy' across the European Union. The strategy laid the foundations for an ambitious free Wi-Fi service to be deployed primarily in parks, squares, libraries, museums, hospitals and other public buildings. The European Parliament, Council and Commission has raised 120 million euros to fund the infrastructure and installation costs.

Municipalities can apply for the funding, and must then support operation fees. Like many services, the drivers for this centre around economic growth, the provision of amenities such as e-government and e-health, and the civic responsibility for including more of the population in a functional internet service. This final aim is evinced in the plans for a geographically balanced roll-out, rather than a focus on high income business areas. Consider Jean-Claude Juncker's State of the Union comment: 'Everyone benefiting from connectivity means that it should not matter where you live or how much you earn. So we propose today to equip every European village and every city with free wireless internet access around the main centres of public life by 2020' (European Union 2016). As we discussed in our original strategy, desires for inclusivity and remedying digital divides do not always succeed, as municipalities often face planning and funding problems. For instance, will municipalities in lower income areas be able to raise revenue to support data demands, and if not how will they enter into partnerships that do not compromise their civic responsibilities?

Entrepreneurial Services

As well as telcos and cities, private companies and entrepreneurs have provided various different forms of wireless service. For example, New York's Link City NYC project has rented old public telephones, fitted them with fibre backhaul, and plans to establish over 7000 hotspots throughout the city (Fung, 2016). The service can be accessed through over 60 kiosks, and requires an email address. It is supported by Qualcomm and Alphabet's subsidiary, Sidewalk Labs. These companies have signalled monetisation strategies such as adds appearing on captive portal splash screens, email data brokering, and the analysis and commodification of city population rhythms, commercial activity and logistics.

Within cities, immense commercial spaces are capitalising on the new business opportunities afforded by the wireless mediatisation of public space. Sports stadiums, for instance, are investing in services that can provide terabytes of instantaneous throughput to upwards of a hundred thousand spectators. This amplifies the commodified experience of spectatorship through multiple screens, including personal smart devices, which can carry targeted pay-per-view, subscription and advertisement based services such as internet-protocol TV, while at the same time affording broadcasters, sponsors, stadium and team owners new forms of fine grained data analytics (Hutchins, 2016). Interestingly, Hutchins (ibid) explores the controversial way that these innovations have been received by many sports fans across the globe. Fans are concerned, sometimes to the point of protest, with the possibility that wireless stadiums may reduce the sensory and emotionally intense experience of presence they associate with live sport, as well as the possibility for spontaneous moments of 'communitas' with fellow fans. Hutchins argues that this related to a concern over the reductive commodification of the sports fan and his or her performance of identity in public space.

Mobile Public Spaces

Given the increased mobility of people and material objects, a diagnosis of the current trends in wirelessness should be cognisant of what Shelly and Urry (2006) call 'mobilities'. This includes the various actors that circulate through regional and global networks, the structures that enable this, and the various critical consequences for social relations, politics, culture, health, and natural and built environments. Wireless technologies of various kinds are a key form of support for mobile people such as tourists, migrants and mobile workers. This has led to international partnerships between telcos to offer cheaper roaming packages. WiMax enabled mobile internet modems are now common in countries such as Japan, and are often offered to travellers within the context of informal economies such as AirBNB house sharing (Lobato & Thomas, 2015). Hotels have long offered Wi-Fi, and businesses target hotels as environments to experiment with new platforms. For example, Ruckus is testing the new 802.11ac wireless standard in 3,700 hotels across 90 countries (Gabriel & Fellah, 2016). For many years it has been a point of contention that many of these traveller-oriented services, especially those in hotels, are not free, and are sometimes onerously expensive (Lambert, McQuire, & Papastergiadis, 2014b). In terms of direct and hidden costs, the landscape of traveller-accessible services remains complex and heterogeneous. Hence, we can anticipate the continuation of a phenomenon that we identified in our original study: a 'mobile digital divide' that jeopardises the capacity for low-income travellers to access the online services they depend on.

Mobile objects are also increasingly converging with wireless technologies, becoming part of the Internet of Things. Cars serve as a useful example. As Goggin (2012) explores, cars and mobile media emerged within the same historical advancement of time-space compression. Early automobiles were equipped with radios, and were one of the first spaces embedded with mobile telephones. First generation cellular phones were predominantly car phones. Satnav systems began to proliferate in cars during the 1990s. Over the course of the last two decades cars have converged with person-to-person mobile communication via smartphone docking systems. Wireless internet enabled vehicles are coming to market that boast multiple screens and internet services. The car has always been a core example of what Williams (1983) calls 'mobile privatisation', creating a bubble of private space within the public realm of city streets. Scholars also argue that mobile media create a form of mobile privatisation where virtual social connections are given preference over co-present ones (Ling, 2008), and digital entertainment is used as a barrier against the sensory complexity of public space (Richardson & Wilken, 2013). It is likely that these 'telecocooning' (Habuchi, 2005) effects of mobile media will amplify this phenomenon for drivers. Cars are increasingly equipped with various internal sensors, many of which are diagnostic, that can communicate with vehicular infrastructure. This innovation is often argued as enhancing safety and efficiency (Goggin, 2012), yet it also of course enables new forms of mobile surveillance.

Web Company Interventions

It is also essential to consider the role of Web giants such as Google and Facebook in the wireless ecosystem. Of course, these companies converge various online practices into their search, mapping and social media platforms, and thus drive a significant amount of internet use. They have also made significant investments in wireless infrastructure. Google Fibre promised super-fast gigabit internet when it began to roll out in 2010. As part of its marketing strategy it sort to provide capped free Wi-Fi to subscribers in certain American Cities. However the tremendous cost of laying fibre to the premises, cooperating and competing with incumbent carriers, and encouraging users to switch subscriptions has stalled its progress, and most probably doomed any future free initiatives of this kind (Watson, 2017).

Instead, Google (2017) has focused its investment in free wireless on its experimental 'Project Loon'. Google is launching hundreds of air balloons into the stratosphere that use directional antennas to pick up signals from hubs connected to the global internet and redirect them to antennas on the ground or directly to google devices. Machine learning algorithms are used to combine the signals of clustered balloons and enhance the internet capacity over a particular area. Google is seemingly driven by a philanthropic vision, to bring internet to remote areas on the unconnected side of the digital divide. A similar vision supposedly drives Facebook's (2017) subsidiary 'Internet.org', also known as 'Free Basics'. This organisation uses low orbit satellites to beam internet to remote areas, and has been successfully deployed in over 60 countries in Africa, the Middle East, Asia Pacific and Latin America. On its web-page, Free Basics promotes its 'success stories', all of which involve governments and non-profits providing health services to needy populations. However one must interrogate the commercial justification for this network. Like Facebook, Free Basics allows third party businesses and developers to access the network, grow their audience and scale their brand in a previously untapped market. Moreover, Facebook itself can expand into a vast new market, creating the potential for a significant increase in advertising revenue. These forms of monetisation rely on the now ubiquitous process of targeting individual users through highly granular surveillance (Trottier & Lyon, 2014). As well as the privacy issues that flow from this, services such as Free Basics have also been critiqued for establishing monopolies in new markets and hence undermining net-neutrality (Surendran, 2015).

Smart Environments

Implicit in all these developments is the construction of what are often called smart cities. Internet-connected devices that enable real time monitoring of environments are quickly proliferating throughout the world (Andrejevic & Burdon, 2014). This Internet of Things relies on the various forms of wirelessness mentioned above. Of course, there is a bias toward urban infrastructures in the very term smart cities, but sensor proliferation also occurs in rural and remote areas. Hence a more apt term may be 'smart environments'. We argue that smart environments should not be understood purely as sensors that monitor municipal services such as transportation. This suggests a centralised relationship between governance and infrastructure. Given the many public and private actors making use of these technologies, a better way to imagine smart environments is as an assemblage of cross-cutting sensor trajectories controlled by different institutions with different aims and responsibilities.

An ongoing critical question is: how can the public engage in the governance and transformation of smart environments and wireless public spaces, and hence in the architecting of particular sociocultural experiences? In her exploration of what she calls 'open source urbanism', Sassen (2011) persuasively argues that we must imagine new ways in which non-expert city dwellers can 'talk back' to the city, transforming its technological evolution. Sassen points to various instances in which technologists and artists 'hack' city spaces. Sometimes this occurs without explicit permission, and sometimes it occurs as a response to municipalities that make their city data and spaces open in planned 'hackathons'. This also involves imagining new forms of what McQuire (2016) calls 'participatory public space'. These are spaces that are not fully co-opted by commercial and surveillance logics, are inclusive and heterogeneous, facilitate serendipitous sociality, and experiment with commons-based forms of ownership, rights, and cultural production. This will be discussed in more detail below, but first we turn our attention to the significant issue of surveillance in the wireless environments just discussed.

SURVEILLANCE

In a study of the security of Wi-Fi use at 20 airports across the globe, Cheng and colleagues (2013) deduced a concerning variety of user-data that can be collected from an unsecure or partially-secure Wi-Fi connection. These include: device MAC address; device IP address; MAC and IP layer flags; the POP3, SMTP and HTTP email protocols and the personal information that comes with these such as age and work affiliation; and DNS and HTTP packets, enabling the surveillance of online behaviour, including connections with third party applications. This illustrates how the behaviour of users both online and offline can be tracked in a highly granular fashion.

Wi-Fi in particular can capture fine grained information about user movements. This has been expertly analysed by Whitworth (In Press) in relation to the multitude of visitor tracking techniques employed by the high-density Queen Victoria Market (QVM) in Melbourne's CBD. Whitworth illustrates the value of understanding visitor frequencies, movement patterns, and dwell times to dense commercial precincts whose stakeholders seek to optimise their businesses. Precinct managers at QVM have adopted a host of different wireless and surveillance technologies to gather this information, including infrared beam counters at entries and exits, overheat infrared lasers that can count pedestrian traffic and rely on Wi-Fi backhaul, CCTV with thermal imaging sensors that are capable of more accurate body counts, and finally free Wi-Fi services that are capable of granular tracking. QVM first employed the now defunct company Path Intelligence to monitor visitor origin and online activities via the free network. Inaccuracies in the system caused it to be replaced by the Blix Traffic System, which can identify when devices connect to Wi-Fi, and measure footfall and dwell periods across visitation times. QVM are deft at combining multiple solutions to the problem of accurate surveillance. The evolution of these solutions illustrates the canny way that organisations seeking to maximise surveillance can respond to data gaps and security measures taken by phone manufacturers, such as MAC address scrambling.

These forms of surveillance can be considered through a framework of security and privacy, as well as a broader political-economic framework that positions Wi-Fi within the global sensor and surveillance society. The Centre for Urban Research's report explores the way that Wi-Fi enable 'man-in-the-middle' attacks in which cybercriminals intervene in a connection, monitor personal information and redirect traffic (McShane, Gregory, & Wilson, 2016). This opens up serious risks, such as identity theft and financial fraud. The report's authors chart the evolution of security technologies such as WEP and WAP encryption, as well as regulatory efforts and government initiatives to educate citizens. Yet their survey of organisations providing free Wi-Fi finds that 1 in 5 hotposts are not adequately secured. Multiple Australian government departments provide advice on Wi-Fi security, yet these are not well known. Also, terms and conditions that lay out risks on splash pages are often ignored due to their length and complexity. Given this, the authors suggest a need for an increase in user agency when it comes to protecting their connections. This would involve a range of strategies, including anti-malware and virus software, as well as virtual private networks. The authors advance earlier work on risk awareness and privacy-enhancement behaviour, arguing that the promotion of the former will lead to the latter. While we do not disagree with this, a simple link between knowledge and agency can tend to overlook the ideological dimension of technology use. The internet and related technologies have a seductive allure that Mosco (2005) famously refers to as the 'digital sublime'. The attraction of the digital sublime undoubtedly inflects the degree to which people will take steps to protect their privacy, even if they are aware of privacy risks.

A security driven discourse is useful, but often focuses on technological affordances, regulations, and user choices. Looking at the broader picture of Wi-Fi roll out involves positioning it within a global political, economic and technological shift to a mediatised society that produces value through ubiquitous forms of surveillance, hence normalising the disclosure of information to companies and states. Technologies such as Wi-Fi act as important elements in what Andrejevic and Burdon (2014) call the 'sensor society'. Through this term the authors seek to capture the simultaneous diffusion of physical and code-based sensor mechanism through offline and online spaces, as well as the way these gather an increasingly vast amount of information on people, objects and environments. Lyon (2014) refers to this as 'actuarial surveillance': a regime that seeks to capture information on whole populations with the anticipation that it will be of use in the future. Here surveillance no longer seeks to discipline or control a known phenomenon. Instead it is predicated on the expectation of contingent opportunities for market expansion or risk control. This is partly driven by commercial imperatives, such as the identification of consumption patterns, advertisement markets, high credit risk consumers, and areas for profit-oriented business innovation, including automation.

It is also driven by national security-state imperatives, including the identification of potential terrorists or so called illegal migrants.

Far from being passive, these techniques enable the algorithmically automated 'social sorting' of people into categories of value and risk, and these in turn often are constructed through sexist, racist, geographic and socio-economic frameworks (Lyon, 2003). This can be applied to the way in which people access and move through public spaces, a problem we investigate in the following section.

WIRELESS PARTICIPATION AND INCLUSION

In the first decade of the 21st century, free Wi-Fi was heralded as a potentially radical technology, capable of stimulating new forms of civic participation and narrowing the digital divide (Crawford, 2013; Gibbons & Ruth, 2006; Sandvig, 2004; Shaffer, 2007). Today, when the uptake of both licensed and unlicensed wireless broadband has greatly increased, it is worth asking whether these original justifications for free Wi-Fi still hold value.

Economists continue to argue that free Wi-Fi services often produce market discrimination, reduce competition and raise rents (Potts, 2014). Yet these arguments often ignore the market failures that only governments have been successful in remedying (Crawford, 2013). Moreover, they ignore the surveillance practices that are now essential to the operations of private operators, and the way that these practices serve to commodify space and hence narrow the value of being in public.

The commodification of public space both attracts those with purchasing power and constructs them as consumers. This socially sorts lower income people out of consumption spaces. It also precludes other modes of 'being-in-public'. For example, less space is afforded for non-commercial cultural activities and informal economies such as unlicensed communal markets and busking performers. These are the strategic modes of resistance to neo-liberal capitalism that de Certeau (1984) argues are essential for preserving a sense of self and community. It is also not difficult to imagine these capacities being used by governments to shape public political action. The identification and quelling of protesters via locative social media has already occurred in the early period of the Egyptian Revolution and Syrian civil war. These problems reaffirm the necessity of thinking wirelessness through the lens of a more participatory concept of public space that focuses on a range of social, cultural and political values.

Issues relating to digital divides, inclusivity and equity also remain pressing. While the Australian mobile market is highly saturated, many developing countries remain largely unconnected to the internet (http://www.itu.int/en/ITU-D/Statistics). Moreover, many regions rely on Wi-Fi to access important services, but suffer from low bandwidth connections. Anayo and Horst (2016) have explored this in relation to the pacific island of Niue, which has evolved a dependence on internet services, yet remains beleaguered by low network quality, and has yet to be connected to the South Pacific Submarine Cable. Witness, then, the all too common issue of a population with a growing dependence on internet services that is unmet by support from key institutions.

Yet even in countries with a high mobile broadband saturation such as Australia, there remain significant digital inclusion and equity issues. Low income citizens, travellers and homeless people would all potential benefit from internet access. This becomes more significant when one considers the movement of government services online at both municipal, state and federal levels. The movement toward data hungry e-government services should parallel a digital inclusion strategy that carefully evaluates the social structures that impinge on wireless access. A key problem here is that Wi-Fi roll outs are generally aimed at higher income areas, even while they make cursory gestures to social inclusion (McShane, I., Gregory, M. A., & Wilson, C. K., 2014). This is also evident in developing and highly socio-economically stratified countries such as China. Wang (2016) and colleagues examine the free service provided to areas in Shanghai as a partnership between the government, China Mobile Communications Cooperation, China Unicom and China Telecom. At the time of writing this service had 450 hotspots active. Using publically available data, the authors find that the overwhelming majority of hotspots cover high income business centres and university campuses, and ignore nearby lower-income areas such as the Pengpu Xingun shantytown. One cannot help but notice the paradoxically elitist nature of free Wi-Fi.

Given these persisting problems, it is important to reaffirm alternative models of public space, participation, and inclusion. We conclude by describing the key ideas and ideals underpinnings these alternatives.

CONCLUSION

Public spaces provide a kind of commons, but this is often cut through with different forms of power and class (Harvey, 2012). Public spaces provide an opportunity for serendipity and cosmopolitan interactions with strangers, yet we often treat the mass of strangers as a kind of spectacle, and dwell instead in private and intimate social groups (Bauman, 1995). As Lambert (2017) explores, both online and in places like cities we often live at the threshold between intimacy and cosmopolitanism. This is a problem, yet being on the threshold also provides an opportunity for reaching out to the other side. That is, for experimenting with cosmopolitanism. As McQuire (2016) argues, the integration of digital technologies into public spaces has ambivalent consequences. On the one hand they commodify and monitor spaces, and potentially widen inequalities while narrowing forms of cultural expression, yet on other they can open up opportunities for diverse and new forms of sociality. The latter requires active intervention in design, planning and policy. It is suited to what Gordon (2013) calls 'designing for the civic web', and to non-technocratic, participatory forms of planning that take on diverse perspectives.

Such interventions can take their cues from a broader conception of public space that exceeds the rational of commodification. An influential theorist in this regard is Henri Lefebvre (1996), who in the 1960s argued for a 'right to the city', and championed play, difference and spontaneity as a challenge to forms of urban planning and governance that script movements toward commercial exchanges. This can produce a more complex and varied collection of urban experiences and values. The right to the city can be thought of politically, and hence brought into a discourse of public policy. It can also be thought of culturally, as part of an endeavour to engage *cultural citizenship* where people are afforded the right to express their cultural identities in public. Stevenson (2003) has argued for this form of cultural citizenship as a necessary aspect of cosmopolitan societies. Interestingly, this cultural dimension is often lacking in 'smart city' planning. Two of the dominant discourse in urban policy are 'smart cities' and 'creative cities', yet they remain strikingly unconnected. Smart city strategies are almost exclusively concerned with services, skills, innovation and business (Kitchin, 2014).

Returning to Wi-Fi, Antoniadis and Apostle expand on the Lefebvre's notional right to the city to identify the key aspects of the 'hybrid' city that connect social, cultural and technological practices. These consist of specific rights, such as:

The right to access the core resources of the city; the right to be represented, to be part of the collective identity; the right to participate in important decisions, such as urban policies and design; the right to ownership of the urban commons, which refers to commonly held property, and use, stewardship and management in common of the available and produced resources (Antoniadis & Apostol, 2014, p. 2).

This complex re-imagining of participatory, hybrid public space requires us to imagine Wi-Fi as more than just a 'fourth utility' that can be licensed and commodified, and hence have its access rights stratified (McShane, I., Gregory, M. A., & Wilson, C. K., 2014).

Where can we find these ideas put into practice today? A variety of art projects have conceptualised alternative forms of wireless life, often making use of smart devices, location-based apps, and different modes of representing and problematising the invisible spectrum (Gordon & de Souza e Silva, 2011). For example, Mark Shepard's 'Sentient City Survival Kit' uses mobile apps to encourage reflection on surveillance mechanisms as well as serendipitous interactions. Cultural institutions have also experimented with using their free Wi-Fi services to encourage serendipity and cosmopolitanism through media events that hybridise online and offline practices (McQuire, 2016). This has been taken to interesting places by Federation Square in Melbourne, which has curated trans-national communication between visitors in the square and visitors in a partner organisation in Soul (Papastergiadis, 2013). While events such as these provide a valuable testing ground for new concepts, and ones which have been shown to influence the service design of mainstream social media (Gordon & de Souza e Silva, 2011), events themselves are ephemeral occasions. While they may engender new perspectives toward difference, the social connections formed will likely be fleeting. Also, through looking at Melbourne's large scale urban art event White Night, Lambert (2015) has shown how such events are often co-opted by big business and the logic of the spectacle.

Remaining within the realm of creative cities and cultural policy, other possibilities for alternative Wi-Fi use are provided by creative precincts that have a long-term footprint in communities and often carry a mandate to perform community outreach to marginalised and disadvantaged people (Grodach, 2010). These precincts are increasingly being equipped with wireless networks and digital media, for example the Watershed Creative Studios in Bristol and the planned Collingwood Arts Precinct in Melbourne. The latter case provides an excellent example of the fusion of creative city and smart city policy, as it is funded by the Victorian State Government's 'creative state' initiative (as well as philanthropic sources), and is also employing digital urban design company Arup to provide a unique digital infrastructure for the precinct. The ideals driving this 'digital offer' include a focus on community building and serendipitous sociality. The precinct lies in the centre of Collingwood, a culturally diverse yet socio-economically stratified suburb in inner Melbourne that reflects a global trend toward multi-cultural, migrant cities that are becoming increasingly unequal due to wealth concentration and gentrification (Harvey, 2009).

It is also important to mention the NGOs and non-profits that continue to fight to serve free wireless to underserved communities in an era of intense spectrum competition. Take for instance the Mexican non-profit Indigenous Communities Telecommunications, which was granted a license concession to install networks in 356 marginalised local governments in Mexico. The project utilised Wi-Fi to keep operational costs comparatively cheap, and has already enhanced communication and services such as healthcare.

Of course, these projects occur at small scales, and seem dwarfed by telco roll outs supported by companies that can bid for large swaths of spectrum, support significant capital investments, and lobby regulators. Also, the discourse around the convergence of licensed and unlicensed spectrum and the coming age of 5G and smart environments often neglects to mention the critical issues just discussed. Hence there is a disconnect between planning and policy at the level of civic initiatives and the level of national and international institutions that mediate industry-government relations such as the Wireless Broadband Alliance and the International Telecommunications Union. This is largely an issue of representation, advocacy, and democracy from below. At a time when both the wired and wireless internet is becoming owned and controlled by fewer media empires, this form of democracy is needed more than ever. That urban infrastructure has, as Easterling argues, become to a certain extent an autogenous platform that exceeds traditional governance, also poses a challenge for this kind of democratic intervention. Hence, it is far from hyperbolic to argue for an almost activist approach to the wireless internet, and for the development of grass roots collectives and institutions that give new energy to movements such as the wireless commons initiative and what Sassen calls 'open source urbanism'.

REFERENCES

Anayo, J., & Horst, H. A. (2016). Technologies of the Nation: Public Wi-Fi and the Demand for More in Niue. *Information Technologies and International Development, 124*, 1–9.

Andrejevic, M., & Burdon, M. (2014). Defining the Sensor Society. *Television and New Media*. doi: 1527476414541552

Antoniadis, P., & Apostol, I. (2014). The Right(s) to the Hybrid City and the Role of DIY Networking. *The Journal of Community Informatics, 10*(3).

Anttiroiko, A.-V. (2012). Urban Planning 2.0. *International Journal of E-Planning Research, 1*(1), 16–30. doi:10.4018/ijepr.2012010103

Bauman, Z. (1995). Life. In *Fragments: Essays in Postmodern Morality*. Wiley.

Carr, S., Francis, M., Rivlin, L. G., & Stone, A. (1992). *Public Space*. Cambridge, UK: Cambridge University Press.

Cheng, N., Wang, X., Cheng, W., Mohapatra, P., & Seneviratne, A. (2013). *Characterizing Privacy Leakage of Public WiFi Networks for Users on Travel*. Paper presented at the INFOCOM, 2013, Turin. 10.1109/INFCOM.2013.6567086

Crawford, S. (2013). *Captive Audience: The Telecom Industry and Monopoly Power in the New Gilded Age*. Yale University Press.

de Certeau, M. (1984). *The Practice of Everyday Life*. Berkeley, CA: University of California Press.

de Souza e Silva, A. (2006). From cyber to hybrid: Mobile technologies as interfaces of hybrid spaces. *Space and Culture*, *3*, 261–278.

Easterling, K. (2016). *Extrastatecraft: The Power of Infrastructure Space*. London: Verso.

European Union. (2015). *A Digital Single Market for Europe: Commission Sets Out 16 Initiatives to Make it Happen* [Press release]. Retrieved on 10 June 2017 from http://europa.eu/rapid/press-release_IP-15-4919_en.htm

European Union. (2016). *WiFi4EU: Free Wi-Fi for Europeans* [Press release]. Retrieved on 10 June 2017 from https://ec.europa.eu/digital-single-market/en/policies/wifi4eu-free-wi-fi-europeans

Facebook. (2017). *Free Basics Website*. Retrieved 10 July 2017 from https://info.internet.org/en

Fortano, L. (2008). Anytime? Anywhere? Reframing Debates Around Municipal Wireless Networking. *The Journal of Community Informatics*, *4*(1).

Frow, J. (1995). *Cultural Studies and Cultural Value*. Oxford, UK: Oxford University Press.

Fung, B. (2016). New York Wants Everyone to Have Free, Fast Wi-Fi. *The Sydney Morning Herald*. Retrieved on 17 June 2017 from http://www.smh.com.au/technology/innovation/new-york-wants-everyone-to-have-free-fast-wifi-20160411-go3dow.html

Gabriel, C., & Fellah, A. (2016). *The Unlicensed Road to 5G: WBA Report 2016*. Retrieved on 10 May 2017 from http://www.wballiance.com/resources/industry-reports/

Gibbons, J., & Ruth, S. (2006). Municipal Wi-Fi: Big Wave or Wipeout? *IEEE Internet Computing*, *10*(3), 66–71. doi:10.1109/MIC.2006.60

Goggin, G. (2012). Driving the Internet: Mobile Internets, Cars, and the Social. *Future Internet*, *4*(4), 306–321. doi:10.3390/fi4010306

Google. (2017). *Project Loon Website*. Retrieved on 10 July 2017 from https://x.company/loon

Gordon, E., & de Souza e Silva, A. (2011). *Net Locality: Why Location Matters in a Networked World*. Malden, MA: Blackwell.

Gordon, E. (2013). Beyond Participation: Designing for the Civic Web. *Journal of Digital and Media Literacy*, *1*(1).

Graham, S. (2005). Software-Sorted Geographies. *Progress in Human Geography, 29*(5), 562–580. doi:10.1191/0309132505ph568oa

Grodach, C. (2010). Art Spaces in Community and Economic Development: Connections to Neighborhoods, Artists, and the Cultural Economy. *Journal of Planning Education and Research, 31*(1), 74–85. doi:10.1177/0739456X10391668

Habuchi, I. (2005). Accelerated Reflexivity. In M. Ito, D. Okabe, & M. Matsuda (Eds.), *Personal, Portable, Pedestrian: Mobile Phones in Japanese Life*. Cambridge, MA: MIT Press.

Harvey, D. (2009). *Cosmopolitanism and the Geographies of Freedom*. New York: Columbia University Press.

Harvey, D. (2012). *Rebel Cities: From the Right to the City to the Urban Revolution*. London: Verso.

Hutchins, B. (2016). 'We Don't Need No Stinking Smartphones!' Live Stadium Sports Events, Mediatization, and the Non-use of Mobile Media. *Media Culture & Society, 38*(3), 420–436. doi:10.1177/0163443716635862

Kitchin, R. (2014). The Real-Time City? Big Data and Smart Urbanism. *GeoJournal, 79*(1), 1–14. doi:10.100710708-013-9516-8

Kitchin, R., & Dodge, M. (2011). *Code/Space: Software and Everyday Life*. Cambridge, MA: MIT Press. doi:10.7551/mitpress/9780262042482.001.0001

Lambert, A. (2015). *False Shadows and Tactical Spaces: Encountering Transgressive Participation at White Night. In White Night: City as Event*. Melbourne: Research Unit In Public Cultures.

Lambert, A. (2017). *Cosmopolitan Intimacy and Filter Bubbles*. Working Paper Series, Culture, Media, Economy Research Unit.

Lambert, A., McQuire, S., & Papastergiadis, N. (2014a). Free public Wi-Fi and E-Planning: The Use of Online Planning to Build Better Networked Public Places. *International Journal of E-Planning Research, 3*(2), 70–85. doi:10.4018/ijepr.2014040105

Lambert, A., McQuire, S. & Papastergiadis, N. (2014b). Public Wi-Fi: Space, Sociality and the Social Good. *Australian Journal of Telecommunications and the Digital Economy, 2*(3), 45.41-45.17.

Lefebvre, H. (1996). The Right to the City. In E. Kofman & E. Lebas (Eds.), *Writings on Cities*. Oxford, UK: Blackwell.

Ling, R. (2008). *New Tech, New Ties: How Mobile Communication is Reshaping Social Cohesion*. Cambridge, MA: MIT Press.

Lobato, R., & Thomas, J. (2015). *The Informal Media Economy*. New York: Wiley.

Lyon, D. (2003). Surveillance as Social Sorting: Computer Codes and Mobile Bodies. In D. Lyon (Ed.), *Surveillance as Social Sorting*. London: Routledge.

Lyon, D. (2014). Surveillance, Snowden, and Big Data: Capacities, Consequences, Critique. *Big Data and Society, 1*(2). doi:2053951714541861

McQuire, S. (2016). *Geomedia: Networked Cities and the Future of Public Space*. Hoboken, NJ: Wiley.

McShane, I., Gregory, M. A., & Wilson, C. K. (2016). *Practicing Safe Wi-Fi: Assessing and Managing Data-Security Risks*. Retrieved on 17 June from http://cur.org.au/project/practicing-safe-public-wi-fi-assessing-and-managing-data-and-security-risks/

McShane, I., Wilson, C., & Meredyth, D. (2014). Broadband as Civic Infrastructure: The Australian Case. *Media International Australia, 151*(1), 127–136. doi:10.1177/1329878X1415100117

Mosco, V. (2005). *The Digital Sublime: Myth, Power and Cyberspace*. Boston: MIT.

Olma, S. (2016). *In Defense of Serendipity*. London: Repeater Books.

Papastergiadis, N. (2012). *Cosmopolitanism and Culture*. Cambridge, UK: Polity Press.

Papastergiadis, N.(2013). Mega Screens for Mega Cities. *Theory, Culture and Society*. doi: 0263276413503691

Potts, J. (2014). Economics of Public Wi-Fi. *Australian Journal of Telecommunications and the Digital Economy, 2*(1), 20.21-20.29.

Richardson, I., & Wilken, R. (2013). Parerga of the Third Screen: Mobile Media, Place, and Presence. In G. Goggin (Ed.), *Mobile Technology and Place*. Hoboken, NJ: Taylor and Francis.

Sandvig, C. (2004). An Initial Assessment of Cooperative Action in Wi-Fi Networking. *Telecommunications Policy, 28*(7-8), 579–602. doi:10.1016/j.telpol.2004.05.006

Sassen, S. (2011). *Talking Back to Your Intelligent City*. Retrieved from http://voices.mckinseyonsociety.com/talking-back-to-your-intelligent-city/

Shaffer, G. (2007). Frame-Up: An Analysis of Arguments For and Against Municipal Wireless Initiatives. *Public Works Management & Policy*, *11*(3), 204–216. doi:10.1177/1087724X06297347

Sheller, M., & Urry, J. (2006). The New Mobilities Paradigm. *Environment & Planning*, *38*(2), 207–226. doi:10.1068/a37268

State of Victoria. (2017). *Wi-Fi Initiative Website*. Retrieved 10 July 2017 from http://www.vic.gov.au/wifi.html

Stevenson, N. (2003). Cultural Citizenship in the 'Cultural' Society: A Cosmopolitan Approach. *Citizenship Studies*, *7*(3), 331–348.

Surendran, V. (2015). *All You Need to Know About Internet.org, Net Neutrality and Facebook Profile Picture Row*. Retrieved on 17 June 2017 from http://indiatoday.intoday.in/story/all-you-need-to-know-about-internetorg-net-neutrality-and-facebook-profile-picture-row/1/485496.html

Telstra. (2017). *Telstra Air Information Site*. Retrieved on 17 June 2017 from https://www.telstra.com.au/telstra-air/about-telstra-air

Thrift, N. (2004). Remembering the Technological Unconscious by Foreground Knowledges of Position. *Environment and Planning. D, Society & Space*, *22*(1), 175–190. doi:10.1068/d321t

Trottier, D., & Lyon, D. (2014). Key Features of Social Media Surveillance. In C. Fuchs (Ed.), *Internet and Surveillance: The Challenges of Web 2.0 and Social Media* (pp. 89–105). New York: Routledge.

Wang, M., Liao, F., Lin, J., Huang, L., Gu, C., & Wei, Y. (2016). The Making of a Sustainable Wireless City? Mapping Public Wi-Fi Access in Shanghai. *Sustainability*, *8*(2), 111–126. doi:10.3390u8020111

Watson, L. (2017). *What Happened to Google Fiber*. Retrieved on 21 July 2017 from https://www.gizmodo.com.au/2017/02/what-happened-to-google-fiber/

Whitworth, F. (in press). *Pedestrian monitoring at Queen Victoria Market*. Research Unit in Public Culture Working Paper Series. Melbourne.

Wi-Fi Alliance. (2016). *Wi-Fi® device shipments to surpass 15 billion by end of 2016* [Press release]. Retrieved on 17 June 2017 from http://www.wi-fi.org/news-events/newsroom/wi-fi-device-shipments-to-surpass-15-billion-by-end-of-2016

Williams, R. (1983). *Towards 2000*. London: Chatto and Windus.

Chapter 11
The Role of Social Media in Public Involvement:
Pushing for Sustainability in International Planning and Development

Tooran Alizadeh
University of Sydney, Australia

Reza Farid
Griffith University, Australia

Laura Willems
Griffith University, Australia

ABSTRACT

This chapter explores social media's potential to enhance public involvement to pursue sustainable practices on an international scale across planning and development projects. Using a case-study approach, the international institutions of the World Bank, UN-Habitat, Unilever, and World Business Council for Sustainable Development are investigated. The relationship between public versus the institutions' intake on sustainability is examined. Findings identify strong public push for increased sustainability in international development and show evidence of the ways in which international institutions respond to the public. Contributing to the social media research field, it offers an alternative application to the planning profession via e-planning. This could contribute to an extended form of public engagement through social media that goes beyond the limiting geographical borders of each local community, and assesses planning and development projects for their broader sustainability implications on an international platform.

DOI: 10.4018/978-1-5225-5999-3.ch011

INTRODUCTION TO ORGANIZATIONAL SUSTANIBALITY

The widely accepted definition of sustainable development/practice by Brundtland (1987) is development/practice that meets the needs of current generations without compromising the ability of future generations to meet their own needs. This has traditionally been interpreted on a local level, by analyzing local impacts and intergenerational equality. However, the rapid onset of globalization via multilateral trade negotiations over the last few decades has brought about an unprecedented increase in the local consumption of globally sourced goods and services. This definition of sustainable development detaches local planning, development, and resourcing from the issues of international inequality (AntonioDuro, 2012; Milanovic, 2011; Willard, 2009).

It is argued that the progress towards sustainable practice at an international level has been infrequent and inconsistent (Aras & David Crowther, 2009; Dyllick & Hockerts, 2002). Marsden (2000) outlines the wide variation in progress may be attributed to three distinct attitudes towards organizational uptake of sustainable practices:

- **Denial:** Placing responsibility on the government;
- **Reaction:** Acceptance as a contributor to the problem and taking guidance on solutions; and
- **Autonomous Action:** Recognition of power and proactive, independent acts.

Organizations in the reaction or autonomous response categories often utilize new forms of business models that have hybridized corporate and organizational strategy with sustainability and gained a competitive advantage in the market. The traditional business models, on the other hand, were mainly ruled by the economic system with only minor tradeoffs for social or environmental capital gain (Stubbs & Cocklin, 2008). Two major concepts arose in the literature in response to balancing this, namely 'corporate social responsibility' and 'corporate environmental responsibility/corporate sustainability' (Fernando, 2012; Montiel, 2008; Svensson & Wagner, 2010; Willard, 2009). In many cases though, the extent of these efforts are dictated by the organizations and vary greatly. Numerous international examples of over-consumption of local environmental resources, tax avoidance, and social injustices (Cerro Santamaria, 2013; Crane & Matten, 2016; Huang & Hsu, 2003) represent a disturbing level of ingenuity in the corporate sustainability agenda (Aras & Crowther, 2009). More recently, a growing line of critical scholarship has focused on the international corporations' philanthropic efforts, especially at the international level (Alizadeh, 2017; Gautier & Pache, 2015; McNeill, 2015; Paroutis, S., Bennett, M., & Heracleous, L., 2014). A particular focus has been on how firms have tried to influence national and urban governments, while engaging in providing philanthropic services (Kitchin, 2015). Critics argue that in many cases firms gain so much out of their involvement that do not fit with a pure pro bono, philanthropic model (Hack, 2013; McNeill, 2015). This line of argument suggests that new tools are required to bring organizations' international involvement with local projects to a global stage in a push for sustainable practice.

This chapter is an exploratory study into the potential of social media, as a new tool with the ability to give local level issues a global audience, in short time periods, to enhance public involvement and sustainable practices on an international scale across planning and development projects. Using a comparative case-study approach, the supply-demand relationship - on social media - for sustainable practice is examined. Building upon the literature, the chapter examines the virtually sourced data to discover how internationally renowned development and manufacturing organizations react to, or align with public request for sustainable practice. The results could open the door for an alternative form of public engagement through social media that goes beyond the limiting borders of each local community, and assesses planning and development projects for their broader environmental impacts on an international platform.

THE POTENTIAL OF SOCIAL MEDIA

Social media can be defined as an umbrella term, for blogs, wikis, social networks, and online communities; with the key feature of allowing users to connect (Carr & Hayes, 2015; Fieseler & Fleck, 2013). Some of the social media platforms are already incorporated in learning environments and currently affecting development policy through peer learning and by transforming social norms (Alizadeh, T., Tomerini, D., & Colbran, S., 2016; Denskus & Papan, 2013; Fieseler & Fleck, 2013; Willard, 2009). Social media are broad reaching and allow dispersed groups and individuals to form connections and share or promote information relating to common ideas, concerns, interests, or causes (LaRiviere, K., Snider, J., Stromberg, A., & O'Meara, K., 2012; Minton, E., Lee, C., Orth, U., Chung-Hyun, K., & Lynn, K., 2012; Walther & Jang, 2012). Social media has even been described as abstract forms of infrastructure for communities (Fieseler & Fleck, 2013), and is thought to be more accessible than traditional publications and discussions (Denskus & Papan, 2013). Social media, in particular, has played an important role in a number of civic uprisings around the word including but not limited to Arab Spring, the Occupy Movement, and recent presidential election campaigns in the US. This has prompted a new line of scholarship focusing on the role of social media in enabling participation, creating collective voice, and facilitating socio-political change (Comunello & Anzera, 2012; Howard & Parks, 2012; Kavada, 2015).

It is acknowledged that social media use is limited in some countries, but growing rapidly as cross-cultural platforms. Social media has also been shown as highly effective for reaching rural communities and developing countries with limited internet connectivity (Minton, E., Lee, C., Orth, U., Chung-Hyun, K., & Lynn, K., 2012; Willard, 2009). Increasing use is allowing local sustainability issues to gain international attention and support (Fieseler & Fleck, 2013; Mithas, S., Costello, T., & Tafti, A., 2011). It has been argued that the growing connections via social media may create relatively weak social ties (Ellison, N. B., Steinfield, C., & Lampe, C., 2007; Wellman, B., Haase, A. Q., Witte, J., & Hampton, K., 2001), but they seem sufficient for large enough networks of individuals with common concerns to gain enough power to challenge global superpowers (Willard, 2009).

The opportunities are possible due to social media platforms' participatory and interactive nature, which is also what separates them from the traditional forms of media (Barker, M. S., Barker, D. L., Bormann, N. F., & Neher, K. E., 2008; Minton, E., Lee, C., Orth, U., Chung-Hyun, K., & Lynn, K., 2012; Walther & Jang, 2012). In social media, the traditional producers of content engage with their receivers; and studies show that this eliminates the passive nature associated with traditional media (Fieseler & Fleck, 2013). Denskus and Papan (2013) found in their study that the active engagement provides opportunities for discussing and giving feedback on development issues and practices. The interactivity and user-generated content dominate social media, and provide avenues for bottom-up social movements and concerns (Dennis Linders, 2012; Juris, 2012; Waters, R. D., Burnett, E., Lamm, A., & Lucas, J., 2009). Fieseler and Fleck (2013) suggest that social media removes the formal hierarchy of society, and companies and organizations can no longer shield themselves from the involvement of stakeholders. Social capital is a means of empowerment, and social media can provide the necessary environment to create capital amongst users, empower stakeholders to raise their concerns through virtual cooperation (Scott & Liew, 2012; Tapscott & Williams, 2006).

In contrast to these possibilities of empowerment, there is also growing skepticism around the quality of online actions. Slacktivism or clicktivism has been recognized for its ability to draw attention to progressive causes; but questions of user validity, motivations for participation, and longevity in support stand and may lead to a reduction in perceived quality and therefore little overall impact on decision-making (McCafferty, 2011). Particularly in relation to the younger generations who are more enveloped in social media, participation in online campaigns has become too convenient and rarely translates into effective offline action (Vitak, J., Zube, P., Smock, A., Carr, C. T., Ellison, N., & Lampe, C., 2011). Moreover, concerns over the extent of cyber bullying across different social media platforms; and their special impact on vulnerable groups are growing (Best, P., Manktelow, R., & Tylor, B., 2014). Last but definitely not least, the growing role of social media in major terrorist organizations' operational strategy has brought the discussions to a new level (Klausen, 2015). The discussions are no longer simply about the power of social media to connect like-minded people; and to be influential. It is far more complicated.

Concerns about the role of social media in spreading fake news and also fake identities (accounts) on social media (Ferrara, E., Varol, O., Davis, C., Menczer, F., & Flammini, A., 2016) have broad implications on the legitimacy of citizen voice captured via the social media platforms; and raises skepticism about the level of trust that could be or should be put in the trends on social media. In response to growing concerns as such, different sectors (including governments, NGOs, and also private companies) have started developing social media strategies to capitalize on the opportunities and manage risks involved (Robin Effing & A.M.Spil, 2016; SergioPicazo-Vela, IsisGutiérrez-MartínezOpens, & Felipe Luna-Reyes, L., 2012).

Nevertheless, the complexity of the social media debates is, partially, due to its growing ability to connect people; and provide an alternative voice. It is recognized in some cases that within the virtual world of social media, corporations and governments – the officials in a sense - may not be the loudest voice, especially compared to the rising public disapproval of socio-political and environmental issues (Fieseler & Fleck, 2013; Walther & Jang, 2012). The dissatisfaction in the public realm by stakeholders and the community are frequently broadcasted over social media, in the direction of the responsible bodies (Jansen, B. J., Zhang, J. M., Sobel, K., & Chowdury, A. 2009). One of the powerful mechanisms of social media for gauging dissatisfaction and demand according to Walther & Jang (2012) is that of incidental aggregate user statistics, such as likes, shares, and retweets which are generally beyond the control of organizations and can represent mass evaluations of organizational reputation and dependability. In principle this can change the normal hierarchy and give the public the upper hand over authorities and decision makers.

However, as in traditional media, there is growing evidence of manipulating the public by green washing initiatives, or spearing false information on social media. In the context of this study – with attention to sustainability discussions - green washing is especially important. Following the Gulf of Mexico oilrig explosion and leak in 2010, the corporation responsible launched a traditional and online media campaign to superficially improve its environmental image. It is noted that while some met the campaign with skepticism, others showed approval and acceptance (Matejek & Gossling, 2014). More recently, in 2014, Nestle was accused of green washing for its customized recycling program, for Nespresso disposable coffee pods, which had only a negligible overall waste reduction impact. Unilever was also subjected to a Twitter campaign alleging that its role in a sustainable living online engagement platform did not accurately reflect the firm's true environmental impacts (Bowen & Aragon-Correa, 2014). Examples like these, show that social media can also act as an avenue for organizational advertisement and in some cases misrepresentation, to gain public approval. It is, however, important to acknowledge that in response to green washing attempts on social media, online activism focused on 'green washing exposed' or 'against green washing' has also increased in the recent years (Lyon & Montgomery, 2013).

The literature also discusses the impacts of the social media on organizational credibility, reputation, and trustworthiness (Minton, E., Lee, C., Orth, U., Chung-Hyun, K., & Lynn, K., 2012; Walther & Jang, 2012). Transparency of an organization is highly related to its credibility, and a growing number of studies agree that the internet and social media are highly transparent and open spaces in comparison to the traditional media (Denskus & Papan, 2013; Fieseler & Fleck, 2013; Waters, R. D., Burnett, E., Lamm, A., & Lucas, J., 2009). Research has found that organizations and businesses can utilize social media in a variety of positive ways, including managing operations (Waters et al., 2009), educating and engaging with volunteers, donors, and stakeholders (Fieseler & Fleck, 2013; Waters, R. D., Burnett, E., Lamm, A., & Lucas, J., 2009), and observing and gathering information from interactions with consumers (Minton, E., Lee, C., Orth, U., Chung-Hyun, K., & Lynn, K., 2012). The research in this area also acknowledges the frequency of cases in which high-profile firms are being accused of misleading communications about environmental activities or performance; and the damaging impact on the credibility of managers and researchers of organizations (Bowen & Aragon-Correa, 2014).

Nevertheless, it is hoped that openness embedded in the way social media operates will eventually reduce the incidence of corporate green wash (Lyon & Montgomery, 2013). Here, there is an emphasis on the dynamic nature of social media which allows for both bottom-up and top-down community involvement; as public recognizes the opportunities for engagement. From a bottom-up perspective, there is growing research on the use of community led groups to organize and coordinate via social media in opposition to planning, policy, and manufacturing or development processes (Evans-Cowley, 2010; Maireder & Schwarzenegger, 2012). Shav-Ami (2013) argues that a new and incredibly powerful type of social movement has started through social media. This type of movement has the potential to strongly impact any activity that angers community members and is not necessarily geographically restricted. In a top-down engagement perspective also, there is a growing line of literature (Afzalan & Evans-Cowley, 2015; Evans-Cowley, 2012) that argues planners can greatly utilize social media based opportunities to mobilize and organize citizens. This is part of a bigger trend in the literature which supports the idea that online spaces, such as social media, can be used as public spheres to deliberative democracy (Hanson & Hogea, 2012).

METHOD

Case Studies

This study uses an exploratory case study approach. The case study organizations of the United Nations (UN) Habitat, World Bank, Unilever, and World Business Council for Sustainable Development (WBCSD) have been selected based on the recognition of their contribution to sustainable development or production, and their mission statement (Table 1). All four case study organizations self-declare sustainability as the core of their international development focus.

More importantly, these diverse four organizations represent different types of non-state actors recognized to have the most power in the international development (SparkNotes, 2014). The UN-Habitat and World Bank as Intergovernmental Organizations (IGOs) perform in partnership with national governments; World Business Council for Sustainable Development as an International Non-government Organization (INGO) operates by building international partnerships between non-

Table 1. Case studies' contribution to sustainable development, mission statement, and presence on Facebook and Twitter

Name	International Development Focus/Mission	Sustainability Reputation in the Literature	Social Media Following – Stage One*		Social Media Following- Stage Two*	
			Facebook	Twitter	Facebook	Twitter
World Bank	Poverty reduction and sustainable development	Stiglitz 1999; Stone 2000	1.34M	791K	2.33M	2.48M
Unilever	Increase quality of life and promote sustainable living	Smith 2009; Fernando 2012	2.77M	61.4K	3.66M	167K
UN-Habitat	Finance socially and environmentally sustainable towns and cities including provision of housing	Cobbett 2001; Mukhija 2006	23.4K	29.2K	61.8K	66.4K
WBCSD	Sustainable global business; balance with society and environmental	Dyllick & Hockerts 2002	3.9K	8.5K	7.6K	24.9K

K= Thousand M=Million *Accurate at time of collection – 24 July 2014, and 16 Aug 201

government actors; and finally Unilever as a Multinational Corporation (MNC) is a business operating in multiple countries. The four case studies will provide great opportunities for this study to explore the potential role of the different international actor types in the promotion of sustainability at the international level. Due to their roles, these institutions have elements of both development and business practices. However, they are generally focused more towards one than the other. For example, IGOs most likely have development as their priority while a MNC focuses on production and business.

Having said this, we acknowledge that the four case study organizations have different levels of presence on social media. It would have been easier for our investigation if all case studies have had similar level of activities (e.g. Facebook and Twitter followers, posts and comments). Yet, the different levels of presence on social media could also be linked to the different levels of resources dedicated by each case study organization to their global outreach, which is per se worthy of investigation.

Data Collection

Data collection for the study was conducted in two stages, using different data sources. Stage one of the study has already been published (Willems & Alizadeh, 2015). Yet, this chapter – including both stages - provides an opportunity to shed light on the social media usage pattern change in the recent years.

Stage One

In 2014, we conducted the first round of data collection mainly based on the institutions' Facebook pages. The focus on Facebook was based on its overall status as the most dominant social media of the time, and also driven by the fact that it was also the major social media outlet used by the case study institutions (see Table 1). To complement the data collected via Facebook, we also analyzed each organization' official website; and sent a questionnaire to their social media administration.

Facebook Pages

The official Facebook pages of the case study organizations were used as primary data source. Measuring reactions to Facebook content is difficult and debated. A study conducted by Cvijikj and Michahelles (2013) examined factors affecting online customer engagement with Facebook brand pages. The sample included posts obtained from 100 sponsored brand pages and utilised a data-mining tool to extract seven elements from each post including: (1) the message; (2) post media type; (3)

number of likes; (4) number of comments; (5) number of shares; (6) creation time; and (7) time of last interaction (Cvijikj & Michahelles, 2013).

Even Facebook itself provides limited guidance on measuring reactions. Relevant examples of how Facebook officially gauges reactions include:

- Positive feedback is calculated based on the number of times audience took a desired action after seeing an ad, such as shared or liked it (Facebook, 2015).
- Negative feedback is based on the number of times audience hid an ad (Facebook, 2015).
- Post engagement is the number of people who have clicked, liked, commented on or shared a posts (Facebook, 2014).

Access to some measurements, such as post reach, hides, and clicks, is restricted to the page owner. As observed also, positive and negative measurement relates primarily to advertisements. Given that this chapter is exploring opportunities for community engagement, and the post reach measurements are not accessible, the quantitative measurement of post engagement including likes, comments and shares, is the most viable option.

This research structure (stage one) is similar to that of Cvijikj and Michahelles (2013), with a focus on assessing the engagement of Facebook users with official pages. We examined the official Facebook pages of the four case study organizations using content analysis. The last one hundred posts, on each Facebook page were cataloged for number of likes, shares, and comments.

Grounded theory was applied to qualitatively code the Facebook posts' content. Grounded theory method was adopted as it is attentive to issues of interpretation and does not bind itself too closely to longstanding assumptions (Suddaby, 2006). The grounded theory method utilizes flexible guidelines for data collection and analysis to construct theories grounded in the data themselves (Charmaz, 2006). Corbin and Holt (2005) discuss the fundamental steps of grounded theory that were taken at the stage one of this research to draw a theory out of data, including the constant flux between data collection and analysis:

- **Concept Identification:** On Facebook as posts are made, they spread throughout networks of people and accumulate feedback. A date was chosen which allowed sufficient time for posts to undergo this spread. Then, the 100 most recent posts from that date, from each case study were simultaneously collected in a one off collection. Due to the mechanics of Facebook, which allow observers to review past posts, this one off collection method is akin to observing and recording posts as they occur over multiple months. However, the method of recording upon occurrence poses the issue that it takes time for each post to receive feedback, as it spreads throughout the network. In contrast, the one off collection of previous posts has allowed sufficient time for posts to be observed by users and collect feedback. The posts underwent content analysis and the theme of each post was described. This theme description was based on open coding and a total of 44 minor themes were identified.

- **Data Reduction:** This followed the initial data collection and involved combining similarities in the 44 minor theme descriptions into 10 manageable categories, called subthemes. The subthemes included promotion, news, sustainable economics, equality, health, governance, community, resource management, sustainable urban planning, and environmental management. The ten subthemes were finally reduced to three major themes of economic, social, and environmental.
- **Theoretical Sampling:** Quantitative analysis of the data attached to the subthemes and major themes were examined in relation to one another. Emerging patterns were followed through new levels of data collection and analysis across the sources. Examples of the data which supported the emerging trends include viral Facebook posts, website information comparison to public Facebook comments, and questionnaire responses. It should also be noted that single posts were classed in multiple categories if there was evidence of several themes. This multiple theme coding denoted score frequency to be irrelevant unless calculated into a percentage over total category frequencies. This allowed each category to be recognized for its occurrence compared to other categories.

In order to address one of the repeated criticism of the grounded theory on the potential for bias based on individual researcher preconceptions of the themes and subthemes (Juliet Corbin & Strauss, 2014), this study went through careful examination of the steps taken. Initially, two individual researchers – both trained in grounded theory research – conducted the concept identification and data reduction steps – independent from each other. They then cross-examined each other's identified themes and subthemes to reduce the personal bias. The final sets of themes and subthemes were also examined by three researchers with extensive experience in grounded theory, to minimize the bias in this study.

Official Websites

The official websites correlating to each institution were also analyzed using content analysis. Once data patterns began to emerge from the primary Facebook analysis through grounded theory, the official websites were more closely examined to collect additional data as evidence of the patterns.

Questionnaire

Due to this being a case study research method, input from the case study organizations was desirable. A questionnaire was sent to the Facebook administrator of each institution's page via the Facebook message system.

The questionnaire was designed to gain insight into intentional institution strategies, attitudes, and responses towards demand or promotion of sustainable international development over social media. Due to the limited scope of the questionnaire distribution to the four institutions, responses could only be analyzed qualitatively. Verbatim quotes were matched to the patterns appearing in the content analysis of the Facebook pages.

Due to the size of these institutions, it was recognized that responses were not guaranteed. Where response was not received within a week, the questionnaire was resent through their official websites or to the personnel identified on their webpage as communication focal points. Despite the numerous attempts made via institutions' Facebook pages and their official websites, the UN- Habitat was the only case study organization that responded to the questionnaire. However, two out of three non-respondent organizations reacted to the questionnaire. The reasons given by Unilever and World Bank (passed up to head office/could not identify appropriate person with overview of all the issues) perhaps suggest a tension between horizontal and vertical management of the public relations in the organizations which could be the topic of investigation for another independent study. The fact that the questionnaire ignited some reactions from three out of four case study organizations suggests that the research was understood, and the questions did not seem problematic. At the end, we cannot really know the real reason behind the low response rate. It could be interpreted as a negative mark against the organizations' public engagement practices over social media; or it could be simply because they did not like our questionnaire enough to make it a priority. Whatever the reason, we decided to include the only response received from the UN-Habitat in our investigation; as it provided a unique window of opportunity to directly hear from one of the organizations. This could add another layer to our understanding of these institutions' approach towards sustainable practice.

Stage Two

Stage two of the data collection was conducted in 2017. The aim was to update the learning from stage one, and to investigate if the patterns observed in 2014 have changed at all. However, due to the fast pace of changes around the social media, an update of the research meant much more than just repeating the first stage of the study. First, the number of studies focused on social media analytics has grown dramatically over this short period of time; and research from a wide range of disciplines has started picking up the importance of analyzing the voices raised via the social media platforms (Batrinca & Treleaven, 2015; Khan, 2017). Second, there was a clear shift in the popularity of social media platforms and in particular the role that Twitter versus Facebook play in the public debates. Without getting into too much details of the elements contributing to the shifted role of Twitter versus Facebook, Table 1 shows a much faster growth rate of Twitter followers versus Facebook followers for the four case study institutions. This perused us to use Twitter as the main source of data collection in the stage two of our investigation in 2017.

In order to keep the two stages of data collection consistent and then comparable, we built on the findings from stage one to design data collection approaches for stage two. Results of stage one are elaborated in the later parts of the chapter. Nevertheless, as explained above, the data collected in stage one was categorized around three major themes of economic, social, and environmental; with special attention paid to how the three themes were promoted by the case study institutions; and then received by the public. We followed the same line of investigation in the stage two of the study, using Twitter data.

In order to do so, we extracted the latest 3200+ Twitter posts made by the four case study organizations (as the maximum number of tweets that, at this point of time, Twitter allows to be extracted for free), in a one off collection. Due to the mechanics of Twitter, which allow observers to review past posts, this one off collection method is akin to observing and recording posts as they occur over a longer period of time. Using data analytics, we then built a bank of all of the hashtags and their frequency used in the latest 1000 extracted Twitter posts per each case study. The hashtags with minimum two frequency then thematically analyzed around the three major themes of social, environmental, and economic – identified in the stage one of the study. The thematic analysis of the hashtags was conducted by two researchers to minimize the level of bias (similar to the technique adopted in the stage one of the study). The three theme based categories of hashtags (social, environmental, and economic) were then used to categorize the Twitter posts on the four case study institutions' official accounts. We, however, soon figured out that quite a sizeable number of Twitter posts did not use hashtags; and were being excluded from our investigation. So, in order to address the shortcoming, we extended our analysis by using hashtags as keywords. This basically resembled the context analysis adopted in stage one of the study; and meant that examples such as #budget vs. 'budget',

#women vs. 'women', and # deforestation vs. 'deforestation' were equally accounted for in our data analysis. Moreover, in order to have a comparable sample size across the four case study institutions we continued the study to reach 1000 Twitter posts accounted for using the above technique. It should also be noted that, similar to stage one, single posts were classed in multiple categories if there was evidence of several themes.

The 1000 twitter posts, categorized under three major themes of social, environmental, and economic, were then further analyzed based on the engagement statistics – the level of reaction they received from the public. The public engagement analysis was based on the number of likes and retweets. At this point, Twitter does not allow for mass collection of comments made on the posts. So, this limited our chance of analyzing the direction of public reaction to the posts; whether it was positive or negative or mix. In order to alleviate this shortcoming, we identified the Twitter post on each institution's account that had gone viral by attracting the biggest public reaction (the largest number of likes and retweet). We then conducted a more in-depth analysis of the viral tweets to further understand the nature of public reaction.

RESULTS

Stage One

Initial data analysis of the subthemes allowed the researchers to identify which particular subthemes occurred most in each institution's Facebook posts (Table 2). It was noted that World Business Council went through sporadic periods of inactivity on Facebook, sometimes for multiple months, and therefore limited data was available for this case.

The following provides a theme overview and description, in the format of major themes, subthemes, and their descriptions:

- Economic Major Theme
  - **Promotion:** Any post, which aimed to acquire money or increase popularity, such as donation collection, events or project advertisement, product marketing, competitions, Twitter movements.
  - **News:** Any post, updating followers on recent institution news, information, or publications such as reports.
  - **Sustainable Business Practice:** Any post relating to making business processes and practices more sustainable (general advice for businesses and not just the case study institutions).

- Social Major Theme
    - **Equality:** Posts relating to increasing equality such as through humanitarian aid, gender and age equality, poverty reduction, education, and unemployment.
    - **Health:** Posts containing information on disease, healthcare, and safety.
    - **Governance:** Posts relating to governmental transparency, policy, war, peace, and reform.
    - **Community Building:** Posts relating to capacity building and community engagement.
- Environmental Major Theme
    - **Resource Management:** Posts relating to the management of resources such as water, natural resources, and food.
    - **Sustainable Urban Planning:** Posts containing information about sustainable transport, infrastructure planning, sustainable development, or land use management.
    - **Environmental Management:** Posts relating to managing and protecting environment including climate change, conservation, disaster and risk management, and pollution and energy waste reduction.

The next step of data analysis involved merging each organization's subthemes into major themes to gain a clearer overview. This was used to measure the supply and demand ratio of the posts, by comparing engagement statistics. The occurrence graph (Figure 1) represents the percentage of posts that related to each major theme

Table 2. Frequency of subthemes in each organization's Facebook posts

Subtheme	World Bank	UN-Habitat	Unilever	WBCSD
Promotion	13	11	9	9
News	16	47	35	9
Sustainable Economics	21	2	0	14
Equality	46	20	12	1
Health	4	0	17	2
Governance	6	5	0	3
Community	2	6	1	0
Resource Management	3	3	18	2
Sustainable Urban Planning	3	36	4	2
Environmental Management	3	15	34	4
TOTAL	117	145	130	46

Note: Frequency total greater than total posts as some posts contain multiple subthemes.

by each organization. The popularity chart (Figure 2) was calculated by summing the number of likes, comments, and shares for each post, and then graphing the median of all summations separated by major theme. Average scores provided less accuracy than medians, as they were prone to distortion from viral posts.

The comparison between major theme occurrence (Figure 1) and popularity (Figure 2) shows that the interests of the organizations generally influence and cause a preference in the type of information presented in their posts. However, the public does not unconditionally support these interests. The public popularity data shows a relative balance for social, environmental, and economic information. There is also an increase in the popularity of social and environmental information when compared with its occurrence in the posts. This redistribution of information demand-supply suggests that the public may have interests and demands that are not being met, particularly in relation to environmental and social sustainability.

The questionnaire responses and official website comparisons provided further evidence of a publicly, and not necessarily an institutionally, driven demand. As described earlier, despite the numerous attempts, the questionnaire received a response from only one case study organization - the UN-Habitat. The UN-Habitat Social Media Focal Point's response to the question '*Does your organization identify as: Sustainable Enough, Good But Could Be Better, Not Very Sustainable, Not Sustainable At All?*', was '*Sustainable Enough*'. Similarly, when asked '*Does your organization aim to promote sustainability more/the same/less through social media than currently?*', the organization responded '*The same*'. These answers

Figure 1. Comparison of case studies' major theme occurrence (via Facebook)

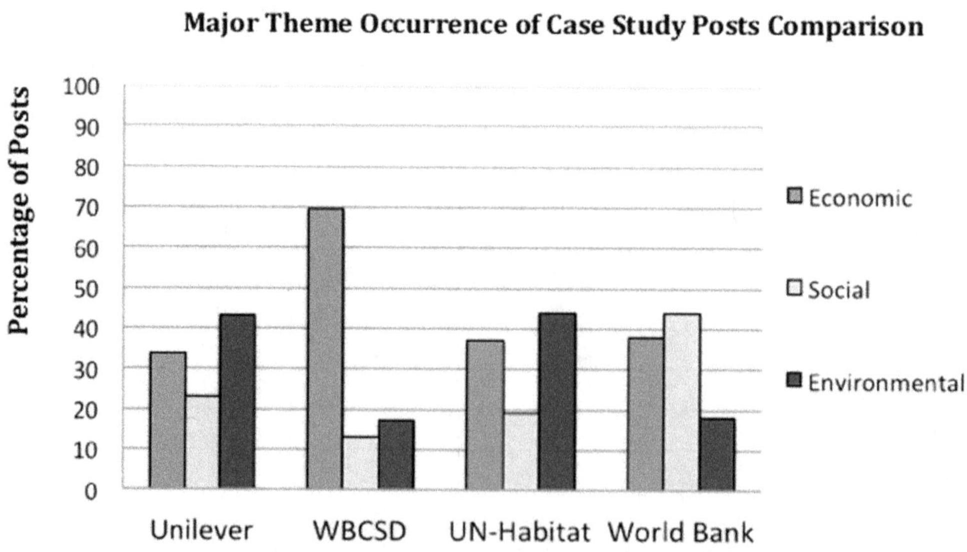

Figure 2. Comparison of case studies' major theme popularity (via Facebook)

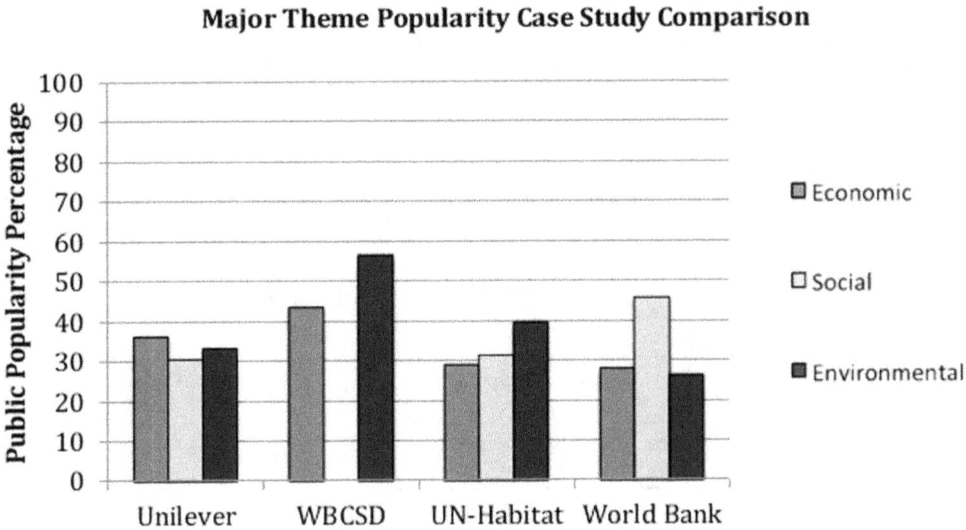

suggest that the organization (or at least its social media manager) is happy with the level of emphasis put on sustainable practice; and considers it a high standard model. This attitude suggests that the international development organization is not necessarily eager to promote sustainability to the public more than it has already been. This does not necessary align with the public demand; especially in respect to environmental and social dimensions of sustainability.

Comparisons between the institutions' official website information and the public comments on Facebook provided further evidence of the emerging pattern described above. A clear example of public demand versus the organizations' contentment is shown in the Unilever case. On their website, Unilever notes they have met their 100% sustainable palm oil sourcing goal by marking it as achieved. However, on their Facebook page a large number of posts were confronted with comments demanding things such as 'stop green-washing' and 'commit to going palm oil free'.

Examining viral posts reveals a tangential aspect of public demand. These are posts originally made by the organizations which attract enourmous attention from the public and receive an amount of likes, comments, and shares that far exceeds the average counts. Viral posts reach a greater number of users through public sharing and newsfeed views. These posts may represent the greatest demands of the public.

Within the total of analyzed posts, eight were considered to be viral outliers in comparison to other posts within each case (Table 4), including:

Table 3. Summary of viral Facebook posts (originally posted by the organizations and turned viral by the public)

Viral Post	Institution	Post Key Words	Popularity*	Trend
1	World Bank	'Empowering citizens'/ '... participate in decision-making process'	2429**L**/11**C**/24**S**	Empowerment Participation
2	UN-Habitat	'...participatory and empowering'	198**L**/1**C**/136**S**	Empowerment Participation
3	World Bank	'...wants to know what you...'/'...leave a comment or post a photo telling us what you...'	741**L**/12**C**/2**S**	Call for participation
4	Unilever	'Have an idea for change? Enter it...'	729**L**/5**C**/3**S**	Call for participation
5	Unilever	'Have a great idea of your own? Enter your big ideas at...'	15095**L**/21**C**/511**S**	Call for participation
6	Unilever	'Do you have a flow of great ideas? Share them with us...'	1580**L**/35**C**/32**S**	Call for participation
7	UN-Habitat	'Thank you to the city of Medellin and all its inhabitants...'	256**L**/10**C**/69**S**	Acknowledgement
8	Unilever	'...big thank you... to our 174,000 amazing employees...'	2546**L**/156**C**/222**S**	Acknowledgement

* L=Likes C=Comments S=Shares

- World Bank and UN-Habitat both experienced two posts each that this research identified as viral outliers.
- Unilever was found to have four posts as viral.
- World Business Council for Sustainable Development did not have any posts that could be considered viral which may be partially attributed to its sporadic periods of inactivity.

The trends in these viral Facebook posts have been identified across the case studies and a summary table (Table 3) shows the following relationships within the content:

- The first trend in the viral posts relates to calls for participation, and empowerment, for example a post calling for the public to engage with an online discussion.
- The next set of viral posts do not contain the specific words of engagement and participation, but ask for the public to submit personal stories or ideas

about improving social or environmental issues. These can be interpreted as calls for engagement.

- The final trend in the viral posts represents a related but slightly different pattern of engagement. In these posts, rather than calling for input or empowerment, the engagement is seen through acknowledgment of community members. It personalizes and thanks regular people for enabling the organizations to conduct their international development activities.

Stage Two

Step two of the study was designed to verify the trends observed in stage one; and to see how the dynamics identified across the social media accounts of the four case study organizations - using Facebook data - in 2014 compares to what is happening in 2017 - using Twitter data. The results are illustrated in Figures 3 and 4. A comparison of the two figures shows the supply and demand ratio based on the presence/occurrence of the three major themes of social, environmental, and economic within the Twitter posts, and the level of reaction that they received form the public (engagement statistics). The occurrence graph (Figure 3) represents the percentage of tweets that relate to each major theme by each organization. The popularity chart (Figure 4) was calculated by summing the number of likes, and

Figure 3. Comparison of case studies' major theme occurrence (via Twitter)

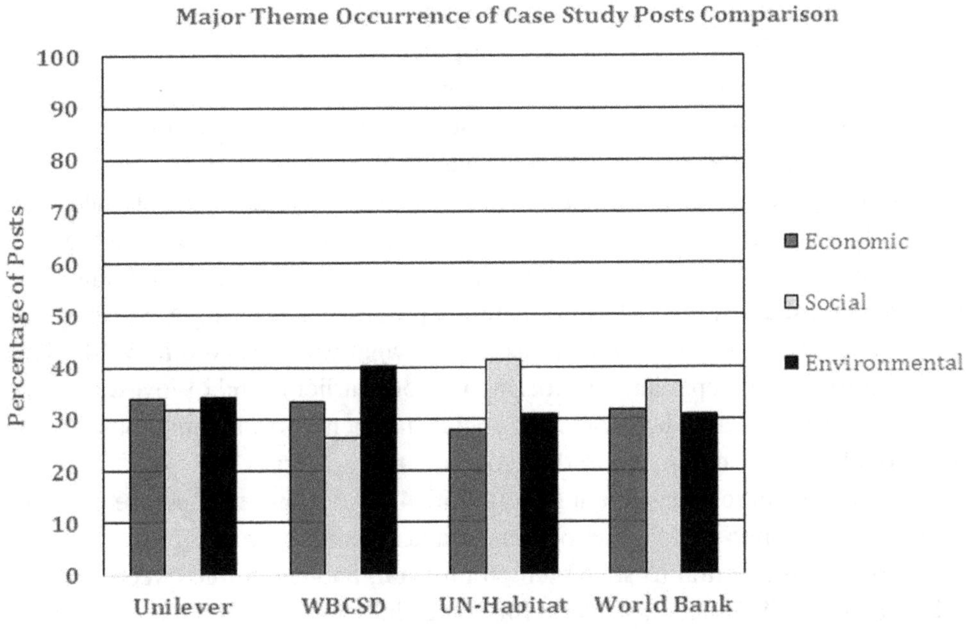

Figure 4. Comparison of case studies' major theme popularity (via Twitter)

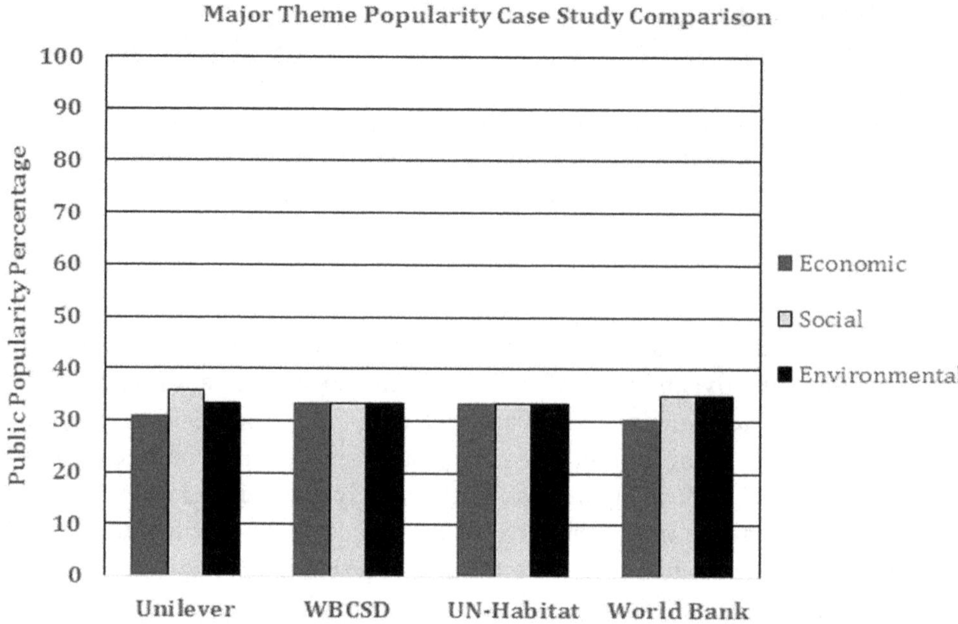

retweets of each tweet; and then graphing the median of all summations separated by major theme. Average scores provided less accuracy than medians, as they were prone to distortion from viral posts.

The trends observed in Figures 3 and 4 are not exactly the same; which suggests that public doesn't necessarily follow the pace set by the organizations. This is similar to the observation made in the stage one of the study focusing on Facebook. However, what is different – compared to stage one – is the level of balance both from organization side and also public's reaction.

First, the patterns observed within each organizations' activities on the social media (Figure 3) seem to have changed towards being more balanced (see Figure 1 for comparison). To be specific, there is an increase of environmental and social posts across all case studies. Interestingly economic posts are not the highest priority for any of the four case study organizations (unlike what was observed in 2014). This is an interesting development, considering that our earlier round of investigation – based on Facebook data - had portrayed social and then environmental posts as the more popular ones – compared to the economic focused posts.

Second, the public engagement data (Figure 4) also shows an increased level of balance in reaction to social, environmental, and economic tweets. There is almost perfect balanced reaction to social, environmental, and economic tweets on UN-Habitat and WBCSD accounts. World Bank and Unilever show just less than perfect balanced reaction with slightly stronger public engagement with their environmental and social posts.

Moreover, examining viral posts reveals a tangential aspect of public demand. These are tweets which attracted enourmous attention from the public and received a large number of likes, and retweets that exceed the average counts for each Twitter acount. Viral posts reach a greater number of users through public sharing (retweets). These posts may represent the greatest demands of the public.

Within the total of analyzed posts, seven were considered to be viral outliers in comparison to other tweets within each case (Table 4), including:

- World Bank has the biggest followers' base among the four case studeis. We have identified two tweets as viral outliers; as the level of public enagement with them was far bigger than other tweets on the organization's account.
- Unilever and UN-Habitat both found to have two tweets as viral. The size of public engagement (the total number of likes and retweets for each viral post) differs across the two companies' accounts which is an indication of different level of social media engagement for the two companies.
- World Business Council for Sustainable Development did not have any tweets that could be considered viral which may be attributed to the fact that it has the smallest follower base on the Twitter among the four cases.

Table 4. Summary of viral Twitter posts (originally posted by the organizations and turned viral by the public)

Viral Post	Institution	Post Key Words	Popularity*	Trend
1	World Bank	Climate change is undeniable; solutions	18102	Environmental Management – Environmental
2	UN-Habitat	Happy Canada Day	13425	Community Building -Social
3	World Bank	Human progress; world betterment	9450	Equality - Social
4	UN-Habitat	Don't stop the refugees; stop the wars that produce them.	8537	Equality - Social
5	World Bank	World's forests; stop deforestation	4752	Resource Management-Environmental
6	Unilever	sustainability goals; growth	2622	Sustainable Urban Planning-Environmental
7	Unilever	young entrepreneurs; world-changing idea	2471	Empowerment – Social

* likes + retweets

The trends in these viral Twitter posts have been identified across the case studies and a summary table (Table 4) of the viral tweets shows interesting patterns. First, none of the viral tweets has an explicit economic focus. This is quite similar to what was observed in the stage one of the study (focusing on the Facebook data). Second, the identified viral posts all focus on bigger picture social and environmental issues (climate change, and refugee crisis for example). Earlier Facebook investigation in stage one also showed an extraordinary level of public response when betterment of major environmental and social issues was the core of a social media post made by the organizations.

DISCUSSION

Our two stage study shows interesting transformation in the dynamic relationship between the four case study organizations in one hand, and the public in the other hand in the short period of 2014-2017. In order to elaborate the transformation observed, we offer the lessons learned in each stage separately before offering the overall learning:

Learning From Stage One

When comparing the relationship between case study organizations' major post themes and public popularity responses, it was found that individual case studies generally favoured one major theme more strongly than the others. In response to this, popularity graph (Figure 2) showed public reaction was relatively balanced across the three major themes of social, environmental, and economic. This trend was evident in all case studies. The conclusion that began to surface was that the public users' level of support was not necessarily dictated by what the institutions posted about. It also showed a demand from the public that is conducive to the balanced triple bottom line for sustainability and showed a public demand for more sustainable information, and potentially outcomes, in development and manufacturing.

It was also noted that the social theme was generally the least common major theme across case study posts, but it received a phenomenal increase in relative popularity across all the case studies except WBSCD which was incalculable due to the low number of social posts. Such a pattern was also corroborated when the viral posts were examined. Emotive and personal connection is one possible explanation why social posts, especially those relating to community empowerment and acknowledgement, were strongly supported. A study by Swani et al. (2013) on the quantitative public popularity of different corporate posts found a trend in which using emotional content in posts generated more likes. This, however, does not completely explain the pattern in our study, as many other emotional posts of both environmental and economic themes were present, but were not as popular. It should be noted that not all social posts were strongly supported, but the trends indicate a generally high distinction between occurrence of social themed posts and their popularity rate.

The World Urban Forum by UN-Habitat is an example of an online public involvement post that received viral proportions of attention. The World Urban Forum is hosted biennially in different locations as an international discussion of the greatest issues facing urban settlements, and is attended by, but not restricted to, government bodies, community groups, professionals, academics, and researchers (UN Habitat, 2014). The forum also has an educational exhibition open to the public and many of the forum's submissions are shared through social media. Turner-Lee (2010) suggested that with the growing technological innovation around the globe, new forms of social media and online platforms could be designed to make virtual public participation a practical revolution whilst being supplemented with strategies to provide internet access and a more equitable engagement process than traditional physical engagement. As technologies advance and become more widely available, planners and development institutions can take advantage of these opportunities and improve delocalized community engagement practices.

Lessons From Stage Two

Our analysis based on the 2017 Twitter data shows a move towards balance. This move is evident when looking at four case study organizations' tweets; and also the reaction received from the public. Interestingly, compared to 2014, organizations seem to have an increased number of social and environmental posts. Yet again, the level of balance observed in public reaction to posts exceeds the one in organizations' posts. Similar to 2014, this could still be interpreted as a sign for public push for a balanced approach to the triple bottom line of sustainability (economic, environmental, and social) in development, planning and manufacturing; a sign that the public is still ahead of what organizations are offering in terms of balanced approach.

The trend observed in viral posts are, however, different and may help better understand the direction of public push. Similar to what was seen in 2014, no economic post has gone viral. However, there seems to be an equal representation of environmental and social tweets among viral posts. This shows an increased attention to environmental issues (such as climate change and deforestation) in comparison to the pattern observed in 2014. In summary, public engagement with social and environmental tweets on the four case study organizations' accounts are far bigger than the level of attraction for economic posts.

Overall Lessons Learned

Comparative analysis of trends observed on the social media accounts of the four case study organizations in stages one and two of the study suggests that the gap between public demand and what organizations are offering in terms of promoting sustainability might be decreasing. This bold observation is based on the move towards balanced offering across the three major themes of economic, social, and environmental on the social media platforms of all case studies. While the level of balance reported in the public engagement (Figure 4) is still ahead of the balance observed in the organizations' posting (Figure 3); the difference is definitely smaller than what observed in 2014 (see Figure 1 and 2).

The golden question, however, is 'What does this mean? '. Is there a real shift across the case study organizations towards sustainability and sustainable practice? Or have they just got better at playing the game? Unfortunately, our study cannot answer this golden question, as far more detail analysis of their practice on the ground – and not just across social media – is needed. Having said this, the change observed in the social media performance of the case studies suggests that the big international organizations have been paying attention and reacting to public. This per se might be an outcome which shows the potential of social media to gain attention of big players.

In a sense, the trend observed in stages one and two of this study may pertain to the growing use of social media as new avenues to instigate bottom-up community engagement. There is growing research on community groups who share information and coordinate via Facebook and Twitter. A growing number of studies (Afzalan & Evans-Cowley, 2015; Evans-Cowley, 2010, 2012) found a significant number of planning-related groups on Facebook and Twitter – many of which were organized to oppose specific developments or plans. Social media has also been recognized as a platform that engages a more targeted youthful audience (Park, N., Kee, K. F., & Valenzuela, S., 2009). There is, however, conflicting information on whether this online subscription to a cause, translates to physical participation.

It is difficult to explain how, why, and if civic and public involvement over social media platforms can be generated into real life actions. However, recent major events have been researched as bottom-up revolutions through social media, such as the Arab Spring Twitter movement, in which four world leaders were ultimately removed from their positions (Bala, 2014). On this severe level of involvement, social media are concluded to have played a major and phenomenal role in facilitation. However, on a less dramatic level, a survey found that administrators of planning issue Facebook groups reported that community members did join the groups, but little physical action resulted (Evans-Cowley, 2010). This suggests that social media may influence large scale, even globally significant issues, more easily than small local issues. This is related to the low rates of participation in traditional engagement techniques. A study by Vogel et al. (2014) found that in local governments, the highest participation rate of residents was 19%, while the majority of local governments recorded comparatively low participation rates.

Research (Boulianne, 2009; Earl & Kimport, 2011; Quan-Haase, A., Wellman, B., Witte, J. C., & Hampton, K. N., 2002) over the last decade or so argues that the internet could supplement engagement but does not necessarily increase level of participation. This, however, could be due to the inequality and inaccessibility issues of traditional community engagement, as it requires extensive commitments from people in time and relocating to be physically present at meetings (Mandarano, L., Meenar, M., & Steins, C., 2010). Nevertheless, social media and the internet have challenged much of the geographical restrictions associated with obtaining local information - as citizens who are not located in close proximity could have access to development and planning information and join the local community.

SIGNIFICANCE OF FINDINGS

Our findings have contributed to the field of research in several ways. Firstly, the chapter has utilised a methodology process and combination that can contribute to more efficient and meaningful social media research. Secondly, the research suggests a role for global citizens to take part – via the social media- in demanding sustainable practice well beyond the geographical borders of local developments and projects. And finally, it has identified potential links between social media and public engagement in the virtual sphere to pursue more sustainable development outcomes. This should be of interest to both planners and policy makers.

The chapter – within the limitations of a comparative case study research - has investigated international development sustainability demands on Facebook and Twitter, and identified that the public push could be a driving factor for increasing organisational sustainability – at least on their social media platforms. This study has also looked at the types of information that the public Facebook and Twitter users seem to respond to and engage with, around the topic of international development. This could be further explored by examining a greater sample of posts, over an extended period of time; and also a questionnaire submitted to the public may strengthen the findings of this research.

The scope could also be scaled down to see if the findings also fit at national, state, or regional levels of impacts. Conducting this research using varied scopes can have the advantage of being able to follow similar research methods. Upon reflection of the research methods and processes, they have allowed the researchers to quickly navigate and analyse the complicated mountain of information sent over Facebook and Twitter. The grounded theory approach – used in stage one - was in particular a key to unlocking a directed and meaningful pattern in the highly variable data. This then guided the structure of investigation in stage two.

Finally, global online public engagement could help planners to alleviate the issues around the traditional engagement and exclusionary issues. A tangential line of research analysing participation and methods in the new online engagement forums will be invaluable to planners. Future research needs also link to the 'corporate social responsibility' and 'corporate sustainability' (Fernando, 2012; Montiel, 2008; Svensson & Wagner, 2010; Willard, 2009) frameworks; in order to actively involve international organizations in the fight for sustainability across the world.

CONCLUSION

This chapter aimed to explore the social media's potential to improve public involvement and sustainable practices on an international scale across planning and development projects. It was based on a multiple case study analysis, by comparing the supply of online information from development and manufacturing organizations, and the public response to this information.

The findings of this research provide insights into the sustainability and consultation demand-supply relationship on social media. The research suggests that such relationship is driven by the public users. Although the development institutions do not necessarily drive sustainability, there is evidence that they are responding to the public demand for sustainability. Nevertheless, this study warns against the danger of green washing in the online domain, and is yet hopeful that the two-sided conversation in the social media provides an opportunity to hold the international organizations accountable; and reduce (and not eliminate) their chance of promoting sustainability hoaxes. This optimistic – and yet pragmatic- view is in line with the previous research (Bowen & Aragon-Correa, 2014; Jacquet, 2015; Lyon & Montgomery, 2013) which suggests public pressure - in the form of public shaming via social media – can challenge corporations and even governments to change policies and behaviors that are detrimental to the environment.

The fundamental premise of this chapter, which targets the role of social media in public engagement with planning and development projects, has been previously discussed in the literature (Afzalan & Evans-Cowley, 2015; Evans-Cowley, 2010, 2012). But previous studies generally took the approach of discovering whether subscriptions to e-planning and development opposing groups, can lead to physical participation. Results were mixed, and it has been more commonly agreed upon that traditional engagement is supplemented by social media activity, but not increased. However, this chapter relates to public involvement from a different angle. It acknowledges the issues with clicktivism and then explores Facebook and Twitter not as just platforms for information sharing and community organization, but for actual online engagement. This extension could potentially offer an alternative venue to address the traditional low engagement rates and accessibility issues. This could simultaneously provide a space for the virtual citizens of the world – regardless of their physical location – to voice their demands for increase in the sustainability of international development.

The framework developed in this research is still in its infancy. In any future attempt to further develop such a framework, the unfolding limits of social media need to be accounted for. It may offer an application to the planning profession via e-planning, which could result in an alternative form of public engagement through social media. This alternative form of public engagement could go beyond the limiting borders of each local community, and assess planning and development projects for their broader environmental impacts on an international platform.

With social media becoming an ever-more influential factor in the operations of international development, and the behaviors of the public, there is a need to further researching and trialing the practical applications of this in the planning profession. Regardless of country, geographical influence level, public institution, or private, research into the potentials of social media in e-planning should be taken up as a unique opportunity to create more equitable and encompassing planning and development outcomes.

REFERENCES

Afzalan, N., & Evans-Cowley, J. (2015). Planning and Social Media: Facebook for Planning at the Neighbourhood Scale. *Planning Practice and Research*, *30*(3), 270–285. doi:10.1080/02697459.2015.1052943

Alizadeh, T. (2017). An investigation of IBM Smarter City Challenge: What do participating cities want? *Cities (London, England)*, *63*, 70–80. doi:10.1016/j.cities.2016.12.009

Alizadeh, T., Tomerini, D., & Colbran, S. (2016). Teaching Planning Studios: An Online Assessment Task to Enhance the First Year Experience. *Journal of Planning Education and Research*. doi:10.1177/0739456X16647162

Aras, G., & Crowther, D. (2009). Corporate Sustainability Reporting: A Study in Disingenuity? *Journal of Business Ethics*, *87*(S1), 279–288. doi:10.100710551-008-9806-0

Aras, G., & Crowther, D. (2009). Making sustainable development sustainable. *Management Decision, 47*(6), 975–988. doi:10.1108/00251740910966686

Bala, M. (2014). Civic engagement in the age of online social networks. *Contemporary Readings in Law and Social Justice, 6*(1), 767–774.

Barker, M. S., Barker, D. L., Bormann, N. F., & Neher, K. E. (2008). *Social Media Marketing: A Strategic Approach*. South-western Cengage Learning.

Batrinca, B., & Treleaven, P. C. (2015). Social media analytics: A survey of techniques, tools and platforms. *AI & Society, 30*(1), 89–116. doi:10.100700146-014-0549-4

Best, P., Manktelow, R., & Tylor, B. (2014). Online communication, social media and adolescent wellbeing: A systematic narrative review. *Children and Youth Services Review, 41*, 27–36. doi:10.1016/j.childyouth.2014.03.001

Boulianne, S. (2009). Does Internet Use Affect Engagement? A Meta-Analysis of Research. *Political Communication, 26*(2), 193–211. doi:10.1080/10584600902854363

Bowen, F., & Aragon-Correa, J. A. (2014). Greenwashing in Corporate Environmentalism Research and Practice: The Importance of What We Say and Do. *Organization & Environment, 27*(2), 107–112. doi:10.1177/1086026614537078

Brundtland, G. H. (1987). *Our Common Future: Report of the World Commission on Environment and Development*. Oxford, UK: Oxford University.

Carr, C. T., & Hayes, R. A. (2015). Social Media: Defining, Developing, and Divining. *Atlantic Journal of Communication, 23*(1), 46–65. doi:10.1080/15456870.2015.972282

Cerro Santamaria, G. (Ed.). (2013). *Urban megaprojects: a worldwide view*. Bingley, UK: Emerald Group Publishing Limited. doi:10.1108/S1047-0042(2013)13

Charmaz, K. (2006). *Constructing Grounded Theory: A Practical Guide Through Qualitative Analysis*. London: Sage Publications.

Comunello, F., & Anzera, G. (2012). Will the revolution be tweeted? A conceptual framework for understanding the social media and the Arab Spring. *Journal of Islam and Christian–Muslim Relations, 23*(4), 453–470. doi:10.1080/09596410.2012.712435

Corbin, J., & Holt, N. (2005). Grounded theory. In B. Somekh & C. Lewin (Eds.), *Research Methods in the Social Sciences*. London: Sage Publications.

Corbin, J., & Strauss, A. (2014). *Basics of qualitative research: Grounded theory procedures and techniques*. London: Sage.

Crane, A., & Matten, D. (2016). *Business ethics: Managing corporate citizenship and sustainability in the age of globalization*. Oxford, UK: Oxford University.

Cvijikj, I. P., & Michahelles, F. (2013). Online engagement factors on Facebook brand pages. *Social Network Analysis and Mining*, *3*(4), 843–861. doi:10.100713278-013-0098-8

Denskus, T., & Papan, A. S. (2013). Reflexive engagements: The international development blogging evolution and its challenges. *Development in Practice*, *23*(4), 455–467. doi:10.1080/09614524.2013.790940

Duro, J. A. (2012). On the automatic application of inequality indexes in the analysis of the international distribution of environmental indicators. *Ecological Economics*, *76*, 1–7. doi:10.1016/j.ecolecon.2011.12.019

Dyllick, T., & Hockerts, K. (2002). Beyond the business case for corporate sustainability. *Business Strategy and the Environment*, *11*(2), 130–141. doi:10.1002/bse.323

Earl, J., & Kimport, K. (2011). *Digitally Enabled Social Change: Activism in the Internet Age*. Cambridge, MA: MIT Press. doi:10.7551/mitpress/9780262015103.001.0001

Effing, R., & Spil, T. A. M. (2016). The social strategy cone: Towards a framework for evaluating social media strategies. *International Journal of Information Management*, *36*(1), 1–8. doi:10.1016/j.ijinfomgt.2015.07.009

Ellison, N. B., Steinfield, C., & Lampe, C. (2007). The benefits of Facebook "friends": Social capital and college students' use of online social network sites. *Journal of Computer-Mediated Communication*, *12*(4), 1143–1168. doi:10.1111/j.1083-6101.2007.00367.x

Evans-Cowley, J. (2010). Planning in the age of Facebook: The role of social networking in planning processes. *GeoJournal*, *75*(5), 407–420. doi:10.100710708-010-9388-0

Evans-Cowley, J. (2012). There's an App for That: Mobile Applications for Urban Planning. *International Journal of E-Planning Research*, *1*(2), 79–87. doi:10.4018/ijepr.2012040105

Facebook. (2014). *Page Post Metrics*. Retrieved 12 May, 2015, from https://www.facebook.com/help/336143376466063/

Facebook. (2015). *Negative Feedback*. Retrieved 12 May, 2015, from https://www.facebook.com/help/401280783409819

Fernando, R. (2012). Sustainable globalization and implications for strategic corporate and national sustainability. *Corporate Governance*, *12*(4), 579–589. doi:10.1108/14720701211267883

Ferrara, E., Varol, O., Davis, C., Menczer, F., & Flammini, A. (2016). The rise of social bots. *Communications of the ACM*, *59*(7), 96–104. doi:10.1145/2818717

Fieseler, C., & Fleck, M. (2013). The pursuit of empowerment through social media: Structural social capital dynamics in CSR-blogging. *Journal of Business Ethics*, *118*(4), 759–775. doi:10.100710551-013-1959-9

Gautier, A., & Pache, A.-C. (2015). Research on Corporate Philanthropy: A Review and Assessment. *Journal of Business Ethics*, *126*(3), 343–369. doi:10.100710551-013-1969-7

Hack, R. (2013). The Google fiber project. *Journal of Computing Sciences in Colleges Archive, 28*(5), 140-140.

Hanson, J., & Hogea, A. (2012). The Internet as the public sphere: Deliberative democracy and civic engagement. In A. Manoharan & M. Holzer (Eds.), *E-Governance and Civic Engagement: Factors and Determinants of E-Democracy* (pp. 467–486). Hershey, PA: IGI Global. doi:10.4018/978-1-61350-083-5.ch023

Howard, P. N., & Parks, M. (2012). Social Media and Political Change: Capacity, Constraint, and Consequence. *Journal of Communication*, *62*(2), 359–362. doi:10.1111/j.1460-2466.2012.01626.x

Huang, S., & Hsu, W. (2003). Materials flow analysis and emergy evaluation of Taipei's urban construction. *Landscape and Urban Planning*, *63*(2), 61–74. doi:10.1016/S0169-2046(02)00152-4

Jacquet, J. (2015). *Is Shame Necessary?: New Uses for an Old Tool*. New York: Pantheon Books.

Jansen, B. J., Zhang, J. M., Sobel, K., & Chowdury, A. (2009). Twitter power: Tweets as electronic word of mouth. *Journal of the American Society for Information Science and Technology*, *60*(11), 2169–2188. doi:10.1002/asi.21149

Juris, J. (2012). Reflections on #Occupy Everywhere: Social media, public space, and emerging logics of aggregation. *Journal of American Ethnological Society*, *39*(2), 259–279.

Kavada, A. (2015). Creating the collective: Social media, the Occupy Movement and its constitution as a collective actor. *Information Communication and Society*, *18*(8), 872–886. doi:10.1080/1369118X.2015.1043318

Khan, G. F. (2017). Social Media Analytics. *Social Media for Government*, 93-118.

Kitchin, R. (2015). Making sense of smart cities: Addressing present shortcomings. *Cambridge Journal of Regions, Economy and Society*, *8*(1), 131–136. doi:10.1093/cjres/rsu027

Klausen, J. (2015). Tweeting the Jihad: Social Media Networks of Western Foreign Fighters in Syria and Iraq. *Studies in Conflict and Terrorism*, *38*(1), 1–22. doi:10.1080/1057610X.2014.974948

LaRiviere, K., Snider, J., Stromberg, A., & O'Meara, K. (2012). Protest: Critical lessons of using digital media for social change. *About Campus: Enriching the Student Learning Experience*, *17*, 10–17. doi:10.1002/abc.21081

Linders, D. (2012). From e-government to we-government: Defining a typology for citizen coproduction in the age of social media. *Government Information Quarterly*, *29*(4), 446–454. doi:10.1016/j.giq.2012.06.003

Lyon, T. P., & Montgomery, A. W. (2013). Tweetjacked: The Impact of Social Media on Corporate Greenwash. *Journal of Business Ethics*, *118*(4), 747–757. doi:10.100710551-013-1958-x

Maireder, A., & Schwarzenegger, C. (2012). A movement of connected individuals. *Information Communication and Society*, *15*(2), 171–195. doi:10.1080/1369118X.2011.589908

Mandarano, L., Meenar, M., & Steins, C. (2010). Building social capital in the digital age of civic engagement. *Journal of Planning Literature*, *25*(2), 123–135. doi:10.1177/0885412210394102

Marsden, C. (2000). The new corpotate citizenship of big business: Part of the solution to sustainability? *Business and Society Review*, *105*(1), 9–25. doi:10.1111/0045-3609.00062

Matejek, S., & Gossling, T. (2014). Beyond legitimacy: A case study in BP's "green lashing. *Journal of Business Ethics*, *120*(4), 571–584. doi:10.100710551-013-2006-6

McCafferty, D. (2011). Activism vs. slacktivism. *Communications of the ACM*, *54*(12), 17–19. doi:10.1145/2043174.2043182

McNeill, D. (2015). Global firms and smart technologies: IBM and the reduction of cities. *Transactions of the Institute of British Geographers*, *40*(4), 562–574. doi:10.1111/tran.12098

Milanovic, B. (2011). *Worlds apart: Measuring international and global inequality*. Princeton, NJ: Princeton University Press. doi:10.1515/9781400840816

Minton, E., Lee, C., Orth, U., Chung-Hyun, K., & Lynn, K. (2012). Sustainable marketing and social media: A cross-country analysis of motives for sustainable behaviors. *Journal of Advertising*, *41*(4), 69–84. doi:10.1080/00913367.2012.10672458

Mithas, S., Costello, T., & Tafti, A. (2011). Social networking in the enterprise. *IT Professional Magazine*, *13*(4), 16–17. doi:10.1109/MITP.2011.69

Montiel, I. (2008). Corporate social responsibility and corporate sustainability: Separate pasts, common futures. *Organization & Environment*, *21*(3), 245–269. doi:10.1177/1086026608321329

Park, N., Kee, K. F., & Valenzuela, S. (2009). Being immersed in social networking environment: Facebook groups, uses and gratifications, and social outcomes. *Cyberpsychology & Behavior*, *12*(6), 729–733. doi:10.1089/cpb.2009.0003 PMID:19619037

Paroutis, S., Bennett, M., & Heracleous, L. (2014). A strategic view on smart city technology: The case of IBM Smarter Cities during a recession. *Technological Forecasting and Social Change*, *89*, 262–272. doi:10.1016/j.techfore.2013.08.041

Picazo-Vela, S., Gutiérrez-Martínez, I., & Luna-Reyes, L. F. (2012). Understanding risks, benefits, and strategic alternatives of social media applications in the public sector. *Government Information Quarterly*, *29*(4), 504–511. doi:10.1016/j.giq.2012.07.002

Quan-Haase, A., Wellman, B., Witte, J. C., & Hampton, K. N. (2002). Capitalizing on the Net: Social contact, civic engagement, and sense of community. In B. Wellman & C. Haythornthwaite (Eds.), *The Internet in everyday life*. Oxford, UK: Blackwell. doi:10.1002/9780470774298.ch10

Scott, K., & Liew, T. (2012). Social networking as a development tool: A critical reflection. *Urban Studies (Edinburgh, Scotland)*, *49*(12), 2751–2767. doi:10.1177/0042098011435279

Shav-Ami, A. (2013). Social protest: The Israeli case. Journal of Enterprising Communities. *People and Places in the Global Economy*, *7*(4), 373–382.

SparkNotes. (2014). *International Organizations*. Retrieved 4 Feb, 2015, from http://www.sparknotes.com/us-government-and-politics/political-science/international-politics/section4.rhtml

Stubbs, W., & Cocklin, C. (2008). Conceptualizing a "Sustainability Business Model". *Organization & Environment, 21*(2), 103–127. doi:10.1177/1086026608318042

Suddaby, R. (2006). From the Editors: What Grounded Theory is Not. Academy of Management Journal, 49(4), 633-642.

Svensson, G., & Wagner, B. (2010). Transformative business sustainability: Multi-layer model and network of e-footprint sources. *European Business Review, 23*(4), 334–352. doi:10.1108/09555341111145735

Swani, K., Milne, G., & Brown, B. P. (2013). Spreading the word through Likes on Facebook: Evalutating the message strategy effectiveness of Fortune 500 companies. *Journal of Research in Interactive Marketing, 7*(4), 269–294. doi:10.1108/JRIM-05-2013-0026

Tapscott, D., & Williams, A. D. (2006). *Wikinomics: How Mass Collaboration Changes Everything*. New York: Penguin.

Turner-Lee, N. (2010). The challenge of increasing civic engagement in the digital age. *Federal Communications Law Journal, 63*(1), 19–32.

UN Habitat. (2014). *The World Urban Forum*. Retrieved 2 March, 2015, from http://wuf7.unhabitat.org/theworldurbanforum

Vitak, J., Zube, P., Smock, A., Carr, C. T., Ellison, N., & Lampe, C. (2011). It's complicated: Facebook users' political participation in the 2008 election. *Cyberpsychology, Behavior, and Social Networking, 14*(3), 107–114. doi:10.1089/cyber.2009.0226 PMID:20649449

Vogel, R., Moulder, R., & Huggins, M. (2014). The extent of public participation. *Public Management, 96*(2), 6–10.

Walther, J. B., & Jang, J. (2012). Communication processes in particapatory websites. *Computer-Mediated Communication, 18*(1), 2–15. doi:10.1111/j.1083-6101.2012.01592.x

Waters, R. D., Burnett, E., Lamm, A., & Lucas, J. (2009). Engaging Stakeholders through Social Networking: How nonprofit organizations are using Facebook. *Public Relations Review, 35*(2), 102–106. doi:10.1016/j.pubrev.2009.01.006

Wellman, B., Haase, A. Q., Witte, J., & Hampton, K. (2001). Does the Internet increase, decrease, or supplement social capital?: Social networks, participation, and community commitment. *The American Behavioral Scientist, 45*(3), 436–455. doi:10.1177/00027640121957286

Willard, T. (2009). Social Networking and Governance for Sustainable Development. Winnipeg, Canada: International Institute for Sustainable Development (IISD).

Willems, L., & Alizadeh, T. (2015). Social Media for Public Involvement and Sustainability in International Planning and Development. *International Journal of E-Planning Research*, *4*(4), 1–17. doi:10.4018/ijepr.2015100101

Compilation of References

@Last Software. (2000). SketchUp. *Trimble Navigation*. Retrieved August 30, 2015, from http://www.sketchup.com

ABS. (2010). 1367.2 - State and Regional Indicators, Victoria, Jun 2010. *Feature article: Water and energy efficiency elements of households in older and newer dwellings*. Retrieved from http://www.abs.gov.au/AUSSTATS/abs@.nsf/Lookup/1367.2Feature+Article1Jun+2010

ABS. (2013). Water Supply and Use 2011–12 — Australia. *4610.0 - Water Account, Australia, 2011–12*. Retrieved 11 August 2014, 2013, from http://www.abs.gov.au/AUSSTATS/abs@.nsf/DetailsPage/4610.02011%E2%80%9312?OpenDocument

Adam, N., J. Eledath, S. Mehrotra, N. Venkatasubramanian. (2012a). Social media alert and response to threats to citizens (SMART-C). *Proceedings of CollaborateCom*, 181-189.

Adam, N., Shafiq, B., & Staffin, R. (2012b). Spatial computing and social media in the context of disaster management. *IEEE Intelligent Systems, Volume*, *27*(6), 90–96. doi:10.1109/MIS.2012.113

Adkins, M., Burgoon, M., & Nunamaker, J. F. Jr. (2003). Using group support systems for strategic planning with the United States Air Force. *Decision Support Systems*, *34*(3), 3, 315–337. doi:10.1016/S0167-9236(02)00124-0

Adla, A., Zarate, P., & Soubie, J. L. (2011). A proposal of toolkit for GDSS facilitators. *Group Decision and Negotiation*, *20*(1), 57–77. doi:10.100710726-010-9204-8

Adua, L. (2010). To cool a sweltering earth: Does energy efficiency improvement offset the climate impacts of lifestyle? *Energy Policy*, *38*(10), 5719–5732. doi:10.1016/j.enpol.2010.05.021

Afzalan, N., & Evans-Cowley, J. (2015). Planning and Social Media: Facebook for Planning at the Neighbourhood Scale. *Planning Practice and Research*, *30*(3), 270–285. doi:10.1080/02697459.2015.1052943

Afzalan, N., Sanchez, T., & Evans-Cowley, J. (2017). Creating smarter cities: Considerations for selecting online participatory. *Cities (London, England)*, *67*, 21–30. doi:10.1016/j.cities.2017.04.002

Aiken, M., Vanjani, M., & Paolillo, J. (1996). A comparison of two electronic idea generation techniques. *Information & Management*, *30*(2), 91–99. doi:10.1016/0378-7206(95)00048-8

Albino, V., Berardi, U., & Dangelico, R. M. (2015). Smart Cities: Definitions, Dimensions, Performance, and Initiatives. *Journal of Urban Technology*, *22*(1), 3–21. doi:10.1080/10630732.2014.942092

Alizadeh, T. (2017). An investigation of IBM Smarter City Challenge: What do participating cities want? *Cities (London, England)*, *63*, 70–80. doi:10.1016/j.cities.2016.12.009

Alizadeh, T., Tomerini, D., & Colbran, S. (2016). Teaching Planning Studios: An Online Assessment Task to Enhance the First Year Experience. *Journal of Planning Education and Research*. doi:10.1177/0739456X16647162

Al-Kodmany, K. (2001). Bridging the gap between technical and local knowledge: Tools for promoting community-based planning and design. *Journal of Architectural and Planning Research*, *18*(2), 110–130.

Al-Kodmany, K. (2009). Planning for the Hajj: Political Power, Pragmatism, and Participatory GIS. *Journal of Urban Technology*, *16*(1), 5–45. doi:10.1080/10630730903090289

Allen, B. (1989). *Blacks in Austin*. Austin, TX: Self-published.

Allison, E. W., & Peters, L. (2011). *Historic Preservation and the Livable City*. Hoboken, NY: John Wiley and Sons.

Al-Qaysi, N., Piroozfar, A., Southall, R., & Farr, E. R. P. (2016) Judgment in architectural competitions as communicative deliberative practice In *Proceedings of the ICC 2016: 6th International Competitions Conference: The Competition Mesh*. Leeds Beckett University.

Amabile, T. M. (1996). *Creativity in context*. Boulder, CO: Westview Press.

Anayo, J., & Horst, H. A. (2016). Technologies of the Nation: Public Wi-Fi and the Demand for More in Niue. *Information Technologies and International Development*, *124*, 1–9.

Anbalahan, E. (2010). *Community Based Flood Management System Using Web 2.0, Master's Project*. NJIT.

Ancona, D. G., Okhuysen, G. A., & Perlow, L. A. (2001). Taking time to integrate temporal research. *Academy of Management Review*, *26*(4), 512–529.

Andersson, J. E. (2016). When all is said and done: An architectural competition, was it a good idea? In ICC 2016. Leeds Beckett University.

Andersson, J. E., Zettersten, G. B., & Rönn, M. (Eds.). (2016). *Architectural Competitions as Institution and Process*. The Royal Institute of Technology and Kulturlandskapet.

Andrejevic, M., & Burdon, M. (2014). Defining the Sensor Society. *Television and New Media*. doi: 1527476414541552

Antoniadis, P., & Apostol, I. (2014). The Right(s) to the Hybrid City and the Role of DIY Networking. *The Journal of Community Informatics*, *10*(3).

Anttiroiko, A.-V. (2012). Urban planning 2.0. *International Journal of E-Planning Research, 1*(1), 16–30. doi:10.4018/ijepr.2012010103

Antunes, P., Sapateiro, C., Zurita, G., & Baloian, N. (2010). Integrating spatial data and decision models in an e-planning tool. In G. Kolfschoten, T. Herrmann, & S. Lukosch (Eds.), Lecture notes in computer science: Vol. 6257. *Collaboration and technology* (pp. 97–112). Berlin: Springer. doi:10.1007/978-3-642-15714-1_8

AOEMA. (2005). *E-government from a user's perspective*. Retrieved July 15, 2017, from http://www.aoema.org/EGovernment/Stages-Phases_of_e-government.htm

Aras, G., & Crowther, D. (2009). Corporate Sustainability Reporting: A Study in Disingenuity? *Journal of Business Ethics*, *87*(S1), 279–288. doi:10.100710551-008-9806-0

Aras, G., & Crowther, D. (2009). Making sustainable development sustainable. *Management Decision*, *47*(6), 975–988. doi:10.1108/00251740910966686

Arbués, F., Villanúa, I., & Barberán, R. (2010). Household size and residential water demand: An empirical approach. *The Australian Journal of Agricultural and Resource Economics*, *54*(1), 61–80. doi:10.1111/j.1467-8489.2009.00479.x

Ardissono, L., Lucenteforte, M., Mauro, N., Savoca, A., Voghera, A., & La Riccia, L. (2016). Exploration of Cultural Heritage Information via Textual Search Queries. *Proceedings. of the 18th International Conference on Human-Computer Interaction with Mobile Devices and Services Adjunct*, 992-1001. 10.1145/2957265.2962648

Arieff, A. (2016, July 20). Mapping Detroit, Inch by Inch. *The New York Times*. Retrieved November 20, 2017 from https://www.nytimes.com/2016/07/21/us/mapping-detroit-inch-by-inch.html?mcubz=1

Arnstein, S. R. (1969). A ladder of citizen participation. *Journal of the American Institute of Planners*, *35*(4), 216–224. doi:10.1080/01944366908977225

Artigas, F., Chun, S., & Sookhu, Y. (2009) Real-time Ocean Surge Warning System, Meadowlands District of New Jersey. *Proceedings of the 10th International Conference on Digital Government Research (dg.o 2009)*, 216-222.

Associated Press. (2012). *Vehicles towed in Seaside Heights by company accused of price gouging to be released*. Retrieved from http://www.nj.com/news/index.ssf/2012/11/vehicles_in_seaside_heights_to.html

Austin Historical Survey Project Team. (2009). *Austin Historical Survey Wiki Phase I: Concept Background and Wiki Design*. Austin, TX: University of Texas at Austin School of Architecture. Retrieved October 9, 2014 from http://www.soa.utexas.edu/files/hp/Austin_Historical_Survey_Web_Tool_PhaseIReport.pdf

Australian Government. (2014). Water Efficiency Labelling and Standards (WELS) scheme. *Water efficiency*. Retrieved 22, February, 2015, from http://www.waterrating.gov.au/consumers/water-efficiency

Axelsson, K., Melin, U., & Lindgren, I. (2013). Public e-services for agency efficiency and citizen benefit—Findings from a stakeholder centered analysis. *Government Information Quarterly*, *30*(1), 10–22. doi:10.1016/j.giq.2012.08.002

Bäcklund, P., & Mäntysalo, R. (2010). Agonism and institutional ambiguity: Ideas on democracy and the role of participation in the development of planning theory and practice – The case of Finland. *Planning Theory*, *9*(4), 333–350. doi:10.1177/1473095210373684

Badland, H. M., Opit, S., Witten, K., Kearns, R. A., & Mavoa, S. (2010). Can virtual streetscape audits reliably replace physical streetscape audits? *Journal of Urban Health*, *87*(6), 1007–1016. doi:10.100711524-010-9505-x PMID:21104331

Bailey, K., & Grossardt, T. (2010). Towards structured public involvement: Justice, geography and collaborative geospatial/geovisual decision support systems. *Annals of the Association of American Geographers*, *100*(1), 57–86. doi:10.1080/00045600903364259

Bala, M. (2014). Civic engagement in the age of online social networks. *Contemporary Readings in Law and Social Justice*, *6*(1), 767–774.

Baller, S., Dutta, S., & Lanvin, B. (Eds.). (2017). *The Global Information Technology Report 2016. Innovating in the Digital Economy*. Geneva: World Economic Forum.

Barber, B. R. (2004). *Strong democracy: Participatory politics for a new age*. University of California Press.

Barker, M. S., Barker, D. L., Bormann, N. F., & Neher, K. E. (2008). *Social Media Marketing: A Strategic Approach*. South-western Cengage Learning.

Bartusch, C., Odlare, M., Wallin, F., & Wester, L. (2012). Exploring variance in residential electricity consumption: Household features and building properties. *Applied Energy*, *92*, 637–643. doi:10.1016/j.apenergy.2011.04.034

Baruah, J., & Paulus, P. B. (2008). Effects of training on idea generation in groups. *Small Group Research*, *39*(5), 523–541. doi:10.1177/1046496408320049

Baruah, J., & Paulus, P. B. (2009). Enhancing creativity in groups: The search for synergy. In M. Neale, B. Mannix, & J. Goncalo (Eds.), Research on managing groups and teams. Oxford, UK: Elsevier Science Press. doi:10.1108/S1534-0856(2009)0000012005

Baruah, J., & Paulus, P. B. (2011). Category assignment and relatedness in the group ideation process. *Journal of Experimental Social Psychology*, *47*(6), 1070–1077. doi:10.1016/j.jesp.2011.04.007

Baruah, J., & Paulus, P. B. (2016). The role of time and category relatedness in electronic brainstorming. *Small Group Research*, *47*(3), 333–342. doi:10.1177/1046496416642296

Batrinca, B., & Treleaven, P. C. (2015). Social media analytics: A survey of techniques, tools and platforms. *AI & Society*, *30*(1), 89–116. doi:10.100700146-014-0549-4

Batty, M. (2017). The Sixth Kondratieff Is the Age of the Smart City, Keynote Speech. 15th International Computers in Urban Planning and Urban Management, Adelaide, Australia.

Baum, C., & DiMaio, A. D. (2000). *Gartner's four phases of e-government model.* Retrieved from http://www.gartner.com

Bauman, Z. (1995). Life. In *Fragments: Essays in Postmodern Morality*. Wiley.

Belanger, F., Carter, L. D., & Schaupp, L. C. (2005). U-government: a framework for the evolution of e-government. *Electronic Government, an International Journal, 2*(4), 426-445.

Bellotti, V., & Bly, S. (1996). *Walking Away from the Desktop Computer: Distributed Collaboration and Mobility in a Product Design Team.* Paper presented at the CSCW, Cambridge, MA. 10.1145/240080.240256

Bell, S. T., Villado, A. J., Lukasik, M. A., Belau, L., & Briggs, A. (2011). Getting specific about demographic diversity variable and team performance relationships: A meta-analysis. *Journal of Management, 37*(3), 709–743. doi:10.1177/0149206310365001

Bennett, L. V., & Manoharan, A. P. (2017). The Use of Social Media Policies by US Municipalities. *International Journal of Public Administration, 40*(4), 317–328. doi:10.1080/01900692.2015.1113182

Berger, J. (2011). Arousal increases social transmission of information. *Psychological Science, 22*(7), 891–893. doi:10.1177/0956797611413294 PMID:21690315

Berger, J., & Milkman, K. L. (2012). What makes online content viral? *JMR, Journal of Marketing Research, 49*(2), 192–205. doi:10.1509/jmr.10.0353

Bernstein, K., Sun, E., & Sucre, R. (2009). *Survey LA Panel Discussion.* Presented at the Association for Preservation Technology: Preservation in the City Without Limits, Los Angeles, CA.

Bernstein, K., & Hansen, J. (2016). SurveyLA: Linking Historic Resources Surveys to Local Planning. *Journal of the American Planning Association, 82*(2), 88–91. doi:10.1080/01944363.2015.1137199

Bertron, C. (2013). Survey Forth! Innovative Survey Methodologies. *Forum Journal, 27*(4), 34–35.

Bertron, C., & Rypkema, D. (2012). *Historic Preservation and Rightsizing Current Practices and Resources Survey.* Washington, DC: PlaceEconomics.

Bessis, N., Brown, A., & Asimakopoulou, E. (2010, July). A Mathematical Analysis of a Disaster Management Data-Grid Push Service. *International Journal of Distributed Systems and Technologies, 1*(3), 56–70. doi:10.4018/jdst.2010070104

Best, P., Manktelow, R., & Tylor, B. (2014). Online communication, social media and adolescent wellbeing: A systematic narrative review. *Children and Youth Services Review, 41*, 27–36. doi:10.1016/j.childyouth.2014.03.001

Beyer, H., & Holtzblatt, K. (1998). *Contextual Design - Defining Customer-Centered Systems*. San Francisco, CA: Morgan Kaufmann Publisher.

Bharosa, N., Lee, J., Janssen, M., & Raghav Rao, H. (2009) A case study of information flows in multi-agency emergency response exercises. *Proceedings of the 10th Annual International Conference on Digital Government Research: Social Networks*, 277-282.

Bilbao-Osorio, B., Dutta, S., & Lanvin, X. B. (2013). *The Global Information Technology Report 2013: Growth and jobs in a hyperconnected world*. Retrieved from http://www3.weforum.org/docs/WEF_GITR_Report_2013.pdf

Bishop, I. D., Eagleson, S., Pettit, C. J., Rajabifard, A., Badland, H., Day, J. E., ... White, M. (2017). Using an Online Data Portal and Prototype Analysis Tools in an Investigation of Spatial Livability Planning. *International Journal of E-Planning Research*, *6*(2), 1–21. doi:10.4018/IJEPR.2017040101

Bobrow, D. G., & Winograd, T. (1976). *An Overview of KRL, a Knowledge Representation Language*. Report n. CS-TR-76-581. Stanford University Press.

Boczkowski, P. J., & Mitchelstein, E. (2012). How users take advantage of different forms of interactivity on online news sites: Clicking, E-Mailing, and commenting. *Human Communication Research*, *38*(1), 1–22. doi:10.1111/j.1468-2958.2011.01418.x

Bollier, D., & Firestone, C. M. (2010). *The Promise and Peril of Big Data*. Washington, DC: Aspen Institute.

Boulianne, S. (2009). Does Internet Use Affect Engagement? A Meta-Analysis of Research. *Political Communication*, *26*(2), 193–211. doi:10.1080/10584600902854363

Bowen, F., & Aragon-Correa, J. A. (2014). Greenwashing in Corporate Environmentalism Research and Practice: The Importance of What We Say and Do. *Organization & Environment*, *27*(2), 107–112. doi:10.1177/1086026614537078

Bowser, A., & Shanley, L. (2013) *New Visions in Citizen Science, Case Study Series, Vol. 3*. Washington, DC: Commons Lab, Science and Technology Innovation Program Woodrow Wilson International Center for Scholars. Retrieved November 20, 2017 from: https://www.wilsoncenter.org/sites/default/files/NewVisionsInCitizenScience.pdf

Boyd, D., & Crawford, K. (2012). Critical Questions for Big Data: Provocations for a Cultural, Technological, and Scholarly Phenomenon. *Information Communication and Society*, *15*(5), 662–679. doi:10.1080/1369118X.2012.678878

Brabham, D. C. (2009). Crowdsourcing the Public Participation Process for Planning Projects. *Planning Theory*, *8*(3), 242–263. doi:10.1177/1473095209104824

Bragg, A. E. (2014, January 8). *Preservation Advocates Launch Critical Survey of Historic Detroit Neighborhoods*. Retrieved July 1, 2017 from Preservation Detroit: http://preservationdetroit.org/call-to-action-major-preservation-survey-of-historic-detroit-neighborhoods/

Brail, R. K., & Klosterman, R. E. (Eds.). (2001). *Planning Support Systems: Integrating Geographic Information Systems, Models, and Visualization Tools*. Redlands, CA: ESRI Press.

Branon, R. F., & Essex, C. (2001). Synchronous and asynchronous communication tools in distance education. *TechTrends*, *45*(1), 36–36. doi:10.1007/BF02763377

Bratton, M. (Ed.). (2013). *Voting and democratic citizenship in Africa*. Boulder, CO: Lynne Rienner Publishers.

Brown, G., & Weber, D. (2012). Measuring change in place values using public participation GIS (PPGIS). *Applied Geography (Sevenoaks, England)*, *34*, 316–324. doi:10.1016/j.apgeog.2011.12.007

Brown, V. R., & Paulus, P. B. (2002). Making group brainstorming more effective: Recommendations from an associative memory perspective. *Current Directions in Psychological Science*, *11*(6), 208–212. doi:10.1111/1467-8721.00202

Brown, V., Tumeo, M., Larey, T. S., & Paulus, P. B. (1998). Modeling cognitive interactions during group brainstorming. *Small Group Research*, *29*(4), 495–526. doi:10.1177/1046496498294005

Brundtland, G. H. (1987). *Our Common Future: Report of the World Commission on Environment and Development*. Oxford, UK: Oxford University.

Bryant, S. L., Forte, A., & Bruckman, A. (2005). Becoming Wikipedian: transformation of participation in a collaborative online encyclopedia. In *GROUP '05 Proceedings of the 2005 International ACM SIGGROUP Conference on Supporting Group Work* (pp. 1–10). New York: ACM Press.

Buccella, A., Cechich, A., Gendarmi, D., Lanubile, F., Semeraro, G., & Colagrossi, A. (2011). Building a global normalized ontology for integrating data sources. *Computers & Geosciences*, *37*(7), 833–916. doi:10.1016/j.cageo.2011.02.022

Buck, N. T., & While, A. (2017). Competitive urbanism and the limits to smart city innovation: The UK Future Cities initiative. *Urban Studies (Edinburgh, Scotland)*, *54*(2), 501–519. doi:10.1177/0042098015597162

Burrough, P. A. (1986). *Principles of geographical information systems for land resource assessment*. Oxford, UK: Clarendon Press.

Bush, A. O., Lafferty, K. D., Lotz, J. M., & Shostak, A. W. (1997). Parasitology meets ecology on its own terms: Margolis et al. revisited. *The Journal of Parasitology*, 575-583.

Calista, D. J., Melitski, J., Holzer, M., & Manoharan, A. (2010). Digitized government in worldwide municipalities between 2003 and 2007. *International Journal of Public Administration*, *33*(12/13), 588–600. doi:10.1080/01900692.2010.513881

Camacho, L. M., & Paulus, P. B. (1995). The role of social anxiousness in group brainstorming. *Journal of Personality and Social Psychology*, *68*(6), 1071–1080. doi:10.1037/0022-3514.68.6.1071

Cambria, E., Rajagopal, D., Olsher, D., & Das, D. (2013). Big social data analysis. In R. Akerkar (Ed.), *Big Data Computer* (pp. 401–414). Boca Raton, FL: CRC Press. doi:10.1201/b16014-19

Capra, C. F. (2016). The Smart City and its Citizens: Governance and Citizen Participation in Amsterdam Smart City. *International Journal of E-Planning Research*, *5*(1), 20–38. doi:10.4018/IJEPR.2016010102

Careem, M., De Silva, C., De Silva, R., Raschid, L., & Weerawarana, S. (2007) Demonstration of Sahana: free and open source disaster management. *Proceedings of the 8th annual international conference on Digital government research: bridging disciplines \& domains*, 266-267.

Carr, C. T., & Hayes, R. A. (2015). Social Media: Defining, Developing, and Divining. *Atlantic Journal of Communication*, *23*(1), 46–65. doi:10.1080/15456870.2015.972282

Carrizales, T. (2008). Functions of e-government: A study of municipal practices. *State & Local Government Review*, *40*(1), 12–26. doi:10.1177/0160323X0804000102

Carr, S., Francis, M., Rivlin, L. G., & Stone, A. (1992). *Public Space*. Cambridge, UK: Cambridge University Press.

Cayirci, E., & Coplu, T. (2007, June). SENDROM: Sensor networks for disaster relief operations management. *Wireless Networks*, *13*(3), 409–423. doi:10.100711276-006-5684-5

Cegarra-Navarro, J. G., Garcia-Perez, A., & Moreno-Cegarra, J. L. (2014). Technology knowledge and governance: Empowering citizen engagement and participation. *Government Information Quarterly*, *31*(4), 660–668. doi:10.1016/j.giq.2014.07.001

Central Bureau of Statistics. (2011). *Local Authorities in Israel 2011*. Retrieved from http://www.cbs.gov.il/publications/p_libud_11.xls

Cerro Santamaria, G. (Ed.). (2013). *Urban megaprojects: a worldwide view*. Bingley, UK: Emerald Group Publishing Limited. doi:10.1108/S1047-0042(2013)13

Chadwick, A. (2003). Bringing E-Democracy Back In Why it Matters for Future Research on E-Governance. *Social Science Computer Review*, *21*(4), 443–455. doi:10.1177/0894439303256372

Chang, H., Parandvash, G. H., & Shandas, V. (2010). Spatial variations of single-family residential water consumption in Portland, Oregon. *Urban Geography*, *31*(7), 953–972. doi:10.2747/0272-3638.31.7.953

Charmaz, K. (2006). *Constructing Grounded Theory: A Practical Guide Through Qualitative Analysis*. London: Sage Publications.

Cheng, N., Wang, X., Cheng, W., Mohapatra, P., & Seneviratne, A. (2013). *Characterizing Privacy Leakage of Public WiFi Networks for Users on Travel*. Paper presented at the INFOCOM, 2013, Turin. 10.1109/INFCOM.2013.6567086

Chen, R., Sharman, R., Rao, H. R., & Upadhyaya, S. J. (2008, May). Coordination in emergency response management. *Communications of the ACM*, *51*(5), 66–73. doi:10.1145/1342327.1342340

Chen, X. (2015). Decentralized computation offloading game for mobile cloud computing. *IEEE Transactions on Parallel and Distributed Systems*, *26*(4), 974–983. doi:10.1109/TPDS.2014.2316834

Chen, Y. (2010). Citizen-centric e-government services: Understanding integrated citizen service information systems. *Social Science Computer Review*, *28*(4), 427–442. doi:10.1177/0894439309359050

Chen, Y., Liu, C., & Wong, R. (2007). The adoption of synchronous and asynchronous media in the teaching of a second language. *Issues in Information Systems*, *3*(1), 217–223.

Choi, Y., Blackburn, J. M., & Seo, H. (2006, November). A connection oriented telecommunications disaster management model using mobile GIS. *International Journal of Mobile Communications*, *4*(1), 63–81. doi:10.1504/IJMC.2006.008198

Christensen, H. S. (2013). Broadening Democratic Participation? An exploratory study of e-democracy in 188 Finnish municipalities. *Scandinavian Journal of Public Administration*, *17*(3).

Chun, S., & Artigas, F. (2012). Sensors and Crowdsourcing for Environmental Awareness and Emergency Planning. *International Journal of E-Planning Research*, *1*(1), 56–74. doi:10.4018/ijepr.2012010106

Chusid, J. (2006). Preservation in the Progressive City: Debating History and Gentrification in Austin. *The Next American City,* (12), 23-27.

Cialdini, R. B., & Goldstein, N. J. (2004). Social influence: Compliance and conformity. *Annual Review of Psychology*, *55*(1), 591–621. doi:10.1146/annurev.psych.55.090902.142015 PMID:14744228

Cioffi-Revilla, C. (2010). Computational Social Science. *Wiley Interdisciplinary Reviews: Computational Statistics*, *2*(3), 259–271. doi:10.1002/wics.95

City of Austin. (2003). *Staff Task Force on Gentrification in East Austin: Findings and Recommendations*. Austin, TX: City of Austin.

City of Austin. (2011). *Downtown Austin Plan.* Austin, TX: City of Austin.

City of Austin. (2012). *Imagine Austin.* Austin, TX: City of Austin.

City of Los Angeles. (2013). *SurveyLA. City of Los Angeles Website*. Retrieved November 20, 2017 from: http://preservation.lacity.org/survey

City of Vaasa. (2017a). *Ak1000: Palosaaren salmen alue, Mansikkasaari ja Salmenpuisto* [Detailed plan for the area of Palosaaren salmi, The island of Mansikkasaari and park Salmenpuisto]. Retrieved July 13, 2017 at: https://www.vaasa.fi/kaavoitus/ak1000

City of Vaasa. (2017b). *Ak1081: Palosaaren Salmen alue, Rahkola-Varisselän puisto* [Detailed plan for the area of Palosaaren salmi, The park of Rahkola-Varisselkä]. Retrieved July 13, 2017 at: https://www.vaasa.fi/kaavoitus/ak1081

Clarke, P., Ailshire, J., Melendez, R., Bader, M., & Morenoff, J. (2010). Using Google Earth to conduct a neighborhood audit: Reliability of a virtual audit instrument. *Health & Place*, *16*(6), 1224–1229. doi:10.1016/j.healthplace.2010.08.007 PMID:20797897

Clift, S. (2004). *E-Government and Democracy. Representation and citizen engagement in the information age*. Retrieved July 15, 2017, from http://www.publicus.net/e-government/

Coleman, S., & Gotze, J. (2001). *Bowling together: Online public engagement in policy deliberation* (pp. 39–50). London: Hansard Society.

ComScore. (2011). *It's a social world: Top 10 need-to-knows about social networking and where it's headed*. Retrieved from www.comscore.com/it_is_a_social_world

Comunello, F., & Anzera, G. (2012). Will the revolution be tweeted? A conceptual framework for understanding the social media and the Arab Spring. *Journal of Islam and Christian–Muslim Relations*, *23*(4), 453–470. doi:10.1080/09596410.2012.712435

Connors, J. P., Lei, S., & Kelly, M. (2012). Citizen Science in the Age of Neogeography: Utilizing Volunteered Geographic Information for Environmental Monitoring. *Annals of the Association of American Geographers*, *102*(6), 1267–1289. doi:10.1080/00045608.2011.627058

Convertino, G., Mentis, H. M., Slavkovic, A., Rosson, M., & Carroll, J. M. (2011, December). Supporting common ground and awareness in emergency management planning: A design research project. *ACM Transactions on Computer-Human Interaction*, *18*(4), 22. doi:10.1145/2063231.2063236

Cooper, A. (2004). *Inmates Are Running the Asylum, The: Why High-Tech Products Drive Us Crazy and How to Restore the Sanity*. Indianapolis, IN: Sams Publishing.

Cooper, W. H., Gallupe, R. B., Pollard, S., & Cadsby, J. (1998). Some liberating effects of anonymous electronic brainstorming. *Small Group Research*, *29*(2), 147–178. doi:10.1177/1046496498292001

Coppola, S. (2012, October 23). Austin's tax breaks for historic properties challenged in lawsuit. *Austin American-Statesman*.

Corbin, J., & Holt, N. (2005). Grounded theory. In B. Somekh & C. Lewin (Eds.), *Research Methods in the Social Sciences*. London: Sage Publications.

Corbin, J., & Strauss, A. (2014). *Basics of qualitative research: Grounded theory procedures and techniques*. London: Sage.

Cornelius, C., & Boos, M. (2003). Enhancing mutual understanding in synchronous computer-mediated communication by training: Trade-offs in judgmental tasks. *Communication Research*, *30*(2), 147–177. doi:10.1177/0093650202250874

Coskun, H. (2000). *The effects of outgroup comparison, social context, intrinsic motivation, and collective identity in brainstorming groups* (Unpublished doctoral dissertation). The University of Texas at Arlington.

Coskun, H., Paulus, P. B., Brown, V., & Sherwood, J. J. (2000). Cognitive stimulation and problem presentation in idea generation groups. *Group Dynamics*, *4*(4), 307–329. doi:10.1037/1089-2699.4.4.307

Coursey, L. E., Kenworthy, J. B., Paulus, P. B., & Doboli, S. (2017). *Diversity and group creativity in an online, asynchronous environment* (Unpublished manuscript). University of Texas at Arlington.

Crabtree, A. (2003). *Designing Collaborative Systems: A Practical Guide to Ethnography*. London: Springer.

Craig, W. J., Harris, T. M., & Weiner, D. (2003). *Community participation and Geographic Information Systems*. London: Taylor and Francis.

Cramton, C. D. (2001). The mutual knowledge problem and its consequences for dispersed collaboration. *Organization Science*, *12*(3), 346–371. doi:10.1287/orsc.12.3.346.10098

Crane, A., & Matten, D. (2016). *Business ethics: Managing corporate citizenship and sustainability in the age of globalization*. Oxford, UK: Oxford University.

Crawford, R., & Fuller, R. (2011). *Energy and greenhouse gas emissions implications of alternative housing types for Australia*. Paper presented at the State of Australian Cities National Conference, Melbourne, Australia.

Crawford, S. (2013). *Captive Audience: The Telecom Industry and Monopoly Power in the New Gilded Age*. Yale University Press.

Creative Commons. (2017). *State of the Commons 2016*. Retrieved November 20, 2017: http://creativecommons.org

Creswell, J. W. (2007). *Qualitative Inquiry and Research Design: Choosing Among Five Approaches*. SAGE Publishing.

Criado, J. I., Sandoval-Almazan, R., & Gil-Garcia, J. R. (2013). Government innovation through social media. *Government Information Quarterly*, *30*(4), 319–326. doi:10.1016/j.giq.2013.10.003

Cunha, E., Allegretti, G., & Matias, M. (2011). Participatory Budgeting and the Use of Information and Communication Technologies: A Virtuous Cycle? *RCCS Annual Review, 3*. Retrieved July 15, 2017, from http://rccsar.revues.org/316

Currion, P., de Silva, C., & Van de Walle, B. (2007, March). Open source software for disaster management. *Communications of the ACM*, *50*(3), 61–65. doi:10.1145/1226736.1226768

Cvijikj, I. P., & Michahelles, F. (2013). Online engagement factors on Facebook brand pages. *Social Network Analysis and Mining*, *3*(4), 843–861. doi:10.100713278-013-0098-8

Dahl, R. A. (1967). The city in the future of democracy. *The American Political Science Review*, *61*(04), 953–970. doi:10.2307/1953398

Dahl, R. A. (1970). *After the revolution? Authority in a good society*. New Haven, CT: Yale University Press.

Dahl, R. A. (2005). *Who governs? Democracy and power in an American city*. Yale University Press.

Daily, B., Whatley, A., Ash, S. R., & Steiner, R. L. (1996). The effects of a group decision support system on culturally diverse and culturally homogeneous group decision making. *Information & Management*, *30*(6), 281–289. doi:10.1016/S0378-7206(96)01062-2

Damanpour, F., & Schneider, M. (2009). Characteristics of innovation and innovation adoption in public organizations: Assessing the role of managers. *Journal of Public Administration: Research and Theory*, *19*(3), 495–522. doi:10.1093/jopart/mun021

Damurski, L. (2016). Smart City, Integrated Planning, and Multilevel Governance: A Conceptual Framework for e-Planning in Europe. *International Journal of E-Planning Research*, *5*(4), 41–53. doi:10.4018/IJEPR.2016100103

Data Driven Detroit. (2017). *Historic Resource Survey*. Retrieved from Data Driven Detroit: http://portal.datadrivendetroit.org/datasets/d464738110584038bfb836f861e566a0_0

Davey, M. (2014, February 17). A Picture of Detroit Ruin, Street by Forlorn Street. *The New York Times*. Retrieved July 21, 2017, from https://www.nytimes.com/2014/02/18/us/detroit-tries-to-get-a-clear-picture-of-its-blight.html

Dawes, S. S. (2008, December). The evolution and continuing challenges of e-governance. *Public Administration Review,* 86-102.

de Certeau, M. (1984). *The Practice of Everyday Life*. Berkeley, CA: University of California Press.

de Souza e Silva, A. (2006). From cyber to hybrid: Mobile technologies as interfaces of hybrid spaces. *Space and Culture*, *3*, 261–278.

de Vreede, G.-J., Briggs, R. O., & Reiter-Palmon, R. (2010). Exploring asynchronous brainstorming in large groups: A field comparison of serial and parallel subgroups. *Human Factors*, *52*(2), 189–202. doi:10.1177/0018720809354748 PMID:20942250

de Vreede, T., de Vreede, G.-J., & Reiter-Palmon, R. (2017, January). Antecedents of engagement in community-based crowdsourcing. *Proceedings of the 50th Hawaii International Conference on System Sciences*. 10.24251/HICSS.2017.027

Dennis, A. R. (1996). Information exchange and use in group decision making: You can lead a group to information, but you can't make it think. *Management Information Systems Quarterly*, *20*(4), 433–457. doi:10.2307/249563

Dennis, A. R., Minus, R. K., & Williams, M. L. (in press). Creativity in computer-mediated virtual roups. In P. B. Paulus & B. A. Nijstad (Eds.), *Handbook of group creativity*. New York: Oxford University Press.

Dennis, A. R., Valacich, J. S., Connolly, T., & Wynne, B. E. (1996). Process structuring in electronic brainstorming. *Information Systems Research*, *7*(2), 268–277. doi:10.1287/isre.7.2.268

Dennis, A. R., & Williams, M. L. (2003). Electronic brainstorming: Theory, research, and future Directions. In P. B. Paulus & B. A. Nijstad (Eds.), *Group creativity: Innovation through collaboration* (pp. 160–180). New York, NY: Oxford University Press. doi:10.1093/acprof:oso/9780195147308.003.0008

Dennis, N. (1970). *People and Planning. Faber*. London: DOE.

Denskus, T., & Papan, A. S. (2013). Reflexive engagements: The international development blogging evolution and its challenges. *Development in Practice*, *23*(4), 455–467. doi:10.1080/09614524.2013.790940

Derosa, D. M., Smith, C. L., & Hantula, D. A. (2007). The medium matters: Mining the long-promised merit of group interaction in creative idea generation tasks in a meta-analysis of the electronic group brainstorming literature. *Computers in Human Behavior*, *23*(3), 1549–1581. doi:10.1016/j.chb.2005.07.003

Detroit Blight Removal Task Force. (n.d.a). *Frequently Asked Questions*. Retrieved July 21, 2017, from The Detroit Blight Removal Task Force: http://www.timetoendblight.com/faq/#1

Detroit Blight Removal Task Force. (n.d.b) *What is the context for our work?* Retrieved July 21, 2017, from Detroit Blight Removal Task Force: https://s3.amazonaws.com/detroit-blight-taskforce/CHAPTER+02.pdf

Deuja, A., Kohn, N. W., Paulus, P. B., & Korde, R. M. (2014). Taking a broad perspective before brainstorming. *Group Dynamics*, *18*(3), 222–236. doi:10.1037/gdn0000008

DHI. (2008). Linking Water, Energy & Climate Change: A proposed water and energy policy initiative for the UN Climate Change Conference, COP15. In *Copenhagen 2009*. Draft Concept Note.

Dia, H. (2017). *When Artificial Intelligence Permeates the Modern City and Transforms Urban Life (Keynote Speech)*. 15th Annual Computers in Urban Planning and Urban Management Conference, Adelaide, Australia.

Dickson, G. W., Lee-Partridge, J. E., Limayem, M., & Desanctis, G. L. (1996). Facilitating computer-supported meetings: A cumulative analysis in a multiple-criteria task environment. *Group Decision and Negotiation*, *5*(1), 51–72. doi:10.1007/BF02404176

Diehl, M., & Stroebe, W. (1987). Productivity loss in brainstorming groups: Toward the solution of a riddle. *Journal of Personality and Social Psychology*, *53*(3), 497–509. doi:10.1037/0022-3514.53.3.497

Domene, E., & Saurí, D. (2006). Urbanisation and water consumption: Influencing factors in the metropolitan region of Barcelona. *Urban Studies (Edinburgh, Scotland)*, *43*(9), 1605–1623. doi:10.1080/00420980600749969

Domene, E., Saurí, D., & Parés, M. (2005). Urbanization and sustainable resource use: The case of garden watering in the metropolitan region of Barcelona. *Urban Geography*, *26*(6), 520–535. doi:10.2747/0272-3638.26.6.520

Dornburg, C. C., Stevens, S. M., Hendrickson, S. M., & Davidson, G. S. (2009). Improving extreme-scale problem solving: Assessing electronic brainstorming effectiveness in an industrial setting. *Human Factors*, *51*(4), 519–527. doi:10.1177/0018720809343587 PMID:19899361

Druckman, A., & Jackson, T. (2008). Household energy consumption in the UK: A highly geographically and socio-economically disaggregated model. *Energy Policy*, *36*(8), 3177–3192. doi:10.1016/j.enpol.2008.03.021

Dugosh, K. L., & Paulus, P. B. (2005). Cognitive and social comparison processes in brainstorming. *Journal of Experimental Social Psychology*, *41*(3), 313–320. doi:10.1016/j.jesp.2004.05.009

Dugosh, K., Paulus, P. B., Roland, E. J., & Yang, H.-C. (2000). Cognitive stimulation in brainstorming. *Journal of Personality and Social Psychology*, *79*(5), 722–735. doi:10.1037/0022-3514.79.5.722 PMID:11079237

Duhl, L. J., & Sanchez, A. K. (1999). *Healthy cities and the city planning process: A background document on links between health and urban planning*. Copenhagen, Denmark: World Health Organization.

Duro, J. A. (2012). On the automatic application of inequality indexes in the analysis of the international distribution of environmental indicators. *Ecological Economics*, *76*, 1–7. doi:10.1016/j.ecolecon.2011.12.019

Dutil, P., Howard, C., Langford, J., & Roy, J. (2007). 'Re-imagining the service recipient: Customers, clients and citizens in the emerging world of e-government. *Journal of Information Technology & Politics*, *4*(1), 77–90. doi:10.1300/J516v04n01_06

Dyllick, T., & Hockerts, K. (2002). Beyond the business case for corporate sustainability. *Business Strategy and the Environment*, *11*(2), 130–141. doi:10.1002/bse.323

Dzindolet, M. T., Paulus, P. B., & Glazer, C. (2012). Brainstorming in virtual teams. In C. N. Silva (Ed.), *Online research methods in urban and planning studies: Design and outcome* (pp. 138–156). Hershey, PA: IGI Global. doi:10.4018/978-1-4666-0074-4.ch009

Earl, J., & Kimport, K. (2011). *Digitally Enabled Social Change: Activism in the Internet Age*. Cambridge, MA: MIT Press. doi:10.7551/mitpress/9780262015103.001.0001

Easterling, K. (2016). *Extrastatecraft: The Power of Infrastructure Space*. London: Verso.

Ebdon, C. (2000). The relationship between citizen involvement in the budget process and city structure and culture. *Public Productivity & Management Review*, 383-393.

Eckler, P., & Bolls, P. (2011). Spreading the virus: Emotional tone of viral advertising and its effect on forwarding intentions and attitudes. *Journal of Interactive Advertising*, *11*(2), 1–11. doi:10.1080/15252019.2011.10722180

Effing, R., & Spil, T. A. M. (2016). The social strategy cone: Towards a framework for evaluating social media strategies. *International Journal of Information Management*, *36*(1), 1–8. doi:10.1016/j.ijinfomgt.2015.07.009

Eger, J. (1997). Cyberspace and cyberplace: Building the smart communities of tomorrow. *San Diego Union-Tribune, 26*. empirical assessment. *Public Administration Review, 68*(3), 523–536.

EIA. (2011). *Residential Energy Consumption Survey (RECS)*. Retrieved 11,August, 2014, from http://www.eia.gov/consumption/residential/data/2009/index.cfm?view=characteristics

Eisner, D. (2006). *What is neogeography anyway?* Retrieved July 15, 2017, from http://platial.typepad.com/news/2006/05/what_is_neogeog.html

Ellison, N. B., Steinfield, C., & Lampe, C. (2007). The benefits of Facebook "friends": Social capital and college students' use of online social network sites. *Journal of Computer-Mediated Communication, 12*(4), 1143–1168. doi:10.1111/j.1083-6101.2007.00367.x

Elwood, S. (2008). Volunteered geographic information: Key questions, concepts and methods to guide emerging research and practice. *GeoJournal, 72*(3/4), 133–135. doi:10.100710708-008-9187-z

Elwood, S., Goodchild, M. F., & Sui, D. Z. (2012). Researching volunteered geographic information: Spatial data, geographic research, and new social practice. *Annals of the Association of American Geographers, 102*(3), 571–590. doi:10.1080/00045608.2011.595657

English, K., & Feaster, L. S. (2003). *Community geography: GIS in action*. Redlands, CA: ESRI Press.

Eräranta, S. (2013). *Situation awareness in urban planning* (Unpublished Master's Thesis). Aalto University, Espoo, Finland.

Eräranta, S., Kahila-Tani, M., & Nummi-Sund, P. (2015). Web-based public participation in urban planning competitions. *International Journal of E-Planning Research, 4*(1), 1–18. doi:10.4018/ijepr.2015010101

Estelles-Arolas, E., & Gonzalez-Ladron-de-Guevara, F. (2012). Towards an Integrated Crowdsourcing Definition. *Journal of Information Science, 38*(2), 189–200. doi:10.1177/0165551512437638

European Commission. (2015). *Digital Single Market strategy*. COM(2015) 192 final. Retrieved July 15, 2017, from https://ec.europa.eu/digital-single-market/en/news/digital-single-market-strategy-europe-com2015-192-final

European Union. (2015). *A Digital Single Market for Europe: Commission Sets Out 16 Initiatives to Make it Happen* [Press release]. Retrieved on 10 June 2017 from http://europa.eu/rapid/press-release_IP-15-4919_en.htm

European Union. (2016). *WiFi4EU: Free Wi-Fi for Europeans* [Press release]. Retrieved on 10 June 2017 from https://ec.europa.eu/digital-single-market/en/policies/wifi4eu-free-wi-fi-europeans

Evaluation of the Land Use and Building Act in practice 2013 [Arviointi maankäyttö- ja rakennuslain toimivuudesta 2013]. (2014). Suomen ympäristö, Ympäristöministeriö, 1/2014. Helsinki.

Evans-Cowley, J. (2010). Planning in the age of Facebook: The role of social networking in planning processes. *GeoJournal, 75*(5), 407–420. doi:10.100710708-010-9388-0

Evans-Cowley, J. (2010). Planning in the Real-Time City: The Future of Mobile Technology. *Journal of Planning Literature, 25*(2), 136–149. doi:10.1177/0885412210394100

Evans-Cowley, J. (2012). There's an App for That: Mobile Applications for Urban Planning. *International Journal of E-Planning Research, 1*(2), 79–87. doi:10.4018/ijepr.2012040105

Evans-Cowley, J., & Hollander, J. (2010). The New Generation of Public Participation: Internet-based Participation Tools. *Planning Practice and Research, 25*(3), 397–408. doi:10.1080/02697459.2010.503432

Evans, E. (2014). Integrity, Character, and Intactness: A preservation survey model for legacy cities. In T. Schwartz (Ed.), *Historic Preservation and Urban Change*. Cleveland, OH: Cleveland Urban Design Collaborative, Kent State University.

Faccioli, P., & Losacco, G. (2010). *Nuovo manuale di sociologia visuale: dall'analogico al digitale*. Milan: Franco Angeli.

Facebook. (2014). *Page Post Metrics*. Retrieved 12 May, 2015, from https://www.facebook.com/help/336143376466063/

Facebook. (2015). *Negative Feedback*. Retrieved 12 May, 2015, from https://www.facebook.com/help/401280783409819

Facebook. (2017). *Free Basics Website*. Retrieved 10 July 2017 from https://info.internet.org/en

Faradani, S., Hartmann, B., & Ipeirotis, P. (2011) What's the Right Price? Pricing Tasks for Finishing on Time. *Human Computation: Papers from the 2011 AAAI Workshop*, 26-31.

Faure, C. (2004). Beyond brainstorming: Effects of different group procedures on selection of ideas and satisfaction with the process. *The Journal of Creative Behavior, 38*(1), 13–34. doi:10.1002/j.2162-6057.2004.tb01229.x

FEMA. (2002). *Preventing a disaster within the disaster: The effective use and management of unaffiliated volunteers*. UPS Foundation. Retrieved from https://www.ncjrs.gov/pdffiles1/Archive/202852NCJRS.pdf

FEMA. (2005). *Managing Spontaneous Volunteers in Times of Disaster: The Synergy of Structure and Good Intentions*. UPS Foundation. Retrieved from https://www.fema.gov/pdf/donations/ManagingSpontaneousVolunteers.pdf

Fernandez, L., Barbera, J., & van Dorp, J. (2006). Spontaneous volunteer response 36 to disasters: The benefits and consequences of good intentions. *Journal of Emergency Management (Weston, Mass.), 4*(5), 57–68.

Fernando, R. (2012). Sustainable globalization and implications for strategic corporate and national sustainability. *Corporate Governance, 12*(4), 579–589. doi:10.1108/14720701211267883

Ferrara, E., Varol, O., Davis, C., Menczer, F., & Flammini, A. (2016). The rise of social bots. *Communications of the ACM*, *59*(7), 96–104. doi:10.1145/2818717

Fieseler, C., & Fleck, M. (2013). The pursuit of empowerment through social media: Structural social capital dynamics in CSR-blogging. *Journal of Business Ethics*, *118*(4), 759–775. doi:10.100710551-013-1959-9

Fietkiewicz, K. J., Mainka, A., & Stock, W. G. (2017). eGovernment in cities of the knowledge society. An empirical investigation of Smart Cities' governmental websites. *Government Information Quarterly*, *34*(1), 75–83. doi:10.1016/j.giq.2016.08.003

Filippi, F. D., Coscia, C., & Guido, R. (2017). How Technologies Can Enhance Open Policy Making and Citizen-Responsive Urban Planning: MiraMap - A Governing Tool for the Mirafiori Sud District in Turin (Italy). *International Journal of E-Planning Research*, *6*(1), 23–42. doi:10.4018/IJEPR.2017010102

Fisher, J., & Ackerman, F. (2011). *The Water-Energy Nexus in the Western States: Projections to 2100 U.S.* Stockholm Environment Institute.

Fitzjohn, M. (2009). The Use of GIS in Landscape Heritage and Attitudes to Place: Digital Deep Maps. In L. S. S. Marie & J. Carman (Eds.), *Heritage Studies: Methods and Approaches* (pp. 237–252). London: Routledge.

Flew, T. (2005). *New media: an introduction*. South Melbourne, Australia: Oxford University Press.

Fonseca, F., Egenhofer, M., Davis, C. Jr, & Borges, K. (2000). Ontologies and knowledge sharing in urban GIS computer. *Environment and Urban Systems*, *24*(3), 232–251. doi:10.1016/S0198-9715(00)00004-1

Forester, J. (1987). *Critical Theory and Public Life*. Cambridge, MA: The MIT Press.

Forester, J. (1989). *Planning in the Face of Power*. Los Angeles, CA: University of California Press.

Fortano, L. (2008). Anytime? Anywhere? Reframing Debates Around Municipal Wireless Networking. *The Journal of Community Informatics*, *4*(1).

Fox, C., McIntosh, B., & Jeffrey, P. (2009). Classifying households for water demand forecasting using physical property characteristics. *Land Use Policy*, *26*(3), 558–568. doi:10.1016/j.landusepol.2008.08.004

Fredericks, Tomitsch, & Stewart. (2017). Design Patterns for Integrating Digitally Augmented Pop-ups with Community Engagement. *International Journal of E-Planning Research*, *6*(3), 19-41.

Frey, P. (2008). *Building Reuse: Finding a place on American climate policy agendas*. Retrieved June 27, 2014, from National Trust for Historic Preservation website: http://www.preservationnation.org/information-center/sustainable-communities/additional-resources/buillding_reuse.pdf

Friedrich, F., & Burghardt, P. (2007, March). Agent-based systems for disaster management. *Communications of the ACM*, *50*(3), 41–42. doi:10.1145/1226736.1226763

Frow, J. (1995). *Cultural Studies and Cultural Value*. Oxford, UK: Oxford University Press.

Fung, B. (2016). New York Wants Everyone to Have Free, Fast Wi-Fi. *The Sydney Morning Herald*. Retrieved on 17 June 2017 from http://www.smh.com.au/technology/innovation/new-york-wants-everyone-to-have-free-fast-wifi-20160411-go3dow.html

Gabriel, C., & Fellah, A. (2016). *The Unlicensed Road to 5G: WBA Report 2016*. Retrieved on 10 May 2017 from http://www.wballiance.com/resources/industry-reports/

Gagliardi, D., Schina, L., Sarcinella, M. L., Mangialardi, G., Niglia, F., & Corallo, A. (2017). Information and communication technologies and public participation: Interactive maps and value added for citizens. *Government Information Quarterly*, *34*(1), 153–166. doi:10.1016/j.giq.2016.09.002

Gallupe, R. B., Bastianutti, L. M., & Cooper, W. H. (1991). Unblocking brainstorms. *The Journal of Applied Psychology*, *76*(1), 137–142. doi:10.1037/0021-9010.76.1.137 PMID:2016214

Ganapati, S. (2010). *Using geographic information systems to increase citizen engagement* (White paper). Retrieved March 9, 2012, from IBM Center for the Business of Government website: http://www.businessofgovernment.org/sites/default/files/GanapatiReport.pdf

Ganpati, S., & Reddick, C. (2016). An ostrich burying its head in the sand? The 2009 NASPAA standards and scope of information technology and e-government curricula. *Journal of Public Affairs Education*, *22*(2), 267–286.

Garde, A. (2014). Citizen participation, design competition and the product in urban design: Insights from the Orange County Great Park. *Journal of Urban Design*, *19*(1), 89–118. doi:10.1080/13574809.2013.854697

Garrett, R. K., & Jensen, M. J. (2011). E-Democracy Writ Small: The impact of the Internet on citizen access to local elected officials. *Information Communication and Society*, *14*(2), 177–197. doi:10.1080/1369118X.2010.490558

Garson, G. D. (2006). *Public information technology and e-governance: Managing the virtual state*. Sudbury, MA: Jones and Bartlett Publishers.

Gascó, M. (2017). Living labs: Implementing open innovation in the public sector. *Government Information Quarterly*, *34*(1), 90–98. doi:10.1016/j.giq.2016.09.003

Gautier, A., & Pache, A.-C. (2015). Research on Corporate Philanthropy: A Review and Assessment. *Journal of Business Ethics*, *126*(3), 343–369. doi:10.100710551-013-1969-7

Gaver, W. W. (1996). Situating Action II: Affordances for Interaction: The Social Is Material for Design. *Ecological Psychology*, *8*(2), 111–129. doi:10.120715326969eco0802_2

Geschka, H. (1993). The development and assessment of creative thinking techniques: A German perspective. In S. G. Isaksen, M. C. Murdock, R. L. Firestien, & D. J. Treffinger (Eds.), *Nurturing and developing creativity: The emergence of a discipline* (pp. 215–236). Norwood, NJ: Ablex.

Getty Foundation. (2017, July 20). *Our Priorities*. Retrieved from The Getty Foundation: http://www.getty.edu/foundation/initiatives/

Gibbons, J., & Ruth, S. (2006). Municipal Wi-Fi: Big Wave or Wipeout? *IEEE Internet Computing*, *10*(3), 66–71. doi:10.1109/MIC.2006.60

Gibson, J. J. (1986). *The Ecological Approach to Visual Perception*. Hillsdale, NJ: Lawrence Erlbaum Associates.

Gil de Zúñiga, H. G., Copeland, L., & Bimber, B. (2014). Political consumerism: Civic engagement and the social media connection. *New Media & Society*, *16*(3), 488–506. doi:10.1177/1461444813487960

Giles, G. (2012). Computational Social Science: Making the Links. *NATNews*, *488*, 448–450. Retrieved from http://www.nature.com/news/computational-social-science-making-the-links-1.11243 PMID:22914149

Gilson, L. L., Maynard, M. T., Jones Young, N. C., Vartiainen, M., & Hakonen, M. (2015). Virtual teams research. *Journal of Management*, *41*(5), 1313–1337. doi:10.1177/0149206314559946

Glasmeier, A., & Christopherson, S. (2015). Thinking about Smart Cities. *Cambridge Journal of Regions, Economy and Society*, *8*(1), 3–12. doi:10.1093/cjres/rsu034

Goggin, G. (2012). Driving the Internet: Mobile Internets, Cars, and the Social. *Future Internet*, *4*(4), 306–321. doi:10.3390/fi4010306

Goldenberg, O., Larson, J. R. Jr, & Wiley, J. (2013). Goal instructions, response format, and idea generation in groups. *Small Group Research*, *44*(3), 227–256. doi:10.1177/1046496413486701

Goodchild, M. (2009). NeoGeography and the nature of geographic expertise. *Journal of Location Based Services*, *3*(2), 82–96. doi:10.1080/17489720902950374

Goodchild, M. F. (2007). Citizens as sensors: The world of volunteered geography. *GeoJournal*, *69*(4), 211–221. doi:10.100710708-007-9111-y

Goodman, G. (1995). Brainwriting: What to do when there's not a cloud in the brainstorming sky. In J. Heilbrunn (Ed.), *Marketing encyclopedia: Issues and trends shaping the future* (pp. 40–46). Lincolnwood, IL: NTC Business Books.

Goodspeed, R., Spanring, C., & Reardon, T. (2012). *Crowdsourcing as Data Sharing: A Regional Web-based Real Estate Development Database*. Presented at the ICEGOV '12, Albany, NY. 10.1145/2463728.2463819

Google. (2017). *Project Loon Website*. Retrieved on 10 July 2017 from https://x.company/loon

Gordon, E., & de Souza e Silva, A. (2011). *Net Locality: Why Location Matters in a Networked World*. Malden, MA: Blackwell.

Gordon, E., & de Souza e Silva, A. (2011). *Net locality: Why location matters in a networked world*. New York, NY: Wiley-Blackwell.

Gordon, E. (2013). Beyond Participation: Designing for the Civic Web. *Journal of Digital and Media Literacy*, *1*(1).

Graham, M., & Avery, E. (2013). Government public relations and social media: An analysis of the perceptions and trends of social media use at the local government level. *The Public Relations Journal*, *7*(4), 1–21.

Graham, S. (2005). Software-Sorted Geographies. *Progress in Human Geography*, *29*(5), 562–580. doi:10.1191/0309132505ph568oa

Gram-Hanssen, K. (2014). New needs for better understanding of household's energy consumption–behaviour, lifestyle or practices? *Architectural Engineering and Design Management*, *10*(1-2), 91–107. doi:10.1080/17452007.2013.837251

Grangaard, S., & Ryhl, C. (2016). Vandhalla – A sport centre and a successful example of first-generation universal design. In *Universal Design 2016: Learning from the Past, Designing for the Future: Proceedings of the 3rd International Conference on Universal Design (UD 2016)* (pp. 243-245). Amsterdam: IOS Press.

Graziano, T. (2017). Citizen e-participation in urban planning: Achievements and future challenges in a Mediterranean city. *International Journal of E-Planning Research*, *3*(6), 1–18. doi:10.4018/IJEPR.2017070101

Grise, M., & Gallupe, R. B. (2000). Information overload: Addressing the productivity paradox in face-to-face electronic meetings. *Journal of Management Information Systems*, *16*(3), 157–185. doi:10.1080/07421222.1999.11518260

Grodach, C. (2010). Art Spaces in Community and Economic Development: Connections to Neighborhoods, Artists, and the Cultural Economy. *Journal of Planning Education and Research*, *31*(1), 74–85. doi:10.1177/0739456X10391668

Gruber, T. R. (1995). Towards Principles for the Design of Ontologies Used for Knowledge Sharing. *International Journal of Human-Computer Studies*, *43*(5/6), 907–928. doi:10.1006/ijhc.1995.1081

GSMA. (2016). *The Mobile Economy 2016*. London: GSMA.

Guarino, N., Oberle, D., & Staab, S. (2009). *What is an ontology? In Handbook on ontologies*. Dordrecht: Springer.

Guerra Santin, O., Itard, L., & Visscher, H. (2009). The effect of occupancy and building characteristics on energy use for space and water heating in Dutch residential stock. *Energy and Building*, *41*(11), 1223–1232. doi:10.1016/j.enbuild.2009.07.002

Habermas, J., & McCarthy, T. (1985). *The Theory of Communicative Action* (Vol. 2). Boston, MA: Beacon Press.

Habuchi, I. (2005). Accelerated Reflexivity. In M. Ito, D. Okabe, & M. Matsuda (Eds.), *Personal, Portable, Pedestrian: Mobile Phones in Japanese Life*. Cambridge, MA: MIT Press.

Hack, R. (2013). The Google fiber project. *Journal of Computing Sciences in Colleges Archive, 28*(5), 140-140.

Hanson, J., & Hogea, A. (2012). The Internet as the public sphere: Deliberative democracy and civic engagement. In A. Manoharan & M. Holzer (Eds.), *E-Governance and Civic Engagement: Factors and Determinants of E-Democracy* (pp. 467–486). Hershey, PA: IGI Global. doi:10.4018/978-1-61350-083-5.ch023

Hanzl, M. (2007). Information technology as a tool for public participation in urban planning: A review of experiments and potentials. *Design Studies, 28*(3), 289–307. doi:10.1016/j.destud.2007.02.003

Harper, D. (1988). Visual sociology: Expanding sociological vision. *The American Sociologist, 19*(1), 54–70. doi:10.1007/BF02692374

Harvey, D. (2009). *Cosmopolitanism and the Geographies of Freedom.* New York: Columbia University Press.

Harvey, D. (2012). *Rebel Cities: From the Right to the City to the Urban Revolution.* London: Verso.

Healey, P. (1997). *Collaborative planning. Shaping places in fragmented societies.* Hampshire, UK: McMillan Press.

Heberlein, T. A. (1976). Some observations on alternative mechanisms for public involvement: The hearing, public opinion poll, the workshop and the quasi-experiment. *Natural Resources Journal, 16*(1), 197–212.

Helbig, N., Gil-García, J. R., & Ferro, E. (2009). Understanding the complexity of electronic government: Implications from the digital divide literature. *Government Information Quarterly, 26*(1), 89–97. doi:10.1016/j.giq.2008.05.004

Hendler, J., Shadbolt, N., Hall, W., Berners-Lee, T., & Weitzner, D. (2008). Web science: An interdisciplinary approach to understanding the web. *Communications of the ACM, 51*(7), 60–69. doi:10.1145/1364782.1364798

Herring, S. C. (2009). Web Content Analysis: Expanding the Paradigm. In J. Hunsinger, L. Klastrup, & M. M. Allen (Eds.), *International Handbook of Internet Research* (pp. 233–249). New York: Springer. doi:10.1007/978-1-4020-9789-8_14

Heslin, P. A. (2009). Better than brainstorming? Potential boundary conditions to brainwriting for idea generation in organizations. *Journal of Occupational and Organizational Psychology, 82*(1), 129–145. doi:10.1348/096317908X285642

Hiller, J., & Bélanger, H. (2001). *Privacy Strategies for Electronic Government. E-Government Series.* Arlington, VA: PricewaterhouseCoopers Endowment for the Business of Government.

Hiltz, S. R., Johnson, K., & Turoff, M. (1991). Group decision support: The effects of designated human leaders and statistical feedback in computerized conferences. *Journal of Management Information Systems, 8*(2), 81–108. doi:10.1080/07421222.1991.11517922

Hinds, J. M., & Payne, S. J. (2016). Collaborative inhibition and semantic recall: Improving collaboration through computer mediated communication. *Applied Cognitive Psychology*, *30*(4), 554–565. doi:10.1002/acp.3228

Ho, A. T., & Ni, A. (2004). Explaining the progressiveness of e-government: A case study of Iowa county treasurers' offices. *American Review of Public Administration*, *34*(2), 164–180. doi:10.1177/0275074004264355

Hoffman, D. L., Novak, T. P., & Schlosser, A. E. (2001). The evolution of the digital divide: Examining the relationship of race to Internet access and usage over time. *The digital divide: Facing a crisis or creating a myth*, 47-97.

Hofmann, S., Beverungen, D., Räckers, M., & Becker, J. (2013). What makes local governments' online communications successful? Insights from a multi-method analysis of Facebook. *Government Information Quarterly*, *30*(4), 387–396. doi:10.1016/j.giq.2013.05.013

Hollands, R. G. (2008). Will the real smart city please stand up? Intelligent, progressive or entrepreneurial? *City*, *12*(3), 303–320. doi:10.1080/13604810802479126

Hollan, J., Hutchins, E., & Kirsh, D. (2008). Distributed Cognition: Toward a New Foundation for Human-Computer Interaction Research. *ACM Transactions on Computer-Human Interaction*, *7*(2), 174–196. doi:10.1145/353485.353487

Horelli, L., & Wallin, S. (2010). The future-making assessment approach as a tool for e-planning and community development – The case of ubiquitous Helsinki. In C. N. Silva (Ed.), *Handbook of research on e-planning: ICTs for urban development and monitoring* (pp. 58–79). Hershey, PA: IGI Global. doi:10.4018/978-1-61520-929-3.ch004

HousePeters, L., Pratt, B., & Chang, H. (2010). Effects of urban spatial structure, sociodemographics, and climate on residential water consumption in Hillsboro, Oregon. The American Water Resource Association, 46(3).

Howard, P. N., & Parks, M. (2012). Social Media and Political Change: Capacity, Constraint, and Consequence. *Journal of Communication*, *62*(2), 359–362. doi:10.1111/j.1460-2466.2012.01626.x

Howe, J. (2006). Crowdsourcing: A Definition. In *Crowdsourcing: Tracking the Rise of the Amateur*. Retrieved from http://crowdsourcing.typepad.com/cs/2006/06/ crowdsourcing_a.htm

Howe, J. (2007). *Crowdsourcing: a definition. Wired Blog Network: Crowdsourcing*. Retrieved July 15, 2017, from http://crowdsourcing.typepad.com/cs/2006/06/crowdsourcing_a.html

Hristidis, V., Chen, S., Li, T., Luis, S., & Deng, Y. (2010, October). Survey of data management and analysis in disaster situations. *Journal of Systems and Software*, *83*(10), 1701–1714. doi:10.1016/j.jss.2010.04.065

Huang, S., & Hsu, W. (2003). Materials flow analysis and emergy evaluation of Taipei's urban construction. *Landscape and Urban Planning*, *63*(2), 61–74. doi:10.1016/S0169-2046(02)00152-4

Huber, J., Landherr, A., Probst, F., & Reisser, C. (2012). Stimulating user activity on company fan pages in online social networks. ECIS 2012 Proceedings, Paper 188.

Hu, G., Pan, W., Lu, M., & Wang, J. (2009). The widely shared definition of e-Government: An exploratory study. *The Electronic Library*, *27*(6), 968–985. doi:10.1108/02640470911004066

Hülsheger, U. R., Anderson, N., & Salgado, J. F. (2009). Team-level predictors of innovation at work: A comprehensive meta-analysis spanning three decades of research. *The Journal of Applied Psychology*, *94*(5), 1128–1145. doi:10.1037/a0015978 PMID:19702361

Hutchins, B. (2016). 'We Don't Need No Stinking Smartphones!' Live Stadium Sports Events, Mediatization, and the Non-use of Mobile Media. *Media Culture & Society*, *38*(3), 420–436. doi:10.1177/0163443716635862

Hutchins, E. (2008). *Distributed Cognition*. International Encyclopedia of the Social and Behavioral Sciences.

Inman, D., & Jeffrey, P. (2006). A review of residential water conservation tool performance and influences on implementation effectiveness. *Urban Water Journal*, *3*(3), 127–143. doi:10.1080/15730620600961288

ITU. (2016). *Measuring the Information Society Report*. Geneva: International Telecommunication Union.

Jacobsen, P. H., Harty, C., & Tryggestad, K. (2016). (Re)constructing the wicked problem through the visual and the verbal: the case of a dialogue based architectural competition. In *Proceedings 32nd Annual ARCOM Conference*. Manchester UK. Association of Researchers in Construction Management.

Jacquet, J. (2015). *Is Shame Necessary?: New Uses for an Old Tool*. New York: Pantheon Books.

Jansen, B. J., Zhang, J. M., Sobel, K., & Chowdury, A. (2009). Twitter power: Tweets as electronic word of mouth. *Journal of the American Society for Information Science and Technology*, *60*(11), 2169–2188. doi:10.1002/asi.21149

Janssen, M., Lee, J., Bharosa, N., & Cresswell, A. (2010, March). Advances in multi-agency disaster management: Key elements in disaster research. *Information Systems Frontiers*, *12*(1), 1–7. doi:10.100710796-009-9176-x

Jarmusch, A. (2011). 2011 National Planning Excellence Award for Public Outreach. *Planning (Chicago, Ill.)*, *77*(4), 35.

Javed, N., & Muralidhara, B. L. (2015). Framework for Participative and Collaborative Governance using Social Media Mining Techniques. *International Journal of Computers and Applications*, *123*(8), 36–41. doi:10.5120/ijca2015905435

Jessup, L. M., Connolly, T., & Galegher, J. (1990). The effects of anonymity on group process in automated group problem solving. *Management Information Systems Quarterly*, *14*(3), 313–321. doi:10.2307/248893

Jin, G., Nicolai, B., Jiang, K., & Winer, C. (2011) Distributed image processing and classification for GIS based disaster management and communication system. In *Proceedings of the 2nd International Conference on Computing for Geospatial Research & Applications.* ACM. 10.1145/1999320.1999347

Jogan, I. (2002). Alcune considerazioni a margine della 'svolta ontologica' della GIS Science. *Archivio di Studi Urbani e Regionali*, *74*, 183–196.

Johnson, G. M. (2006). Synchronous and asynchronous text-based CMC in educational contexts: A review of recent research. *TechTrends*, *50*(4), 46–53. doi:10.100711528-006-0046-9

Johnson, G. M., Howell, A. J., & Code, J. R. (2005). Online discussion and college student learning: Toward a model of influence. *Technology, Pedagogy and Education*, *14*(1), 61–76. doi:10.1080/14759390500200193

Johnson, P. A. (2016). Reflecting on the Success of Open Data: How Municipal Government Evaluates their Open Data Programs. *International Journal of E-Planning Research*, *5*(3), 1–12. doi:10.4018/IJEPR.2016070101

Johnson, P., & Sieber, R. (2013). Situating the Adoption of VGI by Government. In D. Sui, S. Elwood, & M. Goodchild (Eds.), *Crowdsourcing Geographic Knowledge: Volunteered Geographic Information (VGI) in Theory and Practice*. Dordrecht, The Netherlands: Springer. doi:10.1007/978-94-007-4587-2_5

Joss, S., Cook, M., & Dayot, Y. (2017). Smart Cities: Towards a New Citizenship Regime? A Discourse Analysis of the British Smart City Standard. *Journal of Urban Technology*, *24*(4), 29–49. doi:10.1080/10630732.2017.1336027

Juris, J. (2012). Reflections on #Occupy Everywhere: Social media, public space, and emerging logics of aggregation. *Journal of American Ethnological Society*, *39*(2), 259–279.

Kahila, M., & Kyttä, M. (2009). SoftGIS as a bridge-builder in collaborative urban planning. In S. Geertman & J. Stillwell (Eds.), *Planning support systems best practice and new methods* (pp. 389–411). Springer. doi:10.1007/978-1-4020-8952-7_19

Kahila-Tani, M. (2015). *Reshaping the planning process using local experiences: Utilising PPGIS in participatory urban planning* (Doctoral dissertation). Aalto University Publication Series, 223/2015.

Kamar, E., & Horvitz, E. (2012) Incentives for truthful reporting in crowdsourcing. *Proceedings of the 11th International Conference on Autonomous Agents and Multiagent Systems*, 3, 1329-1330.

Kanawattanachai, P., & Yoo, Y. (2007). The impact of knowledge coordination on virtual team performance over time. *Management Information Systems Quarterly*, *31*(4), 783–808. doi:10.2307/25148820

Kansas State Historic Preservation Office & Kansas Historical Society. (2014). *Kansas Historic Resources Inventory*. Retrieved June 27, 2014, from http://khri.kansasgis.org/

Kaufman, J. L. (1963). *Illustrating the Zoning Ordinance*. Chicago: American Planning Association.

Kaufman, N. (2009). *Place, Race and Story: Essays on the Past and Future of Historic Preservation*. New York: Routledge.

Kavada, A. (2015). Creating the collective: Social media, the Occupy Movement and its constitution as a collective actor. *Information Communication and Society*, *18*(8), 872–886. doi:10.1080/1369118X.2015.1043318

Kay, G. (1995). Effective meetings through electronic brainstorming. *Journal of Management Development*, *14*(6), 4–25. doi:10.1108/02621719510086147

Kazemian, R. (2009). Design interactivity and communicative quality judgment versus urban design competition – A design methodology statement. *Nordic Journal of Architectural Research*, *21*(2/3), 68–78.

Kazemian, R., Rönn, M., & Svensson, C. (2006). Arkitekturtävlingar i Finland. *TRITA-ARK-Forskningspublikationer*, *2006*, 11.

Kelly, C. M., Wilson, J. S., Baker, E. A., Miller, D. K., & Schootman, M. (2013). Using Google Street View to audit the built environment: Inter-rater reliability results. *Annals of Behavioral Medicine*, *45*(1), 108–112. doi:10.100712160-012-9419-9 PMID:23054943

Kenney, D. S., Goemans, C., Klein, R., Lowrey, J., & Reidy, K. (2008). Residential Water Demand Management: Lessons from Aurora, Colorado. JAWRA Journal of the American Water Resources Association, 44(1), 192-207.

Kerr, D. S., & Murthy, U. S. (2004). Divergent and convergent idea generation in teams: A comparison of computer-mediated and face-to-face communication. *Group Decision and Negotiation*, *13*(4), 381–399. doi:10.1023/B:GRUP.0000042960.38411.52

Kersten, G. E. (2003). E-democracy and participatory decision processes: Lessons from eNegotiation experiments. *Journal of Multi-Criteria Decision Analysis*, *12*(2-3), 127–143. doi:10.1002/mcda.352

Khan, G. F. (2017). Social Media Analytics. *Social Media for Government*, 93-118.

Kiesler, S., Siegel, J., & McGuire, T. W. (1984). Social psychological aspects of computer-mediated communication. *The American Psychologist*, *39*(10), 1123–1134. doi:10.1037/0003-066X.39.10.1123

Kim, B. J., Kavanaugh, A. L., & Hult, K. M. (2011). Civic engagement and Internet use in local governance: Hierarchical linear models for understanding the role of local community groups. *Administration & Society*, *43*(7), 807–835. doi:10.1177/0095399711413873

Kim, S., & Lee, J. (2012). E-Participation, Transparency, and Trust in Local Government. *Public Administration Review*, *72*(6), 819–828. doi:10.1111/j.1540-6210.2012.02593.x

Kim, Y., & Chen, H. T. (2015). Discussion network heterogeneity matters: Examining a moderated mediation model of social media use and civic engagement. *International Journal of Communication*, *9*, 22.

Kingston, J. (2002). Merging top level ontologies for scientific knowledge management. *Proceedings of the AAAI workshop on ontologies and the semantic web.*

Kitchin, R. (2014). The Real-Time City? Big Data and Smart Urbanism. *GeoJournal*, *79*(1), 1–14. doi:10.100710708-013-9516-8

Kitchin, R. (2015). Making sense of smart cities: Addressing present shortcomings. *Cambridge Journal of Regions, Economy and Society*, *8*(1), 131–136. doi:10.1093/cjres/rsu027

Kitchin, R., & Dodge, M. (2011). *Code/Space: Software and Everyday Life*. Cambridge, MA: MIT Press. doi:10.7551/mitpress/9780262042482.001.0001

Klabbers, J. H. (1999). A paradigm shift in planning: information technology and self-steering in complex systems. *International conference; 6th, Computers in urban planning and urban management.*

Klausen, J. (2015). Tweeting the Jihad: Social Media Networks of Western Foreign Fighters in Syria and Iraq. *Studies in Conflict and Terrorism*, *38*(1), 1–22. doi:10.1080/1057610X.2014.974948

Kock, N. (2005). Media Richness and Media Naturalness? The Evolution of Our Biological Communication Apparatus and Its Influence on Our Behvaiour Toward E-Communication Tools. *IEEE Transactions on Professional Communication*, *48*(2), 117–130. doi:10.1109/TPC.2005.849649

Kock, N., & Nosek, J. (2005). Expanding the Boundaries of E-Collaboration. *IEEE Transactions on Professional Communication*, *48*(1), 1–9. doi:10.1109/TPC.2004.843272

Kohn, N. W., Paulus, P. B., & Choi, Y. (2011). Building on the ideas of others: An examination of the idea combination process. *Journal of Experimental Social Psychology*, *47*(3), 554–561. doi:10.1016/j.jesp.2011.01.004

Kohn, N., & Smith, S. M. (2009). Partly versus completely out of your mind: Effects of incubation and distraction on resolving fixation. *The Journal of Creative Behavior*, *43*(2), 102–118. doi:10.1002/j.2162-6057.2009.tb01309.x

Korde, R., & Paulus, P. B. (2017). Alternating individual and group idea generation: Finding the elusive synergy. *Journal of Experimental Social Psychology*, *70*, 177–190. doi:10.1016/j.jesp.2016.11.002

Kramer, M. W., Fleming, G. P., & Mannis, S. M. (2001). Improving face-to-face brainstorming through modeling and facilitation. *Small Group Research*, *32*(5), 533–557. doi:10.1177/104649640103200502

Krebs, T. B., & Pelissero, J. P. (2010). Urban managers and public policy: Do institutional arrangements influence decisions to initiate policy? *Urban Affairs Review*, *45*(3), 393–411. doi:10.1177/1078087409342693

Kreiner, K., Jacobsen, P. H., & Jensen, D. T. (2011). Dialogues and the problems of knowing: Reinventing the architectural competition. *Scandinavian Journal of Management*, *27*(1), 160–166. doi:10.1016/j.scaman.2010.12.003

Kumar, M., & Sinha, O. P. (2007). M-government–mobile technology for e-government. In *International conference on e-government, India* (pp. 294-301). Academic Press.

Kummitha, R. K. R., & Crutzen, N. (2017). How do we understand smart cities? An evolutionary perspective. *Cities (London, England)*, *67*(July), 43–52. doi:10.1016/j.cities.2017.04.010

Kushchu, I., & Kuscu, H. (2003). From E-government to M-government: Facing the Inevitable. In *the 3rd European Conference on e-Government* (pp. 253-260). MCIL Trinity College Dublin Ireland.

Kyttä, M. (2011). SoftGIS methods in planning evaluation. In A. Hull, E. R. Alexander, A. Khakee, & J. Woltjer (Eds.), *Evaluation for participatory and sustainable planning* (pp. 334–354). London: Routledge.

Kyttä, M., Broberg, A., Tzoulas, T., & Snabb, K. (2013, May). Towards contextually sensitive urban densification: Location-based softGIS knowledge revealing perceived residential environmental quality. *Landscape and Urban Planning*, *113*, 30–46. doi:10.1016/j.landurbplan.2013.01.008

Kyttä, M., & Kahila, M. (2011). SoftGIS methodology – Building bridges in urban planning. *GIM International*, *25*(3), 37–41.

Lambert, A. (2017). *Cosmopolitan Intimacy and Filter Bubbles.* Working Paper Series, Culture, Media, Economy Research Unit.

Lambert, A., McQuire, S. & Papastergiadis, N. (2014b). Public Wi-Fi: Space, Sociality and the Social Good. *Australian Journal of Telecommunications and the Digital Economy, 2*(3), 45.41-45.17.

Lambert, A. (2015). *False Shadows and Tactical Spaces: Encountering Transgressive Participation at White Night. In White Night: City as Event.* Melbourne: Research Unit In Public Cultures.

Lambert, A., McQuire, S., & Papastergiadis, N. (2014a). Free public Wi-Fi and E-Planning: The Use of Online Planning to Build Better Networked Public Places. *International Journal of E-Planning Research*, *3*(2), 70–85. doi:10.4018/ijepr.2014040105

Land Use and Building Act, Finlex. (1999). *Maankäyttö- ja rakennuslaki.* Retrieved November 15, 2013 from http://www.finlex.fi/fi/laki/ajantasa/1999/19990132

Larey, T. S., & Paulus, P. B. (1995). Social comparison and goal setting in brainstorming groups. *Journal of Applied Social Psychology*, *26*(18), 1579–1596. doi:10.1111/j.1559-1816.1995.tb02634.x

Larey, T. S., & Paulus, P. B. (1999). Group preference and convergent tendencies in groups: A content analysis of group brainstorming performance. *Creativity Research Journal*, *12*(3), 175–184. doi:10.120715326934crj1203_2

LaRiviere, K., Snider, J., Stromberg, A., & O'Meara, K. (2012). Protest: Critical lessons of using digital media for social change. *About Campus: Enriching the Student Learning Experience*, *17*, 10–17. doi:10.1002/abc.21081

Las Casas, G., & Scorza, F. (2013). Interoperabilità e condivisione della conoscenza supporto della sviluppo regionale: nuove istanze e prospettive. Proceedings of Conferenza Italiana di Scienze Regionali XXXIV.

Lathrop, D., & Ruma, L. (Eds.). (2010). *Open government*. Beijing: O'Reilly.

Laurie, J. (2008). Historic preservation and cluster based economic development. *Economic Development Journal*, *7*(1), 38–46.

Laurini, R. (2016). The geometric and semantic interoperability of the territorial databases. *Spatial data interoperability for a geographic intelligence, Seminar at Politecnico di Torino.*

Laurini, R. (2001). *Information systems for urban planning: a hypermedia cooperative approach*. London: Taylor & Francis.

Laurini, R. (2007). Pre-consensus ontologies in urban databases. In J. Teller, R. Lee, & C. Roussey (Eds.), *Ontolgies for urban development*. Berlin: Springer Verlag. doi:10.1007/978-3-540-71976-2_3

Laurini, R., & Thompson, D. (1992). *Fundamentals of spatial information systems*. London: Academic Press.

Layne, K., & Lee, J. (2001). Developing fully functional e-government–A four stage model. *Government Information Quarterly*, *18*(2), 118–136. doi:10.1016/S0740-624X(01)00066-1

Lazer, D., Pentland, A., Adamic, L., Aral, S., Barabási, A., Brewer, D., ... Van Alstyne, M. (2009). Life in the Network: The Coming Age of Computational Social Science. *Science*, *323*, 721–723. doi:10.1126cience.1167742 PMID:19197046

Lee, S. M., Tan, X., & Trimi, S. (2006). M-government, from rhetoric to reality: learning from leading countries. *Electronic Government, an International Journal, 3*(2), 113-126.

Lee, H., Kwak, N., Campbell, S. W., & Ling, R. (2014). Mobile communication and political participation in South Korea: Examining the intersections between informational and relational uses. *Computers in Human Behavior*, *38*, 85–92. doi:10.1016/j.chb.2014.05.017

Lee, J. (2010). 10year retrospect on stage models of e-Government: A qualitative meta-synthesis. *Government Information Quarterly*, *27*(3), 220–230. doi:10.1016/j.giq.2009.12.009

Lee, J., Bharosa, N., Yang, J., Janssen, M., & Rao, H. R. (2011, January). Group value and intention to use - A study of multi-agency disaster management information systems for public safety. *Decision Support Systems*, *50*(2), 404–414. doi:10.1016/j.dss.2010.10.002

Lee, S., Lee, S., Kim, K., & Park, J. (2012) Bursty event detection from text streams for disaster management. *Proceedings of the 21st international conference companion on World Wide Web*, 679-682. 10.1145/2187980.2188179

Lefebvre, E., Lefebvre, L. A., Hen, G. L., & Mendgen, R. (2006). Cross-Border E-Collaboration for New Product Development in the Automotive Industry. *International Conference on System Science*.

Lefebvre, H. (1968). *Le Droit à la ville. Suivi de 'Espace et politique*. Paris: Anthropos.

Lefebvre, H. (1996). The Right to the City. In E. Kofman & E. Lebas (Eds.), *Writings on Cities*. Oxford, UK: Blackwell.

Lenk, K. (1999). Electronic support of citizen participation in the planning process. In B. N. Hague & B. D. Loader (Eds.), *Digital democracy: discourse and decision making in the information age* (pp. 87–95). London: Routledge.

Lev-On, A., & Steinfeld, N. (Under Review). *MPs on Facebook: An analysis of politicians' Facebook pages*.

Lev-On, A. (2012). YouTube usage in low-visibility political campaigns. *Journal of Information Technology & Politics*, *9*(2), 205–216. doi:10.1080/19331681.2011.645360

Lev-On, A. (2014). Campaigning online, locally. *International Journal of E-Politics*, *5*(3), 14–28. doi:10.4018/ijep.2014070102

Lev-On, A., & Steinfeld, N. (2015). Local engagement online: Municipal Facebook pages as hubs of interaction. *Government Information Quarterly*, *32*(3), 299–307. doi:10.1016/j.giq.2015.05.007

Liao, H.-C., & Chang, T.-F. (2002). Space-heating and water-heating energy demands of the aged in the US. *Energy Economics*, *24*(3), 267–284. doi:10.1016/S0140-9883(02)00014-2

Li, J. P., Chen, R., Lee, J., & Rao, H. R. (2013, April). A case study of private-public collaboration for humanitarian free and open source disaster management software deployment. *Decision Support Systems*, *55*(1), 1–11. doi:10.1016/j.dss.2012.10.030

Li, L., Wang, D., Shen, C., & Li, T. (2010) Ontology-enriched multi-document summarization in disaster management. *Proceedings of the 33rd international ACM SIGIR conference on Research and development in information retrieval*, 819-820.

Limayem, M., Banerjee, P., & Ma, L. (2006). Impact of GDSS: Opening the black box. *Decision Support Systems*, *42*(2), 945–957. doi:10.1016/j.dss.2005.08.004

Linders, D. (2012). From e-government to we-government: Defining a typology for citizen coproduction in the age of social media. *Government Information Quarterly*, *29*(4), 446–454. doi:10.1016/j.giq.2012.06.003

Ling, R. (2008). *New Tech, New Ties: How Mobile Communication is Reshaping Social Cohesion*. Cambridge, MA: MIT Press.

Lobato, R., & Thomas, J. (2015). *The Informal Media Economy*. New York: Wiley.

Long, J. (2010). *Weird city: sense of place and creative resistance in Austin, Texas* (1st ed.). Austin, TX: University of Texas Press.

Lorenzi, D., Chun, S., Vaidya, J., Shafiq, B., Atluri, V., & Adam, N. (2015). PEER: A Framework for Public Engagement in Emergency Response. *International Journal of E-Planning Research*, *4*(3), 30–47. doi:10.4018/IJEPR.2015070102

Lorenzi, D., Vaidya, J., Chun, S., Shafiq, B., Naik, V., Atluri, V., & Adam, N. (2013). Community Based Emergency Response. *Proceedings of The 14th International Conference of Digital Government Research*, 82-91. 10.1145/2479724.2479739

Lorenzi, D., Vaidya, J., Shafiq, B., Chun, S., Vegesna, N., Alzamil, Z., ... Atluri, V. (2014) Utilizing social media to improve local government responsiveness. *Proceedings of the 15th Annual International Conference on Digital Government Research*, 236-244. 10.1145/2612733.2612773

Lovari, A., & Parisi, L. (2012). Public administrations and citizens 2.0: Exploring digital public communication strategies and civic interaction within Italian municipality pages on Facebook. In F. Comunello (Ed.), *Networked sociability and individualism: Technology for personal and professional relationships* (pp. 238–263). Hershey, PA: Information Science Reference. doi:10.4018/978-1-61350-338-6.ch012

Luff, P., & Heath, C. (1998). *Mobility in Collaboration*. Paper presented at the CSCW, Seattle, WA. 10.1145/289444.289505

Lynch, K. (1972). *What Time is This Place?* London: The MIT Press.

Lyon, D. (2014). Surveillance, Snowden, and Big Data: Capacities, Consequences, Critique. *Big Data and Society, 1*(2). doi:2053951714541861

Lyon, D. (2003). Surveillance as Social Sorting: Computer Codes and Mobile Bodies. In D. Lyon (Ed.), *Surveillance as Social Sorting*. London: Routledge.

Lyon, T. P., & Montgomery, A. W. (2013). Tweetjacked: The Impact of Social Media on Corporate Greenwash. *Journal of Business Ethics*, *118*(4), 747–757. doi:10.100710551-013-1958-x

MacDonald, P., Amparo, G.-C., & José Miguel, C.-S. (2013). Computer learner corpora: Analysing interlanguage errors in synchronous and asynchronous communication. *Language Learning & Technology*, *17*(2), 36–56.

Macintosh, A. (2006). eParticipation in Policy-making: The research and the challenges. In Exploiting the knowledge economy: Issues, applications and case studies (pp. 364–369). Amsterdam, The Netherlands: IOS Press.

Macintosh, A., & Whyte, A. (2008). Towards an evaluation framework for eParticipation. *Transforming Government: People. Process and Policy*, *2*(1), 16–30.

Magnusson, M., Bellström, P., & Magnusson, M. (2012). Facebook usage in government: A case study of information content. In *Proceedings of the 18th Americas Conference on Information Systems 2012* (pp. 353–362). Red Hook, NY: Curran Associates.

Maier, E., & Reimer, U. (2010). Process Support for Increasing Participation in eParticipation. *eJournal of eDemocracy and Open Government*, *2*(1), 46-55. Retrieved July 15, 2017, from http://www.jedem.org/article/view/16/53

Mainka, A., Hartmann, S., Meschede, C., & Stock, W. G. (2015). Mobile application services based upon open urban government data. *iConference 2015 Proceedings.*

Maireder, A., & Schwarzenegger, C. (2012). A movement of connected individuals. *Information Communication and Society*, *15*(2), 171–195. doi:10.1080/1369118X.2011.589908

Mako. (2010). *Coming soon: A Facebook page for all MKs*. Retrieved from: http://www.mako.co.il/news-money/tech/Article-f89c1d2216b4721004.htm

Mallach, A. (2011, July/August). Demolition and preservation in shrinking US industrial cities. *Building Research and Information*, *39*(4), 380–394. doi:10.1080/09613218.2011.573743

Mandarano, L., Meenar, M., & Steins, C. (2010). Building social capital in the digital age of civic engagement. *Journal of Planning Literature*, *25*(2), 123–135. doi:10.1177/0885412210394102

Mannarini, T., Fedi, A., & Trippetti, S. (2010). Public involvement: How to encourage citizen participation. *Journal of Community & Applied Social Psychology*, *20*(4), 262–274. doi:10.1002/casp.1030

Mann, R., & Lev-On, A. (2017). *Annual Report: The Media in Israel 2016. Agendas, Uses and Trends. Ariel.* Ariel University. (in Hebrew)

Manoharan, A. P., Bennett, L. V., & Carrizales, T. J. (2012). M-Government: An opportunity for addressing the digital divide. *Citizen 2.0: Public and Governmental Interaction through Web 2.0 Technologies: Public and Governmental Interaction Through Web 2.0 Technologies*, 87.

Manoharan, A. (2013). A three dimensional assessment of US county e-government. *State & Local Government Review*, *45*(3), 153–162. doi:10.1177/0160323X13494858

Manoharan, A. P., & McQuiston, J. (2016). Technology and pedagogy: Information technology competencies in public administration and public policy programs. *Journal of Public Affairs Education*, *22*(2), 175–186.

Manoharan, A., & Carrizales, T. (2011). Recent trends in e-government: States' and local governments' utilization of Websites. *International Journal of Electronic Governance*, *4*(4), 283–303. doi:10.1504/IJEG.2011.046014

Manosevitch, I., Steinfeld, N., & Lev-On, A. (2014). Promoting online deliberation quality: Cognitive cues matter. *Information Communication and Society*, *17*(10), 1177–1195. doi:10.1080/1369118X.2014.899610

Manovich, L. (2011). Trending: the promises and the challenges of big social data. In M. K. Gold (Ed.), *Debates in Digital Humanity* (pp. 460–475). Minneapolis, MN: University of Minnesota Press.

March, H., & Saurí, D. (2010). The suburbanization of water scarcity in the Barcelona metropolitan region: Sociodemographic and urban changes influencing domestic water consumption. *The Professional Geographer*, *62*(1), 32–45. doi:10.1080/00330120903375860

Marsden, C. (2000). The new corpotate citizenship of big business: Part of the solution to sustainability? *Business and Society Review*, *105*(1), 9–25. doi:10.1111/0045-3609.00062

Marsh, D. M. (2008). *The Water-Energy Nexus: A comprehensive analysis in the context of New South Wales (PhD)*. Sydney, Australia: University of Technology.

Martins, L. L., & Shalley, C. E. (2011). Creativity in virtual work: Effects of demographic differences. *Small Group Research*, *42*(5), 536–561. doi:10.1177/1046496410397382

Mason, R. (2006). Theoretical and Practical Arguments for Values-Centered Preservation. *CRM: The Journal of Heritage Stewardship*, *3*(2), 21–47.

Matejek, S., & Gossling, T. (2014). Beyond legitimacy: A case study in BP's "green lashing. *Journal of Business Ethics*, *120*(4), 571–584. doi:10.100710551-013-2006-6

Matsusaka, J. G. (2005). Direct democracy works. *The Journal of Economic Perspectives*, *19*(2), 185–206. doi:10.1257/0895330054048713

McCafferty, D. (2011). Activism vs. slacktivism. *Communications of the ACM*, *54*(12), 17–19. doi:10.1145/2043174.2043182

McCall, M. K. (2017). Applying Participatory GIS and Participatory Mapping to Participatory Spatial Planning and Local–level Land & Resources Management utilizing Local & Indigenous Spatial Knowledge. Morelia: UNAM, CIGA.

McGhee, F. L. (2015, October 15). *Gentrification, Historic Preservation, Public Housing and Human Rights*. Retrieved July 21, 2017, from City of Austin website: http://www.austintexas.gov/edims/document.cfm?id=242870

McLeod, P. L., & Liker, J. K. (1992). Electronic meeting systems: Evidence from a low structure environment. *Information Systems Research*, *3*(3), 195–223. doi:10.1287/isre.3.3.195

McNeill, D. (2015). Global firms and smart technologies: IBM and the reduction of cities. *Transactions of the Institute of British Geographers*, *40*(4), 562–574. doi:10.1111/tran.12098

McQuire, S. (2016). *Geomedia: Networked Cities and the Future of Public Space*. Hoboken, NJ: Wiley.

McShane, I., Gregory, M. A., & Wilson, C. K. (2016). *Practicing Safe Wi-Fi: Assessing and Managing Data-Security Risks*. Retrieved on 17 June from http://cur.org.au/project/practicing-safe-public-wi-fi-assessing-and-managing-data-and-security-risks/

McShane, I., Wilson, C., & Meredyth, D. (2014). Broadband as Civic Infrastructure: The Australian Case. *Media International Australia*, *151*(1), 127–136. doi:10.1177/1329878X1415100117

Medaglia, R. (2007). The diffusion of eParticipation in English and Italian local government. In Understanding eParticipation: Contemporary PhD eParticipation Research in Europe (pp. 131–151). Örebro: University Library.

Melbourne Water. (2012). *Permanent water use rules and targets*, from http://www.melbournewater.com.au/getinvolved/saveandreusewater/Pages/Water-restrictions.aspx

Melitski, J. (2003). Capacity and e-government performance: An analysis based on early adopters of Internet technologies in New Jersey. *Public Performance & Management Review*, *26*(4), 376–390. doi:10.1177/1530957603026004005

Mergel, I. (2013). A framework for interpreting social media interactions in the public sector. *Government Information Quarterly*, *30*(4), 327–334. doi:10.1016/j.giq.2013.05.015

Merikoski, T., Eräranta, S., & Staffans, A. (2012). *Sibbesborg – Kestävän yhdyskunnan suunnittelukilpailu. Raportti kilpailuhankkeen valmistelusta ja toteutuksesta 2010–2011*. Espoo: Aalto University.

Michinov, N. (2012). Is electronic brainstorming or brainwriting the best way to improve creative performance in groups? An overlooked comparison of two idea generation techniques. *Journal of Applied Social Psychology*, *42*, 222–243. doi:10.1111/j.1559-1816.2012.01024.x

Michinov, N., & Primois, C. (2005). Improving productivity and creativity in online groups through social comparison process: New evidence for asynchronous electronic brainstorming. *Computers in Human Behavior*, *21*(1), 11–28. doi:10.1016/j.chb.2004.02.004

Mikhaylov, S., Laver, M., & Benoit, K. R. (2012). Coder Reliability and Misclassification in the Human Coding of Party Manifestos. *Political Analysis*, *20*(01), 78–91. doi:10.1093/pan/mpr047

Milakovich, M. E. (2010). The Internet and increased citizen participation in government. *JeDEM-eJournal of eDemocracy and Open Government, 2*(1), 1-9.

Milanovic, B. (2011). *Worlds apart: Measuring international and global inequality*. Princeton, NJ: Princeton University Press. doi:10.1515/9781400840816

Minner, J. (2013) *Landscapes of Thrift and Choreographies of Change: Reinvestment and Adaptation along Austin's Commercial Strips* (Dissertation). Austin, TX: University of Texas at Austin.

Minner, J. (2015). IMHO to IOHO: A Call for Crowdsourcing Creative Possibilities for Urban Adaptation. *Forum*, 50–62.

Minner, J. (2016). Revealing Synergies, Tensions, and Silences Between Preservation and Planning. *Journal of the American Planning Association*, *82*(2), 72–87. doi:10.1080/01944363.2016.1147976

Minner, J., & Holleran, M. (2016). Introduction to the Special Issue. *Journal of the American Planning Association*, *82*(2), 69–17. doi:10.1080/01944363.2016.1142391

Minton, E., Lee, C., Orth, U., Chung-Hyun, K., & Lynn, K. (2012). Sustainable marketing and social media: A cross-country analysis of motives for sustainable behaviors. *Journal of Advertising*, *41*(4), 69–84. doi:10.1080/00913367.2012.10672458

Mithas, S., Costello, T., & Tafti, A. (2011). Social networking in the enterprise. *IT Professional Magazine*, *13*(4), 16–17. doi:10.1109/MITP.2011.69

Montiel, I. (2008). Corporate social responsibility and corporate sustainability: Separate pasts, common futures. *Organization & Environment*, *21*(3), 245–269. doi:10.1177/1086026608321329

Moon, J. M. (2002). The evolution of e-government among municipalities: Rhetoric or reality? *Public Administration Review*, *62*(4), 434–433. doi:10.1111/0033-3352.00196

Mora, L., Bolici, R., & Deakin, M. (2017). The First Two Decades of Smart-City Research: A Bibliometric Analysis. *Journal of Urban Technology*, *24*(1), 1, 3–27. doi:10.1080/10630732.2017.1285123

Morgeson, F. V., VanAmburg, D., & Mithas, S. (2011). Misplaced trust? Exploring the structure of the e-government-citizen trust relationship. *Journal of Public Administration: Research and Theory*, *21*(2), 257–283. doi:10.1093/jopart/muq006

Mosco, V. (2005). *The Digital Sublime: Myth, Power and Cyberspace*. Boston: MIT.

Mossberger, K., Wu, Y., & Crawford, J. (2013). Connecting citizens and local governments? Social media and interactivity in major US cities. *Government Information Quarterly*, *30*(4), 351–358. doi:10.1016/j.giq.2013.05.016

Mossey, S., & Manoharan, A. P. (2017). Seasteads and E-Democracy: The Potential of Mobile the Potential of Mobile Technology. *Seasteads: Opportunities and Challenges for Small New Societies*, 81.

Municipality of Sipoo. (2015). *N 40, Sipoon Jokilaakso, selostus* [Report for the detailed plan of River Valley of Sipoo]. Retrieved July 13, 2017 at: https://www.sipoo.fi/fi/jokilaakso/asemakaava

Munkes, J., & Diehl, M. (2003). Matching or competition? Performance comparison processes in an idea generation task. *Group Processes & Intergroup Relations*, *6*(3), 305–320. doi:10.1177/13684302030063006

Murray, J. H. (1997). *Hamlet on the Holodeck – The Future of Narrative in Cyberspace*. Cambridge, MA: MIT Press.

Murray, J. H. (2012). *Inventing the medium: principles of interaction design as a cultural practice*. Cambridge, MA: MIT Press.

Myers, K. K., & Sadaghiani, K. (2010). Millennials in the workplace: A communication perspective on millennials' organizational relationships and performance. *Journal of Business and Psychology*, *25*(2), 225–238. doi:10.100710869-010-9172-7 PMID:20502509

Nagasundaram, M., & Dennis, A. R. (1993). When a group is not a group: The cognitive foundation of group idea generation. *Small Group Research*, *24*(4), 463–489. doi:10.1177/1046496493244003

Ndlovu, M., & Mbenga, C. (2013). Facebook, the public sphere and political youth leagues in South Africa. *Journal of African Media Studies*, *5*(2), 169–186. doi:10.1386/jams.5.2.169_1

Nelson-Field, K., Riebe, E., & Newstead, K. (2013). The emotions that drive viral video. *Australasian Marketing Journal*, *21*(4), 205–211. doi:10.1016/j.ausmj.2013.07.003

Nemeth, C. J., & Nemeth-Brown, B. (2003). Better than individuals? The potential benefits of dissent and diversity for group creativity. In P. B. Paulus & B. A. Nijstad (Eds.), *Group creativity: Innovation through collaboration* (pp. 63–84). New York: Oxford University Press. doi:10.1093/acprof:oso/9780195147308.003.0004

Nemeth, C. J., & Ormiston, M. (2007). Creative idea generation: Harmony versus stimulation. *European Journal of Social Psychology*, *37*(3), 524–535. doi:10.1002/ejsp.373

Nemeth, C. J., Personnaz, B., Personnaz, M., & Goncalo, J. A. (2004). The liberating role of conflict in group creativity: A study in two countries. *European Journal of Social Psychology*, *34*(4), 365–374. doi:10.1002/ejsp.210

Newman, G., Zimmerman, D., Crall, A., Laituri, M., Graham, J., & Stapel, L. (2010). User-Friendly Web Mapping: Lessons from a Citizen Science Website. *International Journal of Geographical Information Science*, *24*(12), 1851–1869. doi:10.1080/13658816.2010.490532

Newton, P., & Meyer, D. (2010). The determinants of urban resource consumption. *Environment and Behavior*, *44*(1), 107–135. doi:10.1177/0013916510390494

Nicolai, B., Jin, G., Jiang, K., & Winer, C. (2010) Disaster assessment with parallel image processing for GIS based local area disaster decision support system. In *Proceedings of the 1st International Conference and Exhibition on Computing for Geospatial Research; Application*. ACM. 10.1145/1823854.1823901

Nijstad, B. A. (2009). *Group performance*. New York: Psychology Press.

Nijstad, B. A., & Stroebe, W. (2006). How the group affects the mind: A cognitive model of idea generation in groups. *Personality and Social Psychology Review*, *10*(3), 186–213. doi:10.120715327957pspr1003_1 PMID:16859437

Nijstad, B. A., Stroebe, W., & Lodewijkx, H. F. M. (2002). Cognitive stimulation and interference in groups: Exposure effects in an idea generation task. *Journal of Experimental Social Psychology*, *38*(6), 535–544. doi:10.1016/S0022-1031(02)00500-0

Norman, D. A. (1993). *Things that Make Us Smart.* Addison-Wesley Publishing Company.

Norman, D. A. (2010). *Living with Complexity*. Cambridge, MA: MIT Press.

Norris, P. (2003). *Deepening Democracy via E-Governance (Draft chapter for the UN World Public Sector Report).* Retrieved from http://ksghome.harvard. edu/~.pnorris.shorenstein.ksg/Acrobat/World%20 Public%20Sector%20Report.pdf

Norris, D. F., & Kraemer, K. L. (1996). Mainframe and PC computing in American local governments: Myths and realities. *Public Administration Review*, *56*(6), 568–576. doi:10.2307/977255

Norris, D. F., & Moon, M. J. (2005). Advancing E-Government at the Grassroots: Tortoise or Hare? *Public Administration Review*, *65*(1), 64–75. doi:10.1111/j.1540-6210.2005.00431.x

Norris, D. F., & Reddick, C. G. (2013). Local e-government in the United States: Transformation or incremental change? *Public Administration Review*, *73*(1), 165–175. doi:10.1111/j.1540-6210.2012.02647.x

Noveck, B. S. (2009). *Wiki government: how technology can make government better, democracy stronger, and citizens more powerful.* Washington, DC: Brookings Institution Press.

Noveck, B. S. (2010). The Single Point of Failure. In D. Lathrop & L. Ruma (Eds.), *Open government: collaboration, transparency, and participation in practice* (pp. 49–69). Beijing: O'Reilly.

Noy, N. F., & McGuinness, D. L. (2001). *Ontology Development 101: A Guide to Creating Your First Ontology*. Retrieved July 15, 2017, from http://www.ksl.stanford.edu/people/dlm/papers/ontology-tutorial-noy-mcguinness.pdf

Nummi, P. (2017). Social Media Data Analysis in Urban e-Planning. *International Journal of E-Planning Research*, *6*(4), 18–31. doi:10.4018/IJEPR.2017100102

Nunamaker, J. F. Jr, Dennis, A. R., Valacich, J. S., Vogel, D. R., & George, J. F. (1991). Electronic meetings to support group work. *Communications of the ACM*, *34*(7), 40–61. doi:10.1145/105783.105793

O'Callaghan, B., Green, H. J., Hyde, R. A., Wadley, D., & Upadhyay, A. (2012). Exploring the influence of housing design and occupant environmental attitudes on energy and water usage. *Architectural Science Review*, *55*(3), 176–185. doi:10.1080/00038628.2012.693813

O'Doherty, K., & Einsiedel, E. (2012). Public Engagement and Emerging Technologies. Vancouver: University of British Columbia Press.

Obendorf, H. (2009). *Minimalism: Designing Simplicity*. London, UK: Springer. doi:10.1007/978-1-84882-371-6

Obermeyer, N. J. (1998). The evolution of public participation GIS. *Cartography and Geographic Information Systems*, *25*(2), 65–66. doi:10.1559/152304098782594599

OECD. (2003). *Promises and problems of E-Democracy: Challenges of online citizen engagement*. Retrieved July 15, 2017, from http://www.oecd.org/gov/digital-government/35176328.pdf

Offner, A. K., Kramer, T. J., & Winter, J. P. (1996). The effects of facilitation, recording, and pauses on group brainstorming. *Small Group Research*, *27*(2), 283–298. doi:10.1177/1046496496272005

Olma, S. (2016). *In Defense of Serendipity*. London: Repeater Books.

Organisciak, P. (2008). *Motivation of Crowds: The Incentives That Make Crowdsourcing Work*. Retrieved at http://crowdstorming.wordpress.com/2008/01/31/motivation-of-crowds-the-incentives-that-make-crowdsourcing-work/#comments

Orgaz-Bello, G., Jung, J., & Camachoa, D. (2016). Social big data: Recent achievements and new challenges. *Information Fusion*, *28*, 45–59. doi:10.1016/j.inffus.2015.08.005

Osborn, A. F. (1953). *Applied imagination: Principles and procedures of creative thinking*. New York, NY: Scribner.

OSF Official Statistics of Finland. (2016). *Use of information and communications technology by individuals*. Retrieved July 11, 2017 from http://www.stat.fi/til/sutivi/index_en.html

Ottinger, G. (2010). Buckets of Resistance: Standards and the Effectiveness of Citizen Science. *Science, Technology & Human Values*, *35*(2), 244–270. doi:10.1177/0162243909337121

Oxley, N. L., Dzindolet, M. T., & Paulus, P. B. (1996). The effects of facilitators on the performance of brainstorming groups. *Journal of Social Behavior and Personality*, *11*, 633–646.

Pak, B., Chua, A., & Vande Moere, A. (2017). FixMyStreet Brussels: Socio-Demographic Inequality in Crowdsourced Civic Participation. *Journal of Urban Technology*, *24*(2), 65–87. doi:10.1080/10630732.2016.1270047

Palen, L., Hiltz, S., & Liu, S. B. (2007). Online forums supporting grassroots participation in emergency preparedness and response. *Communications of the ACM*, *50*(3), 54–58. doi:10.1145/1226736.1226766

Papadopoulou, C.-A., & Giaoutzi, M. (2017). Crowdsourcing and Living Labs in Support of Smart Cities' Development. *International Journal of E-Planning Research*, *6*(2), 22–38. doi:10.4018/IJEPR.2017040102

Papastergiadis, N.(2013). Mega Screens for Mega Cities. *Theory, Culture and Society*. doi: 0263276413503691

Papastergiadis, N. (2012). *Cosmopolitanism and Culture*. Cambridge, UK: Polity Press.

Parker, B. (2006). Constructing Community Through Maps? Power and Praxis in Community Mapping. *The Professional Geographer*, *58*(4), 470–484. doi:10.1111/j.1467-9272.2006.00583.x

Park, N., Kee, K. F., & Valenzuela, S. (2009). Being immersed in social networking environment: Facebook groups, uses and gratifications, and social outcomes. *Cyberpsychology & Behavior*, *12*(6), 729–733. doi:10.1089/cpb.2009.0003 PMID:19619037

Parnes, S. J., & Meadow, A. (1959). Effects of" brainstorming" instructions on creative problem solving by trained and untrained subjects. *Journal of Educational Psychology*, *50*(4), 171–176. doi:10.1037/h0047223

Paroutis, S., Bennett, M., & Heracleous, L. (2014). A strategic view on smart city technology: The case of IBM Smarter Cities during a recession. *Technological Forecasting and Social Change*, *89*, 262–272. doi:10.1016/j.techfore.2013.08.041

Pateman, C. (1970). *Participation and democratic theory*. London: Cambridge University Press. doi:10.1017/CBO9780511720444

Paulus, P. B., Korde, R. M., Dickson, J. J., Carmeli, A., & Cohen-Meitar, R. (2015). Asynchronous brainstorming in an industrial setting; Exploratory studies. *Human Factors, 57*(6), 1076–1094.

Paulus, P. B., & Brown, V. (2003). Ideational creativity in groups: Lessons from research on brainstorming. In P. B. Paulus & B. A. Nijstad (Eds.), *Group creativity: Innovation through collaboration* (pp. 110–136). New York, NY: Oxford University Press. doi:10.1093/acprof:oso/9780195147308.003.0006

Paulus, P. B., & Brown, V. R. (2007). Toward more creative and innovative group idea generation: A cognitive-social-motivational perspective of brainstorming. *Social and Personality Psychology Compass*, *1*(1), 248–265. doi:10.1111/j.1751-9004.2007.00006.x

Paulus, P. B., & Coskun, H. (2012). Group creativity. In J. M. Levine (Ed.), *Group Processes* (pp. 215–239). Amsterdam: Elsevier.

Paulus, P. B., Dugosh, K. L., Dzindolet, M. T., Coskun, H., & Putman, V. L. (2002). Social and cognitive influences in group brainstorming. Predicting production gains and losses. *European Review of Social Psychology*, *12*(1), 299–325. doi:10.1080/14792772143000094

Paulus, P. B., & Dzindolet, M. T. (1993). Social influence processes in group brainstorming. *Journal of Personality and Social Psychology*, *64*(4), 575–586. doi:10.1037/0022-3514.64.4.575

Paulus, P. B., & Dzindolet, M. T. (2008). Social influence, creativity and innovation. *Social Influence*, *3*(4), 228–247. doi:10.1080/15534510802341082

Paulus, P. B., Dzindolet, M. T., & Kohn, N. W. (2011). Collaborative creativity-Group creativity and team innovation. In M. D. Mumford (Ed.), *Handbook of organizational creativity* (pp. 327–357). New York, NY: Elsevier.

Paulus, P. B., Kohn, N. W., & Arditti, L. E. (2011). Effects of quantity and quality instructions on brainstorming. *The Journal of Creative Behavior*, *45*(1), 38–46. doi:10.1002/j.2162-6057.2011.tb01083.x

Paulus, P. B., Kohn, N. W., Arditti, L., & Korde, R. (2013). Understanding the group size effect in electronic brainstorming. *Small Group Research*, *44*(3), 332–352. doi:10.1177/1046496413479674

Paulus, P. B., Larey, T. S., Putman, V. L., Leggett, K. L., & Roland, E. J. (1996). Social influence processes in computer brainstorming. *Basic and Applied Social Psychology*, *18*(1), 3–4. doi:10.120715324834basp1801_2

Paulus, P. B., Nakui, T., Putman, V. L., & Brown, V. R. (2006). Effects of task instructions and brief breaks on brainstorming. *Group Dynamics*, *10*(3), 206–219. doi:10.1037/1089-2699.10.3.206

Paulus, P. B., & Nijstad, B. A. (Eds.). (2003). *Group creativity: Innovation through collaboration*. New York: Oxford University Press. doi:10.1093/acprof:oso/9780195147308.001.0001

Paulus, P. B., & van der Zee, K. I. (2015). Creative processes in culturally diverse teams. In S. Otten, K. I. van der Zee, & M. Brewer (Eds.), *Towards inclusive organizations: Determinants of successful diversity management at work* (pp. 108–131). New York: Psychology Press.

Paulus, P. B., & Yang, H. C. (2000). Idea generation in groups: A basis for creativity in organizations. *Organizational Behavior and Human Decision Processes*, *82*(1), 76–87. doi:10.1006/obhd.2000.2888

Peet, R. K. (1974). The measurement of species diversity. *Annual Review of Ecology and Systematics*, *5*(1), 285–307. doi:10.1146/annurev.es.05.110174.001441

Peng, Y., & Choi, S. Y. (2013). Mobile phone use among migrant factory workers in south China: Technologies of power and resistance. *The China Quarterly*, *215*, 553–571. doi:10.1017/S0305741013000738

Petchey, R. (2010). *End use energy intensity in the Australian economy. Australian Bureau of Agricultural and Resource Economics – Bureau of Rural Sciences (ABARE-BRS)*. Australian Government.

Picazo-Vela, S., Gutiérrez-Martínez, I., & Luna-Reyes, L. F. (2012). Understanding risks, benefits, and strategic alternatives of social media applications in the public sector. *Government Information Quarterly*, *29*(4), 504–511. doi:10.1016/j.giq.2012.07.002

Picken, F. (2013). From designed spaces to designer savvy societies: The potential of ideas competitions in willing participation. *Environment & Planning A*, *45*(8), 1963–1976. doi:10.1068/a45460

Pierskalla, J. H., & Hollenbach, F. M. (2013). Technology and collective action: The effect of cell phone coverage on political violence in Africa. *The American Political Science Review*, *107*(2), 207–224. doi:10.1017/S0003055413000075

Pohl, D., Bouchachia, A., & Hellwagner, H. (2012). Automatic sub-event detection in emergency management using social media. In *Proceedings of the 21st international conference companion on World Wide Web*, 683-686. 10.1145/2187980.2188180

Potts, J. (2014). Economics of Public Wi-Fi. *Australian Journal of Telecommunications and the Digital Economy, 2*(1), 20.21-20.29.

Preservation Green Lab. (2011). *The greenest building: Quantifying the environmental value of building reuse*. Retrieved June 27, 2014, from http://www.preservationnation.org/information-center/sustainable-communities/sustainability/green-lab/lca/The_Greenest_Building_lowres.pdf

Prieto-Martín P., de Marcos L., & Martínez J.J. (2012). The e-(R)evolution will not be funded: A transdisciplinary and critical analysis of the developments and troubles of EU-funded eParticipation. *European Journal of ePractice*, *15*, 62-89.

Proust, K., Dovers, S., Foran, B., Newell, B., Steffen, W., & Troy, P. (2007). *Climate, energy and water: Accounting for the links.* The Australian National University, Occasional Paper. Retrieved from http://www.crdc.com.au/uploaded/File/E-Library/E-ENVIRO/Climate_Energy_and_Water_May_2007.pdf

Purian-Lukach, R. (2011). Municipal e-government: Comparative index, Success factors and models for management and evaluation. In E. Cohen & A. Lev-On (Eds.), *Connected: politics, Technology and society in Israel.* Israeli Political Science Association Press. (in Hebrew)

Putman, V. L., & Paulus, P. B. (2009). Brainstorming, brainstorming rules and decision making. *The Journal of Creative Behavior*, *43*(1), 23–39. doi:10.1002/j.2162-6057.2009.tb01304.x

Quan-Haase, A., Wellman, B., Witte, J. C., & Hampton, K. N. (2002). Capitalizing on the Net: Social contact, civic engagement, and sense of community. In B. Wellman & C. Haythornthwaite (Eds.), *The Internet in everyday life*. Oxford, UK: Blackwell. doi:10.1002/9780470774298.ch10

Ramchurn, S., Rogers, A., Macarthur, K., Farinelli, A., Vytelingum, P., Vetsikas, I., & Jennings, N. R. (2008). Agent-based coordination technologies in disaster management. *Proceedings of the 7th international joint conference on Autonomous agents and multiagent systems: demo papers*, 1651-1652.

Randolph, B., & Troy, P. (2008). Attitudes to conservation and water consumption. *Environmental Science & Policy*, *11*(5), 441–455. doi:10.1016/j.envsci.2008.03.003

Randolph, J. (2012). *Environmental land use planning and management* (2nd ed.). Washington, DC: Island press.

Rantanen, H., & Kahila, M. (2008). The SoftGIS approach to local knowledge. *Journal of Environmental Management*, *90*(6), 1981–1990. doi:10.1016/j.jenvman.2007.08.025 PMID:18556113

Raschid, L. (2007). Information integration and dissemination for disaster data management. *Proceedings of the 8th annual international conference on Digital government research: bridging disciplines \& domains*, 333-335.

Ravitsch, P. (2005). *Difference between Israeli municipal websites, 2005: To what extent the inventory of online services fulfills the potentials of the medium and meets public expectations?* (MA thesis). Bar-Ilan University, Department of Political Science. (in Hebrew)

Ray, N. K., & Turuk, A. (2011). A framework for disaster management using wireless ad hoc networks. *Proceedings of the 2011 International Conference on Communication, Computing & Security*, 138-141. 10.1145/1947940.1947970

Reddick, C. G., & Norris, D. F. (2013). Social media adoption at the American grass roots: Web 2.0 or 1.5? *Government Information Quarterly*, *30*(4), 498–507. doi:10.1016/j.giq.2013.05.011

Reddick, C. G., Takeoka Chatfield, A., & Ojo, A. (2016). A social media text analytics framework for double-loop learning for citizen-centric public services: A case study of a local government Facebook use. *Government Information Quarterly*, *34*(1), 110–125. doi:10.1016/j.giq.2016.11.001

Reiter-Palmon, R., Wigert, B., & de Vreede, T. (2012). Team creativity and innovation: The effect of group composition, social processes, and cognition. In M. D. Mumford (Ed.), *Handbook of organizational creativity* (pp. 295–326). Amsterdam: Elsevier. doi:10.1016/B978-0-12-374714-3.00013-6

Richardson, I., & Wilken, R. (2013). Parerga of the Third Screen: Mobile Media, Place, and Presence. In G. Goggin (Ed.), *Mobile Technology and Place*. Hoboken, NJ: Taylor and Francis.

Riesch, H., & Potter, C. (2014). Citizen science as seen by scientists: Methodological, epistemological and ethical dimensions. *Public Understanding of Science (Bristol, England)*, *23*(1), 107–120. doi:10.1177/0963662513497324 PMID:23982281

Rietzschel, E. F., Nijstad, B. A., & Stroebe, W. (2006). Productivity is not enough: A comparison of interactive and nominal brainstorming groups on idea generation and selection. *Journal of Experimental Social Psychology*, *42*(2), 244–251. doi:10.1016/j.jesp.2005.04.005

Rietzschel, E. F., Nijstad, B. A., & Stroebe, W. (2007). Relative accessibility of domain knowledge and creativity: The effects of knowledge activation on the quantity and originality of generated ideas. *Journal of Experimental Social Psychology*, *43*(6), 933–946. doi:10.1016/j.jesp.2006.10.014

Rietzschel, E. F., Nijstad, B. A., & Stroebe, W. (2010). The selection of creative ideas after individual idea generation: Choosing between creativity and impact. *British Journal of Psychology*, *101*(1), 47–68. doi:10.1348/000712609X414204 PMID:19267959

Rogers, R. (2009). *The End of the Virtual – Digital Methods.* Text prepared for the Inaugural Speech, Chair, New Media & Digital Culture, University of Amsterdam. Retrieved from http://www.govcom.org/rogers_oratie.pdf

Rogers, Y., & Lindley, S. (2004). Collaborating around vertical and horizontal large interactive displays: Which way is best? *Interacting with Computers*, *16*(6), 1133–1152. doi:10.1016/j.intcom.2004.07.008

Rogers, Y., Sharp, H., & Preece, J. (2011). *Interaction Design: Beyond Human - Computer Interaction.* Sussex, UK: Willey.

Roininen, J., Horelli, L., & Wallin, S. (2003). *Osallistuminen ja vuorovaikutus kaavoituksessa. Seurannan ja arvioinnin viitekehys ja menetelmät. Suomen ympäristö 664*. Helsinki: Ympäristöministeriö.

Rönn, M. (2009). Judgment in the architectural competition – Rules, policies and dilemmas. *Nordic Journal of Architectural Research, 21*(2/3), 52–67.

Rönn, M. (2011). Architectural quality in competitions. A dialogue-based assessment of design proposals. *FORMakademisk, 4*(1), 100–115.

Rossel, P., & Finger, M. (2007). Conceptualizing E-Governance. In T. Janowaki, & T.A. Pardo (Eds.), *Proceedings of the 1st International Conference on Theory and Practice of Electronic Governance* (pp. 399-407). New York: The Association for Computing Machinery.

Rotberg, R. I., & Aker, J. C. (2013). Mobile phones: Uplifting weak and failed states. *The Washington Quarterly, 36*(1), 111–125. doi:10.1080/0163660X.2013.751651

Rotem, E. (2007). *Municipal e-government in Israel: A comparative perspective.* Hartog School for Governance and Policy. (in Hebrew)

Roy, M. C., Gauvin, S., & Limayem, M. (1996). Electronic group brainstorming: The role of feedback on productivity. *Small Group Research, 27*(2), 215–247. doi:10.1177/1046496496272002

Rundle, A. G., Bader, M. D., Richards, C. A., Neckerman, K. M., & Teitler, J. O. (2011). Using Google Street View to audit neighborhood environments. *American Journal of Preventive Medicine, 40*(1), 94–100. doi:10.1016/j.amepre.2010.09.034 PMID:21146773

Sæbø, Ø., Rose, J., & Flak, L. S. (2008). The shape of eParticipation: Characterizing an emerging research area. *Government Information Quarterly, 25*(3), 400–428. doi:10.1016/j.giq.2007.04.007

SAFA. (2008). *SAFA Competition Rules 2008*. SAFA.

SAFA. (2011). *Competitions 1972-2011.* Retrieved November 12, 2013 from http://safa.fi/document.php?DOC_ID=1133&SEC=03702b3ae4d10944d643485c927250eb&SID=1#competitions_2011.pdf

Salas, E., DiazGranados, D., Klein, C., Burke, C. S., Stagl, K. C., Goodwin, G. F., & Halpin, S. M. (2008). Does team training improve team performance? A meta-analysis. *Human Factors, 50*(6), 903–933. doi:10.1518/001872008X375009 PMID:19292013

Samet, H. (1990). *The design and analysis of spatial data structures*. Boston: Addison- Wesley.

Sandvig, C. (2004). An Initial Assessment of Cooperative Action in Wi-Fi Networking. *Telecommunications Policy, 28*(7-8), 579–602. doi:10.1016/j.telpol.2004.05.006

Santanen, E. L., Briggs, R. O., & de Vreede, G. (2004). Causal relationship in creative problem solving: Comparing facilitation interventions for ideation. *Journal of Management Information Systems, 20*(4), 167–197. doi:10.1080/07421222.2004.11045783

Sanvig-Knudsen, A. M., & Kahila, M. (2012). The role of volunteered geographic information in participatory planning: Examples from Denmark and Finland. *Perspektiv*, *21*, 35–46.

Sassen, S. (2011). *Talking Back to Your Intelligent City*. Retrieved from http://voices.mckinseyonsociety.com/talking-back-to-your-intelligent-city/

Savas, E. S. (2000). *Privatization and public-private partnerships*. Ithaca, NY: Cornell University Press.

Saydi, M. (2015). *An integrated spatial model for understanding residential energy and water consumption* (PhD Thesis). The University of Melbourne.

Saydi, M., Bishop, I. D., & Rajabifard, A. (2015). Virtual identification of dwelling characteristics online for analysis of urban resource consumption. *International Journal of E-Planning Research*, *4*(3), 1–28. doi:10.4018/IJEPR.2015070101

Scassa, T. (2016). Police Service Crime Mapping as Civic Technology: A Critical Assessment. *International Journal of E-Planning Research*, *5*(3), 13–26. doi:10.4018/IJEPR.2016070102

Schlossberg, E., & Shuford, E. (2005). Delineating "public" and "participation" in PPGIS. *URISA Journal*, *16*(2), 15–26.

Schroeder, P. (1996). Report on Public Participation GIS Workshop. In *GIS and Society: The Social Implications of How People, Space and Environment are Represented in GIS*. NCGIA Technical Report n. 96-97, Scientific Report for Initiative 19 Specialist Meeting, South Haven, MN.

Schuler, D. A., Rehbein, K., & Green, C. D. (2016, November). Is Corporate Political Activity a Field? *Business & Society*.

Schwester, R. (2009). Examining the barriers to e-government adoption. *Electronic. Journal of E-Government*, *7*(1), 113–122.

Schwester, R. W. (2010). Socio-Demographic Determinants of E-government Adoption: An Examination of Major US Cities. *Journal of Public Management & Social Policy*, *16*(2).

Scola, N. (2014, February 26). *Why Preservationists Are Mapping Detroit's Historic Layer*. Retrieved from Next City: https://nextcity.org/daily/entry/why-preservationists-are-mapping-detroits-historic-layer

Scott, G., Leritz, L. E., & Mumford, M. D. (2004). The effectiveness of creativity training: A quantitative review. *Creativity Research Journal*, *16*(4), 361–388. doi:10.1080/10400410409534549

Scott, J. K. (2006). "E" the people: Do US municipal government Web sites support public involvement? *Public Administration Review*, *66*(3), 341–353. doi:10.1111/j.1540-6210.2006.00593.x

Scott, K., & Liew, T. (2012). Social networking as a development tool: A critical reflection. *Urban Studies (Edinburgh, Scotland)*, *49*(12), 2751–2767. doi:10.1177/0042098011435279

Seltzer, E., & Mahmoudi, D. (2013). Citizen Participation, Open Innovation, and Crowdsourcing: Challenges and Opportunities for Planning. *Journal of Planning Literature*, *28*(1), 3–18. doi:10.1177/0885412212469112

Senbel, M., & Church, S. P. (2011). Design Empowerment: The Limits of Accessible Visualization Media in Neighborhood Densification. *Journal of Planning Education and Research*, *31*(4), 423–437. doi:10.1177/0739456X11417830

Shaffer, G. (2007). Frame-Up: An Analysis of Arguments For and Against Municipal Wireless Initiatives. *Public Works Management & Policy*, *11*(3), 204–216. doi:10.1177/1087724X06297347

Shafiq, B., Chun, S., Atluri, V., Vaidya, J., & Nabi, G. (2012). Resource Sharing using UICDS Framework for Incident Management *Transforming Government: People Process and Policy*, *6*(1), 41–61.

Shafiq, B., Vaidya, J., Atluri, V., Adam, N., Chun, S., & Lieb, A. (2008). Ontology Driven Resource Management for Emergency Response. *Workshop on Secure Knowledge Management (SKM 2008).*

Shafiq, B., Vaidya, J., Atluri, V., & Chun, S. (2010). UICDS compliant resource management system for emergency response. *Proceedings of the 11th Annual International Digital Government Research Conference on Public Administration Online: Challenges and Opportunities (DGO '10)*, 23-31.

Shapiro, J. M. (2006). Smart cities: Quality of life, productivity, and the growth effects of human capital. *The Review of Economics and Statistics*, *88*(2), 324–335. doi:10.1162/rest.88.2.324

Shapiro, K. (2007). From modernism to McDonald's: Ideology, controversy, and the movement to preserve the recent past. *Journal of Architectural Education*, *61*(2), 6–14. doi:10.1111/j.1531-314X.2007.00145.x

Shav-Ami, A. (2013). Social protest: The Israeli case. Journal of Enterprising Communities. *People and Places in the Global Economy*, *7*(4), 373–382.

H. Shearer (Ed.). (2009). Using Geographical Information Systems to explore the determinants of urban household water consumption and response to the Queensland government demand-side policy measure imposed during the drought of 2006-2008. *Proceedings of the 4th National Conference on the State of Australian Cities.*

Sheller, M., & Urry, J. (2006). The New Mobilities Paradigm. *Environment & Planning*, *38*(2), 207–226. doi:10.1068/a37268

Shen, Z., Lyytinen, K., & Yoo, Y. (2015). Time and information technology in teams: A review of empirical research and future research directions. *European Journal of Information Systems*, *24*(5), 492–518. doi:10.1057/ejis.2014.8

Shepherd, M. M., Briggs, R. O., Reinig, B. A., Yen, J., & Nunamaker, J. F. Jr. (1995). Invoking social comparison to improve electronic brainstorming: Beyond anonymity. *Journal of Management Information Systems*, *12*(3), 155–170. doi:10.1080/07421222.1995.11518095

Shiode, N. (2000). Urban Planning, Information Technology, and Cyberspace. *Journal of Urban Technology*, *7*(2), 105–126. doi:10.1080/713684111

Shirani, A. I., Tafti, M. H., & Affisco, J. F. (1999). Task and technology fit: A comparison of two technologies for synchronous and asynchronous group communication. *Information & Management*, *36*(3), 139–150. doi:10.1016/S0378-7206(99)00015-4

Shkabatur, J. (2010). Cities at crossroads: Digital technology and local democracy in America. *Brooklyn Law Review*, *76*, 1413–1485.

Short, J., Williams, E., & Christie, B. (1976). *The Social Psychology of Telecommunications.* London: Wiley.

Sieber, R. (2006). Public Participation Geographic Information Systems: A Literature Review and Framework. *Annals of the Association of American Geographers*, *96*(3), 491–507. doi:10.1111/j.1467-8306.2006.00702.x

Siebrits, R. (2012). Swimming pools and intra-city climates: Influences on residential water consumption in Cape Town. *Water S.A.*, *38*(1), 133–144. doi:10.4314/wsa.v38i1.16

Siegel, J., Dubrovsky, V., Kiesler, S., & McGuire, T. W. (1986). Group processes in computer-mediated communication. *Organizational Behavior and Human Decision Processes*, *37*(2), 157–187. doi:10.1016/0749-5978(86)90050-6

Silcook, R. (2001). What is e-Government? *Parliamentary Affairs*, *54*(1), 88–101. doi:10.1093/pa/54.1.88

Silva, C. N. (2010). *Handbook of Research on E-Planning. ICTs for Urban Development and Monitoring*. Hershey, PA: IGI Global. doi:10.4018/978-1-61520-929-3

Silva, C. N. (2010). The *e*-planning paradigm - Theory, methods and tools: An overview. In C. N. Silva (Ed.), *Handbook of research on e-planning: ICTs for urban development and monitoring* (pp. 1–14). Hershey, PA: IGI Global. doi:10.4018/978-1-61520-929-3.ch001

Silva, C. N. (2012). *Online Research Methods in Urban and Planning Studies. Design and Outcomes*. Hershey, PA: IGI Global. doi:10.4018/978-1-4666-0074-4

Silva, C. N. (2013). *Citizen E-Participation in Urban Governance. Crowdsourcing and Collaborative Creativity*. Hershey, PA: IGI Global. doi:10.4018/978-1-4666-4169-3

Silva, C. N. (2013). Foreword. In L. Horelli (Ed.), *New approaches to urban planning, insights from participatory communities* (pp. 7–8). Helsinki: Aalto University.

Silva, C. N. (2015). *Emerging Issues, Challenges, and Opportunities in Urban E-Planning*. Hershey, PA: IGI Global. doi:10.4018/978-1-4666-8150-7

Smith, S. (2003). The constraining effects of initial ideas. In P. B. Paulus & B. A. Nijstad (Eds.), *Group creativity: Innovation through collaboration* (pp. 15–31). New York: Oxford University Press. doi:10.1093/acprof:oso/9780195147308.003.0002

Snijders, C., Matzat, U., & Reips, U. D. (2012). "Big Data": Big Knowledge Gaps in the Field of Internet Science. *International Journal of Internet Science*, *7*(1), 1–5.

Solnit, R. (2009). *A Paradise Built in Hell*. New York: Penguin Books.

SparkNotes. (2014). *International Organizations*. Retrieved 4 Feb, 2015, from http://www.sparknotes.com/us-government-and-politics/political-science/international-politics/section4.rhtml

Srinivasan, P. K. (2011). An emergency GSM satellite portable terminal for disaster management. *Proceedings of the 1st International Conference on Wireless Technologies for Humanitarian Relief*, 455-460. 10.1145/2185216.2185333

Staffans, A. (2004). *Vaikuttavat asukkaat. Vuorovaikutus ja paikallinen tieto kaupunkisuunnittelun haasteina. Yhdyskuntasuunnittelun tutkimus- ja koulutuskeskuksen julkaisuja A 29* (Doctoral dissertation). Helsinki: Yliopistopaino.

Stasser, G., Abele, S. P., & Parsons, V. S. (2012). Information flow and influence in collective choice. *Group Processes & Intergroup Relations*, *15*(5), 619–635. doi:10.1177/1368430212453631

State of Victoria. (2017). *Wi-Fi Initiative Website*. Retrieved 10 July 2017 from http://www.vic.gov.au/wifi.html

Stein, C. J. (2010). *Greening modernism: Preservation, sustainability, and the Modern Movement*. New York, NY: W.W. Norton & Company.

Steinfeld, N., & Lev-On, A. (2015). 'Well-Done, Mr. Mayor!': Linguistic Analysis of Municipal Facebook Pages. *International Journal of E-Planning Research*, *4*(2), 26–38. doi:10.4018/ijepr.2015040102

Steiniger, S., & Hunter, A. J. S. (2013). The 2012 free and open source GIS software map – A guide to facilitate research, development, and adoption. *Computers, Environment and Urban Systems*, *39*, 136–150. doi:10.1016/j.compenvurbsys.2012.10.003

Stern, E., Gudes, O., & Svoray, T. (2009). Web-based and traditional public participation in comprehensive planning: A comparative study. *Environment and Planning. B, Planning & Design*, *36*(6), 1067–1085. doi:10.1068/b34113

Stevenson, N. (2003). Cultural Citizenship in the 'Cultural' Society: A Cosmopolitan Approach. *Citizenship Studies*, *7*(3), 331–348.

Stieglitz, S., & Dang-Xuan, L. (2013). Emotions and information diffusion in social media: Sentiment of microblogs and sharing behavior. *Journal of Management Information Systems*, *29*(4), 217–248. doi:10.2753/MIS0742-1222290408

Stillwell, A. S., King, C. W., Webber, M. E., Duncan, I. J., & Hardberger, A. (2011). The Energy-Water Nexus in Texas. *Ecology and Society*, *16*(1), art2. doi:10.5751/ES-03781-160102

Stratigea, A., Leka, A., & Panagiotopoulou, M. (2017). In Search of Indicators for Assessing Smart and Sustainable Cities and Communities' Performance. *International Journal of E-Planning Research*, *6*(1), 43–73. doi:10.4018/IJEPR.2017010103

Stubbs, W., & Cocklin, C. (2008). Conceptualizing a "Sustainability Business Model". *Organization & Environment*, *21*(2), 103–127. doi:10.1177/1086026608318042

Suchman, L. A. (2007). *Human-Machine Reconfigurations: Plans and Situated Actions*. Cambridge, UK: Cambridge University Press.

Suchocki, S. S. (1995). The impact of electronic tools for support of groups engaged in land use planning. *Dissertation Abstracts International Section A: Humanities and Social Sciences*, *56*(3-A), 0742

Suddaby, R. (2006). From the Editors: What Grounded Theory is Not. Academy of Management Journal, 49(4), 633-642.

Surendran, V. (2015). *All You Need to Know About Internet.org, Net Neutrality and Facebook Profile Picture Row*. Retrieved on 17 June 2017 from http://indiatoday.intoday.in/story/all-you-need-to-know-about-internetorg-net-neutrality-and-facebook-profile-picture-row/1/485496.html

Surowiecky, J. M. (2004). *The wisdom of crowds*. New York: Random House.

Susha, I., & Grönlund, A. (2012). eParticipation research: Systematizing the field. *Government Information Quarterly*, *29*(3), 373–382. doi:10.1016/j.giq.2011.11.005

Svensson, C. (2009). Speaking of architecture. A study of the jury's assessment in an invited competition. *Nordic Journal of Architectural Research*, *21*(2/3), 94–107.

Svensson, G., & Wagner, B. (2010). Transformative business sustainability: Multi-layer model and network of e-footprint sources. *European Business Review*, *23*(4), 334–352. doi:10.1108/09555341111145735

Swani, K., Milne, G., & Brown, B. P. (2013). Spreading the word through Likes on Facebook: Evalutating the message strategy effectiveness of Fortune 500 companies. *Journal of Research in Interactive Marketing*, *7*(4), 269–294. doi:10.1108/JRIM-05-2013-0026

Syed, A., Melanie, J., Thorpe, S., & Penney, K. (2007). *Australian energy national and state projections to 2029-30, ABARE research report 07. 24*. Australian Bureau of Agricultural and Resource Economics (ABARE). Retrieved from http://data.daff.gov.au/brs/data/warehouse/pe_abare99001416/rr07.24_energy_projections_full_report.pdf

Syme, G. J., Shao, Q., Po, M., & Campbell, E. (2004). Predicting and understanding home garden water use. *Landscape and Urban Planning*, *68*(1), 121–128. doi:10.1016/j.landurbplan.2003.08.002

Talen, E. (2000). BottomUp GIS. A new tool for individual and group expression in participatory planning. *Journal of the American Planning Association*, *66*, 279–294. doi:10.1080/01944360008976107

Tapscott, D., & Williams, A. D. (2006). *Wikinomics: How Mass Collaboration Changes Everything*. New York: Penguin.

Taylor, B. T., Fernando, P., Bauman, A. E., Williamson, A., Craig, J. C., & Redman, S. (2011). Measuring the quality of public open space using Google Earth. *American Journal of Preventive Medicine*, *40*(2), 105–112. doi:10.1016/j.amepre.2010.10.024 PMID:21238857

Taylor, N. (1998). *Urban Planning Theory Since 1945*. London: SAGE Publications Ltd.

Technologies, L. (2017, June 21). *Detroit*. Retrieved from Loveland: https://detroit.makeloveland.com/

Technologies, L. (n.d.). Retrieved July 21, 2017, from Loveland: https://makeloveland.com/#t=search&b=none

Technologies, L., & the Data Driven Detroit. (n.d.). Retrieved July 21, 2017, from Motor City Mapping: https://www.motorcitymapping.org/#t=overview&s=detroit&f=all

Telstra. (2017). *Telstra Air Information Site*. Retrieved on 17 June 2017 from https://www.telstra.com.au/telstra-air/about-telstra-air

Tenenboim, O., & Cohen, A. A. (2013). What prompts users to click and comment: A longitudinal study of online news. *Journalism*, *20*(10), 1–20.

Teo, T., & Tan, M. (1998). An empirical study of adopters and non-adopters of the Internet in Singapore. *Information & Management*, *34*(6), 339–345. doi:10.1016/S0378-7206(98)00068-8

Terchek, R. J., & Conte, T. C. (2001). *Theories of Democracy; A Reader*. Lanham, MD: Rowman and Littlefield.

Texasinvasives.org. (2011). *Texas Invasives*. Retrieved October 4, 2014, from http://www.texasinvasives.org/

The Getty Conservation Institute. (2013, January). *Arches Project*. Retrieved October 4, 2014, from http://www.getty.edu/conservation/our_projects/field_projects/arches/

The Offwell Woodland & Wildlife Trust. (2008). *Simpson's Diversity Index*. Retrieved 1st,July, 2014, from http://www.countrysideinfo.co.uk/simpsons.htm

Therese, J., Fajardo, B., & Oppus, C. M. (2010, June). A mobile disaster management system using the android technology. *WTOC*, *9*(6), 343–353.

Thiel, S. K., Reisinger, M., Röderer, K., & Fröhlich, P. (2016). Playing (with) Democracy: A Review of Gamified Participation Approaches. *JeDEM-eJournal of eDemocracy and Open Government*, *8*(3), 32-60.

Thrift, N. (2004). Remembering the Technological Unconscious by Foreground Knowledges of Position. *Environment and Planning. D, Society & Space*, *22*(1), 175–190. doi:10.1068/d321t

Tolbert, C. J., & Mossberger, K. (2006). The effects of e-government on trust and confidence in government. *Public Administration Review*, *66*(3), 354–369. doi:10.1111/j.1540-6210.2006.00594.x

Tolbert, C. J., Mossberger, K., & McNeal, R. (2008). Institutions, policy innovation, and e-government in the American states. *Public Administration Review*, *68*(3), 549–563. doi:10.1111/j.1540-6210.2008.00890.x

Townsend, A. M. (2013). *Smart cities: Big data, civic hackers, and the quest for a new utopia* (1st ed.). New York: W.W. Norton & Company.

Tree Folks and Lady Bird Johnson Wildflower Center. (2014). *The Great Texas Tree Roundup*. Retrieved October 4, 2014, from http://treeroundup.org/

Trottier, D., & Lyon, D. (2014). Key Features of Social Media Surveillance. In C. Fuchs (Ed.), *Internet and Surveillance: The Challenges of Web 2.0 and Social Media* (pp. 89–105). New York: Routledge.

Troy, P., & Holloway, D. (2004). The use of residential water consumption as an urban planning tool: A pilot study in Adelaide. *Journal of Environmental Planning and Management*, *47*(1), 97–114. doi:10.1080/0964056042000189826

Truong, H., Manzoor, A., & Dustdar, S. (2009). On modeling, collecting and utilizing context information for disaster responses in pervasive environments. *Proceedings of the first international workshop on Context-aware software technology and applications*, 25-28. 10.1145/1595768.1595779

Tseng, P. T., Yen, D. C., Hung, Y. C., & Wang, N. C. (2008). To explore managerial issues and their implications on e-Government deployment in the public sector: Lessons from Taiwan's Bureau of Foreign Trade. *Government Information Quarterly*, *25*(4), 734–756. doi:10.1016/j.giq.2007.06.003

Tufte, T., & Mefalopulos, P. (2009). *Participatory Communication: A Practical Guide*. Herndon, VA: World Bank Publications.

Tulloch, D. L. (2008). Is VGI participation? From vernal pools to video games. *GeoJournal*, *72*(3-4), 161–171. doi:10.100710708-008-9185-1

Turkucu, A., & Roche, S. (2008). Classification fonctionnelle des Public Participation GIS. *Revue Internationale de Géomatique*, *18*(4), 429–442. doi:10.3166/geo.18.429-442

Turner-Lee, N. (2010). The challenge of increasing civic engagement in the digital age. *Federal Communications Law Journal*, *63*(1), 19–32.

U.S. Department of Energy. (2006). *Energy Demands on Water Resources: Report to Congress on the interdependency of energy and water*. U.S. Department of Energy.

U.S. Department of the Interior, National Park Service. (1985). *Guidelines for local surveys: A basis for preservation planning bulletin* (Revised Version from 1985). Retrieved June 27, 2014, from http://www.nps.gov/nr/publications/bulletins/nrb24/iNDEX.htm

U.S. Department of the Interior, National Parks Service. (2017). *Become a CLG. National Parks Service website*. Retrieved June 28, 2017, from https://grantsdev.cr.nps.gov/CLG_Review/Get_All_CLG.cfm

UN Habitat. (2014). *The World Urban Forum*. Retrieved 2 March, 2015, from http://wuf7.unhabitat.org/theworldurbanforum

UN. (2016). *UN E-Government Survey 2016: E-Government in Support of Sustainable Development*. New York: United Nations.

United Nations. (2014). *E-Government Survey 2014: E-government for the future we want*. Retrieved July 15, 2017, from https://publicadministration.un.org/egovkb/portals/egovkb/documents/un/2014-survey/e-gov_complete_survey-2014.pdf

Våland, M. S. (2009). End user participation as an input to shape the brief in architectural competitions. A threefold translation process. *Nordic Journal of Architectural Research*, *21*(2/3), 108–122.

van Dijk, H., van Engen, M. L., & van Knippenberg, D. (2012). Defying conventional wisdom: A meta-analytical examination of the differences between demographic and job-related diversity relationships with performance. *Organizational Behavior and Human Decision Processes*, *119*(1), 38–53. doi:10.1016/j.obhdp.2012.06.003

van Knippenberg, D., De Dreu, C. K. W., & Homan, A. C. (2004). Work group diversity and group performance: An integrative model and research agenda. *The Journal of Applied Psychology*, *89*(6), 1008–1022. doi:10.1037/0021-9010.89.6.1008 PMID:15584838

van Winden, W., & van den Buuse, D. (2017). Smart City Pilot Projects: Exploring the Dimensions and Conditions of Scaling Up. *Journal of Urban Technology*, *24*(4), 51–72. doi:10.1080/10630732.2017.1348884

VanGundy, A. B. (1981). *Techniques of structured problem solving*. New York: Van. Nostrand–Reinhold.

Vishkaie, R. S. (2014). *SketchBoard: Conceptualizing Interactive Communication Media for the Design Review Process* (Doctoral dissertation). Available from ProQuest Dissertation and Theses database. (URI No. 11023/1426)

Vishkaie, R. S., & Levy, R. M. (2015). Making Design Review Interactive. In G. Mura (Ed.), *Analyzing Art, Culture, and Design in the Digital Age* (pp. 234–257). Hershey, PA: IGI Global. doi:10.4018/978-1-4666-8679-3.ch018

Vitak, J., Zube, P., Smock, A., Carr, C. T., Ellison, N., & Lampe, C. (2011). It's complicated: Facebook users' political participation in the 2008 election. *Cyberpsychology, Behavior, and Social Networking*, *14*(3), 107–114. doi:10.1089/cyber.2009.0226 PMID:20649449

Vogel, R., Moulder, R., & Huggins, M. (2014). The extent of public participation. *Public Management*, *96*(2), 6–10.

Voghera, A., & Crivello, R. (2015). Semantic Spatial Representation, an Experimental Proposal in the Framework of eParticipation. *International Journal of E-Planning Research*, *4*(4), 18–35. doi:10.4018/ijepr.2015100102

Volker, L. (2010). *Designing a design competition: The client perspective*. Paper at the 2010 Design Research Society (DRS) International Conference Design & Complexity, Montreal, Canada.

Walden, D. (2011). Interviews: Ward Cunningham. *IEEE Annals of the History of Computing*, *33*(4), 62–67. doi:10.1109/MAHC.2011.81

Wallhagen, M., Malmqvist, T., & Eriksson, O. (2017). Professionals' knowledge and use of environmental assessment in an architectural competition. *Building Research and Information*, *45*(4), 426–442. doi:10.1080/09613218.2015.1118264

Wallin, S., Amati, M., Saad-Sulonen, J., & Horelli, L. (2012). Exploring *e*-planning practices in different contexts: Similarities and differences between Helsinki and Sydney. *International Journal of E-Planning Research*, *1*(3), 17–39. doi:10.4018/ijepr.2012070102

Walther, J. B., & Jang, J. (2012). Communication processes in particapatory websites. *Computer-Mediated Communication*, *18*(1), 2–15. doi:10.1111/j.1083-6101.2012.01592.x

Wang, H., Song, Y., Hamilton, A., & Curwell, S. (2007). Urban information integration for advanced e-planning in Europe. *Government Information Quarterly*, *24*(4), 736–754. doi:10.1016/j.giq.2007.04.002

Wang, M., Liao, F., Lin, J., Huang, L., Gu, C., & Wei, Y. (2016). The Making of a Sustainable Wireless City? Mapping Public Wi-Fi Access in Shanghai. *Sustainability*, *8*(2), 111–126. doi:10.3390u8020111

Warfel, T. Z. (2009). *Prototyping: A Practitioner's Guide*. Brooklyn, NY: Rosenfeld Media.

Warner, B. E. (2001). John Stuart Mill's theory of bureaucracy within representative government: Balancing competence and participation. *Public Administration Review*, *61*(4), 403–413. doi:10.1111/0033-3352.00044

Waters, R. D., Burnett, E., Lamm, A., & Lucas, J. (2009). Engaging Stakeholders through Social Networking: How nonprofit organizations are using Facebook. *Public Relations Review*, *35*(2), 102–106. doi:10.1016/j.pubrev.2009.01.006

Watson, L. (2017). *What Happened to Google Fiber*. Retrieved on 21 July 2017 from https://www.gizmodo.com.au/2017/02/what-happened-to-google-fiber/

Wellman, B., Haase, A. Q., Witte, J., & Hampton, K. (2001). Does the Internet increase, decrease, or supplement social capital?: Social networks, participation, and community commitment. *The American Behavioral Scientist*, *45*(3), 436–455. doi:10.1177/00027640121957286

Wentz, E. A., & Gober, P. (2007). Determinants of small-area water consumption for the city of Phoenix, Arizona. *Water Resources Management*, *21*(11), 1849–1863. doi:10.100711269-006-9133-0

Wescott, C. (2001). E-Government in the Asia-Pacific region. *Asian Journal of Political Science*, *9*(2), 1–24. doi:10.1080/02185370108434189

Whitworth, F. (in press). *Pedestrian monitoring at Queen Victoria Market.* Research Unit in Public Culture Working Paper Series. Melbourne.

Wi-Fi Alliance. (2016). *Wi-Fi® device shipments to surpass 15 billion by end of 2016* [Press release]. Retrieved on 17 June 2017 from http://www.wi-fi.org/news-events/newsroom/wi-fi-device-shipments-to-surpass-15-billion-by-end-of-2016

WikiMedia Foundation. (2017). *MediaWiki.* Retrieved November 20, 2017, from https://www.mediawiki.org/wiki/MediaWiki

Willard, T. (2009). Social Networking and Governance for Sustainable Development. Winnipeg, Canada: International Institute for Sustainable Development (IISD).

Willems, L., & Alizadeh, T. (2015). Social Media for Public Involvement and Sustainability in International Planning and Development. *International Journal of E-Planning Research*, *4*(4), 1–17. doi:10.4018/ijepr.2015100101

Williams, R. (1983). *Towards 2000.* London: Chatto and Windus.

Wilson, J. S., Kelly, C. M., Schootman, M., Baker, E. A., Banerjee, A., Clennin, M., & Miller, D. K. (2012). Assessing the built environment using omnidirectional imagery. *American Journal of Preventive Medicine*, *42*(2), 193–199. doi:10.1016/j.amepre.2011.09.029 PMID:22261217

Worthington, A. C., & Hoffman, M. (2008). An empirical survey of residential water demand modelling. *Economic Surveys*, *22*(5), 842–871. doi:10.1111/j.1467-6419.2008.00551.x

Wu, A., Yan, X., & Zhang, X. (2011). Geo-tagged mobile photo sharing in collaborative emergency management. *Proceedings of the 2011 Visual Information Communication - International Symposium.* 10.1145/2016656.2016663

Yates, D., & Paquette, S. (2010). Emergency knowledge management and social media technologies: a case study of the 2010 Haitian earthquake. *Proceedings of the 73rd ASIS&T Annual Meeting on Navigating Streams in an Information Ecosystem, 47.*

Yuan, Y., Raubal, M., & Liu, Y. (2012). Correlating mobile phone usage and travel behavior–A case study of Harbin, China. *Computers, Environment and Urban Systems*, *36*(2), 118–130. doi:10.1016/j.compenvurbsys.2011.07.003

Zhang, Y., & Yang, K. (2009). Citizen participation in the budget process: The effect of city managers. *Journal of Public Budgeting, Accounting &. Financial Management*, *21*(2), 289.

Zheng, L., Shen, C., Tang, L., Li, T., Luis, S., & Chen, S. (2011). Applying data mining techniques to address disaster information management challenges on mobile devices. *Proceedings of the 17th ACM SIGKDD international conference on Knowledge discovery and data mining*, 283-291. 10.1145/2020408.2020457

Zheng, L., Shen, C., Tang, L., Li, T., Luis, S., Chen, S., & Hristidis, V. (2010). Using data mining techniques to address critical information exchange needs in disaster affected public-private networks. *Proceedings of the 16th ACM SIGKDD international conference on Knowledge discovery and data mining*, 125-134. 10.1145/1835804.1835823

Zickuhr, K. (2013). *Who's not online and why*. Washington, DC: Pew Research Center. Retrieved June 27, 2014, from http://www.pewinternet.org/files/old-media//Files/Reports/2013/PIP_Offline%20adults_092513_PDF.pdf

Ziegler, R., Diehl, M., & Zijlstra, G. (2000). Idea production in nominal and virtual groups: Does computer mediated communication improve group brainstorming? *Group Processes & Intergroup Relations*, *3*(2), 141–158. doi:10.1177/1368430200032003

About the Contributors

Carlos Nunes Silva, PhD, Professor Auxiliar at the Institute of Geography and Spatial Planning, University of Lisbon, Portugal, holds a degree in Geography (University of Coimbra), a post-graduation in European Studies (University of Coimbra - Faculty of Law), a Master degree in Human Geography: Regional and Local Planning (University of Lisbon), and a PhD in Geography: Regional and Local Planning (University of Lisbon). His research interests focus mainly on urban and metropolitan governance, the history and theory of urban planning, urban planning in Africa, urban e-planning, urban planning ethics, local government policies, local e-government, and research methods. His publications include 'Governing Urban Africa', 'Local Government and Urban Governance in Europe' (co-edited); 'Urban Planning in North Africa'; 'Urban Planning in Lusophone African Countries'; 'Emerging Issues, Challenges, and Opportunities in Urban E-Planning'; 'Urban Planning in Sub-Saharan Africa: Colonial and Postcolonial Planning Cultures'; 'Fiscal Austerity and Innovation in Local Governance in Europe' (co-edited); 'Citizen e-Participation in Urban Governance: Crowdsourcing and Collaborative Creativity'; 'Online Research Methods in Urban and Planning Studies: Design and Outcomes'; 'Handbook of Research on E-Planning: ICT for Urban Development and Monitoring'; 'Portugal: Sistema de Govern Local'; and 'Política Urbana em Lisboa, 1926-1974'. He is the Chair of the International Geographical Union Commission on 'Geography of Governance' (2016-2020), and the founding editor-in-chief of the 'International Journal of E-Planning Research' (IJEPR).

* * *

Nabil Adam is serving as the Vice Chancellor for Research & Collaborations at Rutgers University-Newark. He is a Distinguished Professor of Computers and Information Systems at Rutgers University; the Founding Director of the Rutgers of the Institute for Data Science, Learning, and Applications (I-DSLA); and the Founding Director of the Rutgers CIMIC Research Center. He is a Co-founder and past Director of the Meadowlands Environmental Research Institute. He was on loan as a Fellow to the US Department of Homeland Security-Science & Technology Directorate where he served as a Senior Program Manager, a Branch Chief and managed the Complex Event Modeling, Simulation, and Analysis program, served as the technical lead for the Unified Incident Command & Decision Support System program, and initiated the Cyber-physical Systems Security initiative and the Social Media Alert and Response to Threats to Citizens (SMART-C) initiative. He served as a Research Fellow at the Center of Excellence in Space Data and Information Science, NASA Goddard Space Flight Center. He was a member of the Science Council of the Research Institute for Advanced Computer Science, NASA Ames His research work has been supported by over $21 million from various federal and state agencies, including: DHS; NSF; NSA; EPA; DoD; NASA; NIH; NOAA; National Academies of Sciences, Engineering, and Medicine; and the NJ Meadowlands Commission.

Tooran Alizadeh is a Senior Lecturer and the Director of Urban Design at the University of Sydney. She is an interdisciplinary academic researching policy and planning implications of telecommunication infrastructure with a focus on the National Broadband Network (NBN) in Australia, smart cities, urban digital strategies, and social media. Her broader research interest, however, include spatial planning and urban design, post-disaster/war planning, spatial justice, and planning education.

Vijayalakshmi Atluri is a Professor of Computer Information Systems in the MSIS Department at Rutgers University. Her research interests include Digital Government, Information Security, Privacy, Databases, and Workflow Management. She has published over 150 papers in peer reviewed journals and conferences. In the past, she served as the Vice-chair for the ACM Special Interest Group on Security Audit and Control (SIGSAC), and the Chair of the IFIP WG11.3 Working Conference on Database Security. She was the recipient of the National Science Foundation CAREER Award, the Rutgers University Research Award for untenured faculty, and the 2014 IFIP WG11.3 Workgroup on Data and Application Security and Privacy outstanding research award.

Jonali Baruah is an assistant professor in the Department of Psychological Sciences at Tarleton State University, USA. She maintains a research interest that incorporates social, cognitive, and diversity aspects in the investigation of creativity and performance in small groups and individuals.

Ian Bishop is a Professorial Fellow in the Department of Infrastructure Engineering at the University of Melbourne. His research includes quantitative and analytical approaches to land use, environmental assessment, planning and design especially using tools such as geographic information systems (GIS), visualization and agent-based modeling. He is committed to expanding the role of technology in the management of natural resources and in facilitation of public participation in planning and design issues.

Soon Ae Chun is a professor of Information Systems and Informatics Program, and a doctoral faculty of Computer Science. She is the director of the Information Security Research and Education Lab. She received 2014 Dolphin Award for Outstanding Scholarly Achievement. She serves as the President for the Digital Government Society. Her research areas include Security and Privacy, Semantic data integration, Social Data Analytics and Workflow, focusing on applied areas in Digital Government and Public Health. Her work includes security ontology engineering, knowledge-based disaster and emergency management process and resource discovery, social media in government, policy-based service workflow generation, public engagement and transparency of policy decision making, Geospatial workflow generation, crowd-based environmental planning, and epidemic and public health analytics. Her research has been sponsored by NSF, NOAA, a New Jersey State Agency, and PSC-CUNY. She is an IEEE senior member, a member of ACM Computing Society, the Digital Government Society, and the Beta Gamma Sigma Honor Society.

Josh Conrad is a Ph.D. student in Architecture under Dr. Michael Holleran and a historic preservation consultant and GIS Data Analyst with the firm HHM & Associates in Austin. He was the technical developer of the Austin Historical Survey Wiki, a collaboration between UT Austin's School of Architecture and School of Information, the City of Austin, and Preservation Austin to build an online historical survey database based on open-source GIS technology and capable of continuous updating by citizen and professional preservationists. He is also the founder and editor of the online architectural drawings database RNDRD (@rndrd).

Susa Eräranta is an architect M.Sc. and M.Sc. (Econ). She is a doctoral candidate at the Department of Built Environment at Aalto University and a practicing strategic urban planning architect at the City of Helsinki (Finland). Her research interest is in the dynamics of planning processes.

Reza Farid is a Machine Learning Researcher and an Adjunct Research Fellow in Institute for Integrated and Intelligent Systems (IIIS), at Griffith University. Reza's work is primarily in the fields of Machine Learning, Artificial Intelligence, Computer Vision, Data Analytics and Software Engineering. He has an extensive industry-based experience in R&D projects, software analysis, design and development, databases and data analytics. His focus is on how data can be analysed and utilised to improve the performance of an eco-system.

Michael Holleran joined the UT School of Architecture faculty in fall 2006. He came from the University of Colorado College of Architecture and Planning, where he served as Associate Dean of Research, Associate Professor of Planning and Design, and Director of the Colorado Center for Preservation Research. He has also taught in Architecture and Landscape Architecture at the Rhode Island School of Design. His book, Boston's "Changeful Times": Origins of Preservation and Planning in America (Johns Hopkins University Press, 1998), puts the early preservation movement into its larger context of accelerating environmental change and emerging controls on urban development. It won the Antoinette Downing Award from the Society of Architectural Historians, the biennial Lewis Mumford Award for the best book in North American planning history from the Society for American City and Regional Planning History, and the Historic New England Book Award. He practiced for twelve years in as a partner in Everett · Clarke · Holleran Associates in Providence, Rhode Island, a planning, architecture, and landscape architecture firm working mainly on preservation projects. His public service has included chairing the Landmarks Preservation Board in Boulder, Colorado; and serving on the board and chairing the Public Policy committee of Colorado Preservation, Inc., a statewide preservation advocacy group. He is a board member of the national Recent Past Preservation Network and of the Heritage Society of Austin.

Maarit Kahila-Tani, D.Sc. (Tech.) (Urban planning), works as a postdoctoral researcher at the Department of Built Environment in Aalto University. In her doctoral thesis she focused on the utilisation of PPGIS tools and use of experiential knowledge in urban planning practices.

Luigi La Riccia is a Ph.D. in Regional Planning and Local Development at the Politecnico di Torino and consultant at the Comitato Promotore Transpadana, conducts research activity on Urban and Regional Planning, Landscape and Parks Planning, Environmental Sustainability and Geographic Information Systems (GIS). He has worked on numerous research activities and initiatives, including: the redaction of Landscape Regional Plan of Piemonte (2008-2011); "Landscape Indicators. Assessing and Monitoring the Landscape Quality" (2008-2011); "Fare Paesaggio. Dalla pianificazione di area vasta all'operatività locale"(Research Program of National Interest 2007); the 5th edition of International Exhibition "Creare Paesaggi. Enjoy the Landscape" (2010); Guidelines of Industrial Landscapes of Sardinia Region (2011-2012); Guidelines for the green network of PTC2" (2015). He has also participated in several EU funded projects (implemented and in progress):

"INTERALPES. Transport Intermodal Transalpin" (Alcotra 2011-2013); "POLY5. Polycentric Planning Models for Local Development in Territories interested by Corridor 5 and its TENT ramifications" (Alpine Space, 2011-2014); "PLAT.F.O.R.M. Platform to Form Opinions Related to Mobility" (Alpine Space, 2013-2014); "IP Citygreening" (Erasmus, 2013-2014); "S.T.R.E.E.T. Sustainable Transport Education for Environment and Tourism" (Erasmus+, 2015-2018); "WeGovNow. Towards #WeGovernment: Collective and participative approaches for addressing local policy challenges" (Horizon 2020, 2016-2019). He is the author of 40 scientific publications and international events and exhibitions on these subjects.

Alex Lambert is a lecturer in Communication and Media studies at Monash University and a fellow of the Research Unit in Public Cultures. Dr. Lambert specialises in the study of digital media, intimacy, and cosmopolitanism. His current work focuses on alternative forms of media and spatiality that enable both intimacy and cosmopolitan sociality.

Azi Lev-On is a faculty member in the School of Communication in Ariel University. His research focuses on the uses and perceived effects of social media, public participation and deliberation online, online communities, collective action and campaigns, and behaviors in computer-mediated environments, employing a variety of methods such as content and link analysis, surveys and laboratory experiments.

Richard Levy is a Professor of Planning and Urban Design at The University of Calgary, where he also serves as the Co-Director of the Computational Media Program and is an Adjunct Professor in the Department of Computer Science. Dr. Levy has conducted research projects with faculty from Archaeology, Criminology, Geomatics Engineering, Kinesiology, and Psychology. Dr. Levy speaks at international and national conferences in the fields of urban planning, GIS, and virtual reality. His published work appears in journals including Journal of Archaeological Science, Brain Research, Computers in Urban Planning and Urban Management, Computer Internet Archaeology, and the Journal of Visual Studies, Environment and Planning, Plan Canada, and IEEE, Journal.

Aroon P. Manoharan is an Associate Professor and Graduate Program Director in the Department of Public Policy and Public Affairs, John W. McCormack Graduate School of Policy and Global Studies, University of Massachusetts Boston. His research focuses on e-government, IT pedagogy, strategic planning, performance measurement and reporting, public management, and global comparative public administration. He holds a PhD from Rutgers University-Newark, and an MPA from Kansas State University.

Scott McQuire is Professor of Media and Communications in the School of Culture and Communication at the University of Melbourne, Australia. He is one of the founders of the Research Unit for Public Cultures which fosters interdisciplinary research at the nexus of digital media, art, urbanism, and social theory. He is also Deptuy-Director of the Melbourne Networked Society Institute. His most recent book Geomedia: Networked cities and the future of public space was published by Polity in 2016, and translated into Chinese by Fudan University Press in 2017. An illustrated monograph Art seen under digital light: photography, the image and the aesthetics of data will be published by the Lyon Housemuseum in 2018.

Jennifer Minner is an assistant professor in the Department of City and Regional Planning at Cornell University, in Ithaca, New York. Her teaching and research is focused at the intersecting domains of land use, sustainability, historic preservation, and technology. She researches reinvestment in and repair of urban fabric; tensions and opportunities in land use planning and preservation; sustainable adaptation and conservation of the built environment; and participatory and analytical technologies used in planning. Minner's experience includes planning, research, and community mapping projects related to land use and sustainability, historic preservation, institutional research and higher education assessment, and economic development. She received a B.A. in anthropology from the University of Washington, a Master in Urban and Regional Planning from Portland State University, and a Ph.D. in community and regional planning from the University of Texas at Austin.

Sean Mossey graduated with BA and MPA from the University of New Hampshire in history and public administration, respectively. He has spent his career as a research and teaching assistant working on projects in the realms of international relations, e-governance, m-governance, education policy, and organizational development. His research interests and competencies also include information security policy, quantitative analysis, global comparative policy, and organizational theory. He is currently pursuing his PhD in Public Policy at the University of Massachusetts in Boston, and is the current student representative for the Northeastern Conference on Public Administration (NECOPA).

Pilvi Nummi architect M Sc. is a doctoral candidate in the Department of Built Environment, School of Engineering, at Aalto University. Her research interest is in e-Planning and e-Participation in urban and land-use planning, especially in the use of social media in urban planning. Currently she is a practising town planning architect in the municipality in Sipoo (Finland).

Nikos Papastergiadis is the Director of the Research Unit in Public Culture, at the University of Melbourne.

Paul B. Paulus is the Distinguished University Professor in the Department of Psychology at the University of Texas at Arlington. For the past 25 years he has been investigating the factors that influence group creativity. His research has been funded by the National Science Foundation for the past 10 years. He co-edited the first scholarly volume on group creativity, and he has published over 80 papers and chapters on that topic and the related issue of team innovation. He has made numerous invited presentations to groups from the academic, business, engineering, and intelligence communities. In addition to his teaching and research career, he has served as Chair of the Department of Psychology and Dean of the College of Science. He has been a visiting professor at Bar Ilan University, the University of Groningen, the University of Sidney, the University of Pittsburgh, Carnegie Mellon University, and the Uniformed Services University of the Health Sciences.

Andrea Roberts is an Assistant Professor of Urban Planning and Faculty Fellow of the Center for Heritage Conservation at Texas A &M University. She is also the founder of the Texas Freedom Colonies Project which is dedicated to bringing attention to black settlement descendants' current community development & historic preservation issues. Her research frames planning and historic preservation as cultural resilience and social justice practices and is more broadly concerned with grassroots planning history and contemporary place preservation practices within settlements founded by formerly enslaved Africans in the Americas. She brings to her scholarship 12 years' experience in community and economic development in Philadelphia and Houston. Her current work reflects her ongoing concern with African American spatial justice, unrecorded planning and placemaking histories, and the role of cultural agency in land retention. Her article about African American community building in the Progressive Era, published in the Journal of Planning History, is one example of this work. She's written about social justice, intersectionality, and historic preservation for the National Trust's Preservation Leadership Forum and about the role of homeplace and belonging in African American placemaking for Places Journal. She is currently developing a statewide black settlement inventory and digital humanities project.

Maryam Saydi was graduated from her PhD at the University of Melbourne at Geographic Information system (GIS). She is keen to do spatial analysis for diversity of fields of studies.

Basit Shafiq received his B.S. degree in Electronic Engineering from GIK Institute of Engineering Sciences and Technology, Pakistan, M.S. and Ph.D. degrees in Electrical and Computer Engineering from Purdue University, USA. He is currently an Associate Professor in the Computer Science Department at Lahore University of Management Sciences (LUMS). Dr. Shafiq's interests include information systems security and privacy, access-control management in distributed systems, Web services composition and verification, ontologies, and distributed multimedia systems. His research work resulted in several publications in well-renowned journals, including, IEEE Transactions on Knowledge and Data Engineering, ACM Transactions on Information and System Security, IEEE Transactions on Multimedia, IEEE Transactions on Service Computing, IEEE Computer, IEEE Communications Magazine, and Journal on Information and Computer Systems.

Nili Steinfeld is the head of the Digital Communication track and the academic manager of the Institute for the Study of New-Media, Politics and Society at the School of Communications, Ariel University. Her research interests are e-participation, e-government and e-politics. She studies privacy and surveillance online, use of social networks as public spheres, citizens-politicians online engagement, virtual contact and online political discourse. She combines various methodological approaches, among them online experiments, surveys, eye tracking and digital methods for scraping and analyzing digital objects.

Jaideep Vaidya is the Dean's Research Professor in the Management Science and Information Systems department at Rutgers University. He received the B.E. degree in Computer Engineering from the University of Mumbai, the M.S. and Ph.D. degree in Computer Science from Purdue University. His general area of research is in data mining, data management, security, privacy, and digital government. He has published over 140 technical papers in peer-reviewed journals and conference proceedings, and has received several best paper awards. He is an ACM distinguished scientist and an IEEE senior member.

Rojin Vishkaie is an Assistant Professor of Emerging Media Design & Development. She received her PhD in Computational Media Design (CMD), with a specialization in Human-Computer Interaction and Interaction Design from the University of Calgary in 2014. Since completing her PhD, she has been a Postdoctoral Fellow that was funded by the Bill and Melinda Gates Foundation and German Federal Ministry of Education and Research, as well as a Lecturer of Industrial Design at the University of Illinois at Urbana-Champaign. She also works as a user experience researcher and interaction design consultant.

Angioletta Voghera is Associate Professor of Urban and Regional Planning, Interuniversity Department of Urban and Regional Studies and Planning (DIST), Politecnico of Torino. In the Politecnico di Torino she represents the "Architecture and Urban and Regional Planning Area" in the Task force on Technology Transfer, she is member of the Scientific Committee of S3+LAB-Urban Sustainability & Security Laboratory for Social Challenges and of the South China-Torino Collaboration Lab (Politecnico - South China University of Technology). In 2014 she was visiting researcher in EPFL of Lausanne. She is the Coordinator of the National Commission on Environment and Landscape Planning of the National Institute of Urban Planning (INU), www.inu.it (2014). She is member of the Editorials Boards of American Journal of Civil Engineering. Town and Country Planning, GeoProgress Journal, Sentieri Urbani. Rivista dell'INU Trentino. Participating in territorial governance and elective organisation, she is member of the since 2013 Council of Architects, planners, landscape architects and conservation experts of Turin Province and since 2016 of the Administration Council of Parchi Reali (Piedmont Region). Resulted in over 160 international and national publications, her main research activities are in the field of Regional and Landscape Planning and Urban Design focusing on analysis, evaluation, planning and design of sensible environments, such as natural and landscape protected areas, fluvial and rural landscapes. The main publications in these fields are the monographs: Culture europee di sostenibilità. Storie e innovazioni nella pianificazione (Gangemi Editore, 2006), After the European Landscape Convention. Policies, Plans and Evaluation (Alinea, 2011) and -with Ingaramo R. - Planning and Architecture. Searching for an Approach (Alinea International, 2012). Besides she is editor of: Progetti per il paesaggio (Inu Edizioni, 2015) and - with Bianchetti C., Cogato Lanza E., Enver Kercuku A. and Sampieri A. - Territories in Crisis. Architecture and Urbanism Facing Changes in Europe (JOVIS Verlag GmbH, 2015)

Laura Grace Willems completed her Honours Degree in Urban and Environmental Planning at Griffith University. She has a strong interest in technologies and their potential to improve to urban, social, and environmental outcomes. She works as a transport planner and policy advisor in areas of autonomous vehicles, congestion, climate change, and disruptive transport technologies.

Index

R

S

T

U

W

www.ingramcontent.com/pod-product-compliance
Lightning Source LLC
LaVergne TN
LVHW080608300926
849162LV00005B/1087